TEETH
Second Edition

Archaeological discoveries of teeth provide remarkable information on humans, animals and the health, hygiene and diet of ancient communities. In this fully revised and updated edition of his seminal text Simon Hillson draws together a mass of material from archaeology, anthropology and related disciplines to provide a comprehensive manual on the study of teeth. The range of mammals examined has been extended to include descriptions and line drawings for 325 mammal genera from Europe, North Africa, western, central and northeastern Asia, and North America. The book also introduces dental anatomy and the microscopic structure of dental tissues, explores how the age or season of death is estimated and looks at variations in tooth size and shape. With its detailed descriptions of the techniques and equipment used and its provision of tables and charts, this book is essential reading for students of archaeology, zoology and dental science.

SIMON HILLSON is Professor of Bioarchaeology at the Institute of Archaeology, University College London. His previous publications include *Teeth* (Cambridge, 1986), *Mammal Bones and Teeth* (1992), and *Dental Anthropology* (Cambridge, 1996).

CAMBRIDGE MANUALS IN ARCHAEOLOGY

General Editor
Graeme Barker, *University of Cambridge*

Advisory Editors
Elizabeth Slater, *University of Liverpool*
Peter Bogucki, *Princeton University*

Cambridge Manuals in Archaeology is a series of reference handbooks designed for an international audience of upper-level undergraduate and graduate students, and professional archaeologists and archaeological scientists in universities, museums, research laboratories and field units. Each book includes a survey of current archaeological practice alongside essential reference material on contemporary techniques and methodology.

Books in the series
Clive Orton, Paul Tyers, and Alan Vince, POTTERY IN ARCHAEOLOGY
R. Lee Lyman, VERTEBRATE TAPHONOMY
Peter G. Dorrell, PHOTOGRAPHY IN ARCHAEOLOGY AND CONSERVATION, 2ND EDN
A. G. Brown, ALLUVIAL GEOARCHAEOLOGY
Cheryl Claasen, SHELLS
Elizabeth J. Reitz and Elizabeth S. Wing, ZOOARCHAEOLOGY
Clive Orton, SAMPLING IN ARCHAEOLOGY
Steve Roskams, EXCAVATION
Simon Hillson, TEETH, 2ND EDN
William Andrefskey Jr., LITHICS, 2ND EDN

TEETH

Second Edition

Simon Hillson

Institute of Archaeology, University College London

CAMBRIDGE UNIVERSITY PRESS
Cambridge, New York, Melbourne, Madrid, Cape Town, Singapore, São Paulo

Cambridge University Press
The Edinburgh Building, Cambridge CB2 2RU, UK

Published in the United States of America by Cambridge University Press, New York

www.cambridge.org
Information on this title: www.cambridge.org/9780521545495

© Cambridge University Press 1986, 2005

This book is in copyright. Subject to statutory exception
and to the provisions of relevant collective licensing agreements,
no reproduction of any part may take place without
the written permission of Cambridge University Press.

First published 1986
Paperback edition 1990
Reprinted 1993, 1996
Second edition 2005

Printed in the United Kingdom at the University Press, Cambridge

A catalogue record for this book is available from the British Library

ISBN-13 978-0-521-83701-9 hardback
ISBN-10 0-521-83701-4 hardback
ISBN-13 978-0-521-54549-5 paperback
ISBN-10 0-521-54549-8 paperback

Cambridge University Press has no responsibility for the persistence or accuracy of URLs for external or third-party internet websites referred to in this book, and does not guarantee that any content on such websites is, or will remain, accurate or appropriate.

CONTENTS

List of figures   page viii
List of tables   xii
Preface   xiii

INTRODUCTION   1

1   TOOTH FORM IN MAMMALS   7
    What is included   7
    General structure   8
    Form, function and identification   13
    Subclasses Eutheria, Prototheria and Metatheria   19
    Order Marsupialia   19
    Order Insectivora   20
    Order Chiroptera   29
    Suborder Microchiroptera   29
    Suborder Megachiroptera   40
    Order Primates   42
    Order Carnivora   45
    Order Pinnipedia   63
    Order Cetacea   69
    Suborder Odontoceti   69
    Order Rodentia   73
    Order Lagomorpha   110
    Order Edentata and Order Pholidota   111
    Order Tubulidentata   117
    Order Proboscidea   117
    Order Sirenia   120
    Order Hyracoidea   122
    Order Perissodactyla   122
    Order Artiodactyla   128
    Suborder Suiformes   128
    Suborder Ruminantia (Pecora)   132
    Suborder Tragulina   143
    Suborder Tylopoda   143
    Conclusions   145

2 DENTAL TISSUES — 146
- The inorganic components of dental tissues — 146
- The organic component of dental tissues — 148
- Chemistry and physics of dental tissues in archaeology — 150
- Dental enamel — 155
- Dentine — 184
- Cement — 193
- Resorption of root and crown — 198
- Preparation techniques — 199
- Cameras and light microscopes — 201
- Scanning electron microscopy — 205
- Conclusion — 206

3 TEETH AND AGE — 207
- Growth — 207
- Tooth wear — 214
- Microwear — 219
- Age estimation from dental development, eruption and wear in different orders of mammals — 223
- Circum-annual layering in cement and dentine — 245
- Other age-related histological changes — 255
- Conclusion — 255

4 SIZE AND SHAPE — 257
- Size, shape and populations — 257
- Measurable variation — 260
- Non-metrical variation — 272
- Occlusion and malocclusion — 281
- Conclusion — 284

5 DENTAL DISEASE — 286
- Dental plaque — 286
- Dental calculus — 288
- Dental caries — 290
- Immunity and inflammation — 303
- Trauma — 314
- Anomalies of eruption, resorptions and abrasions — 315
- Cysts, odontomes and tumours — 316
- Conclusion – palaeoepidemiology and recording — 317

APPENDIX A THE GRANT DENTAL ATTRITION AGE ESTIMATION METHOD	319
References	323
Index	364

FIGURES

1.1.	Tooth and periodontium.	*page* 9
1.2.	The dental arcade, using *Talpa* as an example.	11
1.3.	Tribosphenic molar crowns.	14
1.4.	Variations on tribosphenic form.	16
1.5.	*Didelphis* permanent dentitions.	21
1.6.	Talpidae (excluding desmans), upper permanent dentitions.	22
1.7.	Talpidae (excluding desmans), lower permanent dentitions.	23
1.8.	Talpidae (desmans), permanent dentitions.	24
1.9.	*Erinaceus*, permanent dentitions.	26
1.10.	Soricidae, permanent dentitions.	27
1.11.	Soricidae, variation in unicuspids.	28
1.12.	*Elephantulus*, permanent dentitions.	29
1.13.	Vespertilionidae, upper permanent dentitions.	31
1.14.	Vespertilionidae, lower permanent dentitions.	32
1.15.	Molossidae and Rhinopomatidae, permanent dentitions.	33
1.16.	Emballonuridae and Mormoopidae, permanent dentitions.	35
1.17.	Nycteridae and Rhinolophidae, permanent dentitions.	36
1.18.	Phyllostomatidae, New World fruit bats permanent dentitions.	38
1.19.	Phyllostomatidae, nectar feeders permanent dentitions.	39
1.20.	Phyllostomatidae, *Macrotus* permanent dentitions.	40
1.21.	Desmodontidae, permanent dentitions.	41
1.22.	Natalidae, permanent dentitions.	42
1.23.	Pteropodidae, Old World fruit bat permanent dentitions.	43
1.24.	Cercopothecidae, permanent dentitions.	45
1.25.	Hominidae, human permanent and deciduous dentitions.	46
1.26.	Carnivore canines.	47
1.27.	Canidae, *Canis* permanent and deciduous dentitions.	48
1.28.	Canidae, upper permanent dentitions.	49
1.29.	Canidae, lower permanent dentitions.	50
1.30.	Viverridae, permanent dentitions.	52
1.31.	Mustelidae, small-sized (weasels, stoats, polecats and martens) permanent dentitions.	53
1.32.	Mustelidae, skunk permanent dentitions.	54
1.33.	Mustelidae, medium-sized (badgers and otter) permanent dentitions.	55

List of figures ix

1.34.	Mustelidae, large-sized (wolverine, badgers and sea otter) permanent dentitions.	56
1.35.	Hyaenidae, permanent dentitions.	58
1.36.	Felidae, *Felis* permanent and deciduous dentitions.	59
1.37.	Felidae, big cat permanent dentitions.	60
1.38.	Ursidae, upper permanent dentitions.	61
1.39.	Ursidae, lower permanent dentitions.	62
1.40.	Ursidae, *Ailuropoda* permanent dentitions.	63
1.41.	Procyonidae, permanent dentitions.	64
1.42.	Pinnipedia, permanent and deciduous dentitions	65
1.43.	Otariidae, permanent dentitions.	66
1.44.	Phocidae, permanent dentitions.	68
1.45.	Odobenus, permanent dentitions.	69
1.46.	Odontoceti, Ziphiidae, lower teeth.	70
1.47.	Odontoceti. Narwhal tusk, large, medium-sized and small whales, dolphins and porpoises.	72
1.48.	Muridae, *Rattus* upper and lower permanent dentitions.	73
1.49.	Muridae, upper permanent cheek teeth.	76
1.50.	Muridae, lower permanent cheek teeth.	77
1.51.	Cricetidae, isometric views of *Cricetus* permanent cheek teeth.	79
1.52.	Cricetidae, occlusal views of permanent cheek teeth.	80
1.53.	Hesperomyidae, isometric views of permanent cheek teeth.	81
1.54.	Hesperomyidae, occlusal views of permanent cheek teeth.	82
1.55.	Gerbillinae, upper permanent cheek teeth.	83
1.56.	Gerbillinae, lower permanent cheek teeth.	84
1.57.	Microtinae, isometric views of upper permanent cheek teeth.	86
1.58.	Large Microtinae, occlusal views of permanent cheek teeth.	87
1.59.	Medium-sized Microtinae, occlusal views of upper permanent cheek teeth.	88
1.60.	Medium-sized Microtinae, occlusal views of lower permanent cheek teeth.	89
1.61.	Small Microtinae, occlusal views of permanent cheek teeth.	90
1.62.	Zapodidae, permanent cheek teeth.	92
1.63.	Dipodidae, upper permanent cheek teeth.	93
1.64.	Dipodidae, lower permanent cheek teeth.	94
1.65.	Heteromyidae, permanent cheek teeth.	96
1.66.	Ctenodactylidae and Geomyidae, permanent cheek teeth.	97
1.67.	Spalacidae, permanent cheek teeth.	99
1.68.	*Erethizon*, permanent cheek teeth.	100
1.69.	Erethezontidae, Castoridae, Hystricidae and Aplodontidae, permanent cheek teeth.	101
1.70.	Sciuridae, small, lower crowned ground and tree squirrels.	103
1.71.	Sciuridae, large, higher crowned ground and tree squirrels.	104

1.72.	Sciuridae, flying squirrels, permanent cheek teeth.	105
1.73.	Sciuridae, high-crowned flying squirrels, permanent cheek teeth.	106
1.74.	Gliridae, upper permanent cheek teeth.	108
1.75.	Gliridae, lower permanent cheek teeth.	109
1.76.	*Hydrochaeris*, permanent cheek teeth.	110
1.77.	Lagomorpha, permanent dentitions.	112
1.78.	Lagomorpha, permanent cheek teeth.	113
1.79.	*Dasypus* and *Glyptotherium* dentitions.	114
1.80.	Ground sloths.	116
1.81.	Elephantidae, lower third molars.	119
1.82.	Sirenia, dentitions.	121
1.83.	*Procavia*, permanent dentitions.	123
1.84.	*Tapirus*, permanent dentitions.	124
1.85.	Rhinocerotidae, permanent dentitions.	125
1.86.	*Equus*, permanent and deciduous dentitions.	127
1.87.	*Sus*, permanent and deciduous dentitions.	129
1.88.	Tayassuidae, permanent dentitions.	130
1.89.	Suidae and Tayassuidae canine tusks.	131
1.90.	*Hippopotamus*, permanent dentitions.	133
1.91.	*Hippopotamus* tusks and deer upper canines.	134
1.92.	*Cervus*, permanent and deciduous dentitions.	136
1.93.	*Bos*, *Cervus* and *Camelus*, isometric views of permanent molars in various states of wear.	137
1.94.	Cervidae, occlusal views of permanent cheek teeth.	139
1.95.	Bovidae, large-sized (with *Camelus* for comparison), occlusal views of permanent cheek teeth.	141
1.96.	Bovidae, small-sized, occlusal views of permanent cheek teeth.	142
1.97.	*Camelus*, permanent and deciduous dentitions.	144
2.1.	Enamel prisms in a human first molar.	156
2.2.	Enamel patterns 1, 2 and 3.	157
2.3.	Brown striae of Retzius and prism cross striations in a human upper second incisor.	160
2.4.	Prism cross striations in a human incisor.	161
2.5.	Pattern of enamel layering in a cattle molar.	162
2.6.	Perikymata on the enamel crown surface of a horse molar.	164
2.7.	Defects of dental enamel (enamel hypoplasia) in a lower first incisor.	170
2.8.	Radial section of the tooth shown in Figure 2.7.	171
2.9.	Formation times for the crown surface in human permanent incisors and canines.	173
2.10.	Rodent incisor enamel measurements.	181
2.11.	Hystricomorph enamel in a porcupine incisor.	182
2.12.	Dentinal tubules in human premolars.	186

List of figures xi

2.13.	Andresen's lines and calcospheritic structure in dentine from a human molar.	188
2.14.	Diagenetic foci in dentine of a sheep molar.	191
2.15.	Cement layering.	197
2.16.	Orientations of section planes, using an incisor tooth as an example.	201
3.1.	Development of tooth germs.	209
3.2.	Human dental development in one year stages.	224
3.3.	A series of jaws of domestic sheep, showing the sequence of eruption and wear.	230
3.4.	Brown and Chapman (1990) wear recording scheme.	236
3.5.	Canine tooth from an elephant seal *Mirounga leonina*.	247
4.1.	Positions of mesiodistal and buccolingual crown diameters in human teeth.	261
4.2.	Human molar cusps and fissures.	276
4.3.	Pronounced cusps of Carabelli in a human deciduous upper fourth premolar and permanent first molar.	278
4.4.	Pig lower third molar variation in cusps.	280
5.1.	Calculus deposits.	289
5.2.	Caries in human teeth.	292
5.3.	Distribution of dental caries in different tooth classes for different age groups of recent rural Kenyan people.	296
5.4.	Bone loss due to periodontal disease.	306
5.5.	Periapical bone loss.	309
A.1.	Tooth wear stages of cattle teeth.	320
A.2.	Tooth wear stages of sheep/goat teeth.	321
A.3.	Tooth wear stages of pig teeth.	322

TABLES

2.1.	Commonly studied isotopes in dental tissues.	*page* 152
3.1.	Age ranges for stages of human tooth development.	226
3.2.	Formulae for calculating age from measurements of developing human permanent teeth.	228
3.3.	Development stages of lower teeth in cattle.	232
3.4.	Dental eruption and wear stages in cattle lower jaws.	233
3.5.	Gingival emergence timing in permanent teeth of pig.	234
3.6.	Dental development and wear scores in lower cheek teeth for *Cervus elaphus* and *Cervus (Dama) dama*.	235
3.7.	Dental development groups for sorting caribou/reindeer mandibles.	238
3.8.	Gingival emergence times for camel lower jaws.	239
3.9.	Eruption through bone (and first signs of wear) in the horse.	240
3.10.	Gingival emergence stages for domestic cat.	242
3.11.	Gingival emergence stages for domestic dog.	242
3.12.	Elephants and mammoths, loph numbers and attrition ages for lower cheek teeth.	244
3.13.	Lower jaw tooth wear stages in elephants.	246
5.1.	Scoring for carious lesions.	298
5.2.	Dental caries in Anglo-Saxon British dentitions.	300

PREFACE

The first edition of *Teeth* was published in 1986. This second edition, which became affectionately known as *Teeth II*, has very largely been rewritten. One of the main changes is an expansion of the taxonomic range. The first edition included 150 genera of mammals from the western Palaearctic (Europe, western Asia and North Africa). This made it possible to keep the size of the job down to manageable proportions, and also kept the book down to the intended size but, as the largest sales were in North America, this approach did not fit well with its main readership. In 1996, I published *Dental Anthropology*, also with Cambridge University Press, which duplicated a good deal of specifically human material in *Teeth*. This made it possible to give less emphasis to the human component in *Teeth II*, leaving space to include a total of 325 genera, representing the Holarctic in its entirety, including Europe, North Africa, Western, Central and North-east Asia, and North America. Humans are still included, but the level of detail is closer to that of the other mammals. The other changes in *Teeth II* are more to do with changing my mind about various issues, and updating references, rather than dramatic developments in the subject. One of the striking things about returning to the text almost 20 years later is how few of the fundamentals have in fact changed.

It would not have been possible to write this book without access to the great zoological collections of the world. I am very lucky that one of these is here in London, at the Natural History Museum, and I am very grateful to Paula Jenkins and Richard Sabin for allowing me to use this magnificent collection, and helping me during weekly visits over a period of a year or more. Another wonderful collection is at the National Museum of Natural History, part of the Smithsonian Institution in Washington DC. There, I must particularly thank Don Ortner for helping me to organise my visit, Linda Gordon and Charlie Potter for access to the Mammals collections, and Bob Purdy for access to the Vertebrate Paleontology collections. The third great collection is at the Field Museum in Chicago, and I thank Michi Schulenberg (Mammals) and Bill Simpson (Fossil Vertebrates) for their help.

The figures in this book have taken me longer than the writing. It is my first attempt to illustrate a book entirely using computer graphics. Many new drawings were needed, and all the figures from *Teeth I* were redrawn, in order to keep the style consistent. Previously, my artwork has all been pen and ink, and I am grateful to Phil Walker, of University of California, Santa Barbara, who first suggested the kind of thing that is possible with computers. I made the original drawings in the museum, with pencil and paper. For large specimens, they were based on measurements, and the axonometric projections were constructed on tracing paper over a grid. Small specimens were drawn using a drawing tube attachment and an excellent Leitz stereomicroscope. For axonometric projections, the specimen was mounted on a small plinth that tilted it into the correct position, and the focus of the microscope had continually to be adjusted whilst tracing along the tooth row. It was very tiring on the eyes – one eye follows the pencil and the other the specimen, and the brain merges the two together! At museums in the USA, photography was permitted, so some specimens were recorded either with a Pentax 35 mm SLR film camera fitted with a macro lens, or with Nikon or Sony digital cameras. It took a lot of experimentation to find a digital drawing technique that produced results like pen and ink. The crucial piece of equipment is a really good pressure-sensitive graphics tablet, in this case a Wacom. It can be a small one and, coupled with a laptop computer, this is very portable. Most of the drawings were done on a large desktop computer, but some were done by the sea on a small island in Greece, and others in a hotel room in Lima. Pencil drawings and photographic negatives were scanned so that, along with digital images, they could be used as the templates over which the final drawings could be made. These drawings were traced as a separate layer in Adobe Photoshop. After the original template layer had been deleted, these digital images were converted into vector graphics with Corel Trace. These were imported into Corel Draw, in which they could be scaled, shaded and labelled. The original drawings were many times larger than their final size, because the reduction 'tightens' them up and makes them much crisper and cleaner.

I had a lot of support from colleagues and family during the writing and drawing of *Teeth II*. In particular, I would like to thank Daniel Antoine, Louise Martin and Tony Waldron from the Institute of Archaeology in University College London for their advice and discussion. I am also grateful to Peter Ucko, Director of the Institute, for allowing sabbatical leave which helped a great deal. He has always been supportive of my research interests, as well as getting me involved in new research directions which have provided a great deal of interest and enjoyment. As always, I gratefully acknowledge my teachers, Don Brothwell, Alan Boyde and Sheila Jones. Other colleagues, part of a loosely defined 'London' group, who have always been there for dental discussions include Chris Dean, Leslie Aiello, Don Reid, Charles FitzGerald, Fred Spoor, Louise Humphrey and Chris Stringer. Most patient of all have been my family, Kate, William, James and Harriet, my father and sister, who have tolerated my eccentric interest in teeth, and have helped in many ways. In particular, my sons helped a great deal with computing, and James even allowed me to include an illustration of his cusps of Carabelli. Finally, I thank my editors at Cambridge University Press, Simon Whitmore and Tracey Sanderson.

INTRODUCTION

Teeth have the great archaeological advantage of being constructed from remarkably tough materials, which can survive a century and more in the harsh environment of the mouth. They also survive in a very wide range of archaeological sites and conditions of burial. Teeth of large animals are part of the carcass which is thrown away early in the butchery process, and so become incorporated quickly into rubbish deposits. They are readily recognised during excavation and routinely recovered in a similar way to artefacts. Often, they are amongst the most numerous finds. At large town sites in Britain, for instance, the number of identifiable bone and tooth fragments frequently exceeds the total of recognisable sherds of pottery.

The importance of recovering such material from excavations has long been recognised. In his *Primeval Antiquities of Denmark* (1849), J. J. A. Worsaae asserted firmly that all objects from archaeological sites, including animal bones, should be preserved. As archaeology developed, finds of the remains of extinct mammals alongside human bones and artefacts came to provide crucial evidence for the antiquity of man. William Pengelly's famous excavations of Brixham Cave in 1858–9 revealed a deposit containing flint tools and extinct animal bones that was sealed by a thick layer of stalagmite, also containing remains of extinct animals (Daniel, 1978).

Most teeth from mammals larger than a cat can be recognised when trowelling on an archaeological site, or quickly recovered by sieving/screening at a coarse mesh (1 cm). Small mammals – traditionally those not tall enough to be seen above long grass – may have very small teeth indeed. A microscope is required to see them properly, and isolated specimens are often missed on site. To recover the small teeth of voles, mice and similar creatures, large samples need to be taken, and sieved at a fine mesh (ideally 0.5 mm, although 1 mm catches many of them). Outside arid lands, dry sieving is difficult, and wet sieving is required. This is time-consuming and samples need to be carefully selected, to maximise return on effort.

Along with such material as shells and insect skeletons, bones and teeth now form a central part of the discipline of archaeozoology (or zooarchaeology). This development has gone hand in hand with the growth of archaeologically based biology in general and, together with work on soils and geomorphology, makes up the wider discipline of environmental archaeology. Specifically anthropological and pathological investigations of human skeletons have been carried out since the early discoveries in European caves and the Egyptian excavations of Flinders Petrie produced large collections of human remains which formed the basis of much work during the 1900s and 1920s. The origins of physical anthropology therefore lie in Europe, but it is in North America that it has expanded most, particularly through the students of Earnest Hooton who spread the subject into American universities during the first half of the twentieth century. Anthropology is a very broad discipline, but specifically dental anthropology has its origins largely in the work of Al Dahlberg in the 1930s and 1940s, with fellow dentists and anatomists. The use of the phrase 'dental anthropology' dates to a meeting in London in 1958 (Brothwell, 1963a). Zooarchaeology, or archaeozoology, developed mainly during the second half of the twentieth century from a number of academic sources, in particular vertebrate palaeontology and zoology, and out of the interests of archaeologists themselves in answering questions about diet, hunting and the origins of farming. The need for reference material for identification purposes has meant that much of its development has gone hand in hand with the building of large collections, not only at major museums, but also in university departments and government organisations.

Teeth themselves are not normally considered a distinct category of finds in archaeological work. They are, however, very different from bones in their biology. Andreas Vesalius first recognised these differences in structure and function as long ago as 1542 and it is now clear that teeth cannot be considered parts of the bony skeleton in a strict sense. Instead, they comprise the dentition, which is connected to the skeleton, but is derived at least partly from tissues akin to the skin and is exposed at the surface of the body. The anatomy, physiology and pathology of teeth are highly specialised subjects with their own long and honourable history. Teeth were included in the general anatomical and medical works of the classical authors. The *Corpus Hippocraticum*, originating in the fifth century BC, mentioned their anatomy and growth, as well as a number of dental diseases and treatments. Teeth were also described by Aristotle and by Galen of Pergamon, physician to four Roman emperors. Galen must have been one of the first to describe human bones from archaeological sites. He was forbidden by law to dissect human bodies and so turned to the study of remains in ancient tombs and monuments (Magner, 1979). Galen's ideas constituted the basis of anatomical science until the great developments of the Renaissance that took place particularly at the University of Padua. Andreas Vesalius gave the first convincing description of dental anatomy in his *De Humani Corporis Fabrica* of 1542 and in 1563 Bartolomeo Eustachi wrote the first known book on teeth, *Libellus de Dentibus*. The first microscopic studies of dental tissues were carried out by Marcello Malphighi and Anthony van Leeuwenhoek during the later seventeenth century, and van Leeuwenhoek was also the first to see micro-organisms in dental plaque (see Chapter 5). Although Pierre Fauchard's great work *Le Chirurgien Dentiste* followed in 1728, the true starting point of modern dental anatomy is usually taken to be John Hunter's *The Natural History of the Human Teeth* first published in 1771. Similarly, a major impetus for comparative dental anatomy came from Sir Richard Owen's *Odontography* of 1840, which included not only mammals, but also reptiles, birds, amphibians and fish. Many of the main features of microscopic dental anatomy were effectively described in the nineteenth century by such workers as Purkinje, Retzius, Preiswerk, Owen, von Ebner and Tomes (father and son). Sir John Tomes in particular is often regarded as the father of modern dentistry.

During the twentieth century, there were such strides in understanding of the anatomy, growth, physiology and diseases of the teeth that many avenues can now be followed in research on archaeological material. This book aims to draw together ideas and techniques from many fields, including archaeology itself, palaeontology, physical anthropology, anatomy, histology, mammalogy, dentistry, pathology and forensic science. Some of the methods described are already widely used in zooarchaeology and anthropology, but others are rarely applied. Several techniques have been developed specifically for forensic work on human remains but may well point the way to alternative approaches for other mammals, and zooarchaeological ideas also have application in anthropology. For this reason, humans are deliberately combined with non-humans. More detail on specifically human dentitions is given in Hillson (1996). Many of the methods may also have applications for zoologists, particularly those working on museum collections or in the field.

Teeth are often found on archaeological sites, not just as waste, but as art objects. Dental tissues make attractive materials – finely grained, tough and beautifully patterned – they were used to make artefacts in antiquity, and are still used. Ivory is the chief of these and, from an anatomical point of view, is just a large mass of dentine (Chapter 2). It is found in all teeth, but workable pieces come from the tusks of elephants, walrus, hippopotamus, pig or whales. These different forms can often be distinguished, either by eye or under the microscope. Much ancient ivory came from elephants, although it is not easy to distinguish the ivory of mammoth and Asian and African elephants. Ivory objects, presumably from mammoth tusks, are quite common in Upper Palaeolithic contexts in Europe. These include, for example, the ivory figurines from Dolní Věstonice in the Czech Republic and the burials of two boys at Sungir, near Moscow, in clothes richly covered with ivory beads. Ivory also occurs in later prehistoric contexts in Europe and was quite common in Roman times. The Romans imported much of their ivory through the Red Sea ports (MacGregor, 1985), and probably ultimately from North Africa or even India. After the fall of the Empire, elephant ivory artefacts became rare in Europe, but gradually reappeared in ecclesiastical and royal contexts. Elephant ivory was in good supply again by the ninth century AD, by which time it probably came mostly through Iraq and Egypt. Walrus ivory from local sources became common in northern Europe during the tenth to twelfth centuries AD, and the most celebrated examples are the Lewis chess men at the British Museum. Boar tusks were used in Mycenean Greece as the protective coating for helmets. The lower tusks were split into curved plates, incorporating dentine (in effect pig ivory) and enamel, making a white outside which shone in the sun. Holes were drilled in the corners of the plates, which were fixed, probably, to a leather framework. Famous examples can be seen in the National Archaeological Museum of Athens.

In addition to yielding useful materials, teeth are attractive objects in themselves. They have pleasing, rounded outlines and were indeed quite frequently used as amulets or in necklaces. Cave bear canines, perforated at one end, are common in the Upper Palaeolithic French cave sites, and the burials at Sungir included headdresses decorated with arctic fox canines. Red deer (elk) canines were often used in the same way, both in Europe and North America. Anglo-Saxon burials in Britain quite often include amulets made from beaver incisors (MacGregor, 1985).

As noted above, teeth form a prominent part of the large collections of mammal remains representing burials, food debris and industrial waste on archaeological sites. The large degree of variation in their shape and form makes them readily identifiable, and teeth and jaws play a major role in the identification of archaeological material. Many fragments of the bony skeleton may yield little precise information, but even a single small tooth can often be assigned to species. This makes comparative dental anatomy an important subject for archaeologists. Relatively few textbooks exist, and few of these provide sufficient detail. Whereas in modern zoology the accent is on physiology, behaviour and ecology, archaeology still badly needs detailed studies of small and obscure anatomical differences if the fragmentary material from excavations is to be identified.

The first aim of this book is therefore to act as a starting point for making identifications. It deals only with mammals, because inclusion of reptiles, amphibians and fish, most of which have teeth too, would make it far too large a book. For similar reasons, it also applies itself particularly to the Holarctic zoogeographical region, which includes mainly Europe, North Africa, Asia and North America. Chapter 1 introduces the dentitions of some 325 genera of mammals which could be found on archaeological sites from approximately the past 100 000 years. These include both common finds and rare specimens, because it is exactly the chance find of the latter that might cause the greatest problem for archaeologists. It introduces mammalian dental anatomy and summarises the distinctive features of the different families of mammals. Much of the necessary information was not available in published works and Chapter 1 is the result of many hours' work with collections in the British Museum (Natural History), the Odontological Museum of the Royal College of Surgeons of England, the Natural History Museum at the Smithsonian Institution in Washington DC and the Field Museum of Natural History in Chicago. These are some of the largest collections of mammal teeth in the world today and, even so, a few mammal genera are today so rare that less than ten specimens were available. Chapter 1 is largely illustrated by axonometric drawings which were specially designed to clarify the three-dimensional arrangement of the diagnostic patterns of cusps, ridges and fissures on tooth crowns. Where specimens are too damaged or, as in isolated rodent incisors, not very diagnostic, it may be possible to use the microscopic structure of dental tissues as a guide. Dental enamel has particular potential for this, and Chapter 2 outlines methods and possibilities. Teeth do not even have to survive on a site to be of interest. Gnawing marks are found on bone, wood and nutshells, and it is often possible to make an identification of the animal responsible (Bang & Dahlstrom, 1972).

Identification is the first task, and it is necessary to identify not only the species present, but also the type of tooth, its position in the dental arcade and whether it is a permanent or milk tooth. The next step is to establish the relative abundance of different species. This is important in the reconstruction of such factors as demography, diet, husbandry and hunting practice. Estimates of abundance are refined by matching up bones and teeth from the same individual, and teeth have a potential advantage for this. Within one species, teeth vary widely in size and shape, and are modified over time by wear and disease. It is frequently possible to match up teeth and jaws by looking for similar patterns, shapes and sizes. If a very reliable match is needed, it may be possible to match the microscopic layered structure of the enamel and dentine (Chapter 2), which is characteristic of that individual.

Once the relative abundance of different species has been established, it is necessary to establish the age at which each individual died, or the season of the year. This is often difficult and teeth are again important because they provide the best methods for age determination. In young individuals, the formation of teeth and their eruption through bone and gums occurs in a regular sequence. Comparison of the state of development between individuals gives an estimate of relative maturity. Comparison with living animals may give some idea of actual age. Chapter 3 therefore outlines the biology of dental development in mammals and shows how it can be used to estimate age. It also deals with the uncertainties and problems involved. In mature individuals, the only guide to age is often the state of wear on the teeth. With continuous use, the teeth gradually wear down and the extent of this yields an estimate of relative age. Chapter 3 outlines various schemes for recording wear and discusses the difficulties in their use. Still other methods are based on the internal microstructure of teeth. These include counting incremental structures formed in enamel during dental development, and the counting of layers formed during adult life in the cement that coats the tooth roots. Such methods are only beginning to be employed in archaeology. The theory, practice and problems are covered in Chapters 2 and 3.

It is also necessary to determine as far as possible the sex of individuals in an archaeological collection of mammal material. A number of bones in the skeleton may show strong enough differences between males and females to be of value for sex determination. Teeth also are often dimorphic, sometimes to the extent of being present in one sex but not the other. In other instances, the size of the teeth can be used in some animals, even in humans, where the difference is small. One particular advantage of teeth is that they are formed full adult size from the start, so they can be measured in young individuals and compared directly with adults. Bones, by contrast, need to reach adult proportions before they can be used to estimate sex. Morphological variation of this kind is included in Chapter 4, which also deals with variation of other kinds within species. Teeth of one species often show a wide range of variations, not just in size, but also in the details of their form. People who see a large collection of teeth for the first time are often surprised by the diversity. Different populations may to some extent be definable by their pattern of dental morphology. This allows, for example, discussion of origins and relationships of ancient mammal populations, evolution of different morphologies, and the effects of domestication.

Dental diseases, injuries and anomalies (Chapter 5) are amongst the most common pathological conditions seen in archaeological remains. Teeth are used to gather and process food, and touch every particle of it that enters the body. They are strongly influenced, not only by the constituents of the food, but also by the behaviour of the animal. One important factor in dental disease is the wear of the teeth, and related changes in the form of the jaws through life. These are discussed in Chapter 4. Dental conditions are strongly age-related, so it is important to consider this aspect, but they can yield much information on the nature of the diet. Archaeological specimens may provide a pattern of diseases, both for humans and non-humans, for which it is hard to find modern analogues.

Growth and development are potentially of great interest, because the rate and pattern is affected by diet, health and the general environment in which an animal grows up. Generally, it is difficult to study growth in archaeology, because of the lack of an independent technique of age determination – age estimation in juveniles is based on growth-related changes, so it is easy to enter a circular argument. Teeth start at an advantage, because they appear to be less affected than the skeleton by environmental factors and, in any case, growth disruptions can be recognised as defects. Also unlike the bones of the skeleton, they are not continually replaced during life, and the dental tissues preserve a detailed record of growth as a series of layers, some of which record daily growth. This built-in clock can be used to calibrate a detailed record of growth during childhood, and its disturbances. In turn, such a sequence can potentially be used to calibrate bone growth in the skeleton of the same individuals. Chapter 2 describes the complex biology of enamel and dentine growth and the various techniques that can be applied to archaeological material to produce such detailed sequences.

Teeth are complex structures and, although this complexity requires a lot of work to learn, it is also what makes the dentition so information dense. As a class of archaeological finds, they usually repay the amount of effort put into study. At sites where bones are not well preserved and the artefacts are unexceptional, the resilient teeth may yield the most interesting results. Most people would not regard them as beautiful from an aesthetic point of view, but they take on a staggering array of different forms, many of them very stylishly sculptured. Who could resist, for example, the elegant upper molars of a rhinoceros, or the fine lines of microchiropteran bat teeth. Not to mention the astonishing intricacy of the complex-toothed squirrel *Trogopterus*, which probably has the most complicated teeth in the mammal world, and the computer-chip-like detail of *Napaeozapus*, which is difficult to believe when seen for the first time under the microscope.

1

TOOTH FORM IN MAMMALS

What is included
This chapter aims to introduce a wide range of tooth forms, to act as a starting point for identification of archaeological remains, items in museum collections, or zoological specimens. It includes only the teeth of mammals and confines itself to the Holarctic zoogeographical region, and the neighbouring oceans. The idea of dividing the main land masses of the world into broad faunal zones including a range of vertebrates and invertebrates is most strongly associated with Alfred Russel Wallace (1876). The Holarctic region runs in a band around the north of the globe, including Europe, North Africa, and the non-tropical parts of Asia, North and Central America. Various definitions are used and, in this book, the southern boundary follows the line given by Corbet (1978). In Africa, this includes only those countries with a North African coastline. In Asia, it includes the whole of Arabia, Iran and most of Afghanistan, the Himalayas and the high Tibetan plateau, lowland China north of the Hwang Ho river and all the Japanese islands except for the southernmost in the Ryuku archipelago. Together, the European, African and Asian part of the Holarctic comprise the Palaearctic region. The Nearctic part of the Holarctic includes the whole of Canada and the United States of America, together with the northern desert region of Mexico (Hall, 1981). In all, there are 59 families of mammals with members living in the Holarctic land area. Marine mammals are not included in the definitions of zoogeographic zones, but this book also covers nine families of sea mammals whose members approach the Holarctic shores near enough to be included in archaeological sites. In this chapter, geographical range within the Holarctic is denoted by codes:

EU	Europe
AF	North Africa
AS	Asia all areas
ASW	Mainland western Asia
ASC	Central Asia
ASE	Mainland north-eastern Asia
JA	Japan
AM	America all areas
AMN	North America
AMC	Confined to the very south of North America and northern Mexico
ARC	Arctic Ocean and neighbouring seas
ATL	Northern Atlantic and neighbouring seas

PAC   Northern Pacific and neighbouring seas
IND   Northern Indian Ocean and neighbouring seas
MED   Confined mostly to the Mediterranean and Black Seas.

Where the range applies to fossils, and the animal is today extinct in a region, the code is enclosed in parentheses [ ].

Why confine this book to the Holarctic? One of the main roles of teeth in archaeology is identification, particularly in small mammals where the bones are difficult to identify. To be useful, the range of animals included must be comprehensive, but a book this size cannot cover the whole world fully so it must concentrate on one region. Many Holarctic mammal families have wide ranges, with strong general trends in tooth form, and they therefore make it a practical proposition to cover comprehensively. Coverage is also confined to the Late Quaternary period, including the present interglacial and previous cold stage (glaciation), or roughly the last 120 000 years. This includes the majority of archaeological sites.

For comparative dental anatomy, the most useful level of taxonomy is the *genus*. In the majority of mammal families, genera can quite clearly be distinguished from one another by their teeth, but it is harder to define unambiguous identifying features for species, particularly as there is often variation at population or subspecies level. Readers are most likely to find the chapter useful as a starting point, to narrow down the possibilities for identification. Some genera are distinctive enough to be recognised straight away without difficulty, particularly if their geographical range is taken into account. Most, however, require consultation of a reference collection to be certain. The largest general collections at the time of writing are at the Natural History Museum in London, the American Museum of Natural History in Washington DC and the Field Museum in Chicago, but there are other important collections as well.

General structure

Teeth consist of two main elements; a *crown* and a *root* (or roots). The crown generally protrudes into the mouth and the roots are firmly held in the bony sockets of the jaws, but many variations on this theme are possible. Teeth with tall crowns may only show their tips, with more being exposed as it gradually wears away. Other teeth may not have a crown, so the root is the part that protrudes. What really defines the crown is its coating by a layer of hard, shiny *enamel*. The root and, in many animals, the enamel of the crown too, is coated with a layer of bone-like tissue called *cement*. Underlying these surface layers and forming the main structure of the tooth is a very tough and resilient tissue called *dentine*, familiar as the precious material ivory. Deep inside the tooth is the *pulp chamber*. In teeth with roots narrower than the crown, an extension from this runs, like a canal, down the centre of the root to emerge in a tiny hole, the apical foramen, at its tip. Within the pulp chamber in a living tooth is the *pulp* – soft tissue which includes the cells of dentine, the blood and nervous supply. Details of enamel, dentine and cement structure are given in Chapter 2.

Tooth form in mammals 9

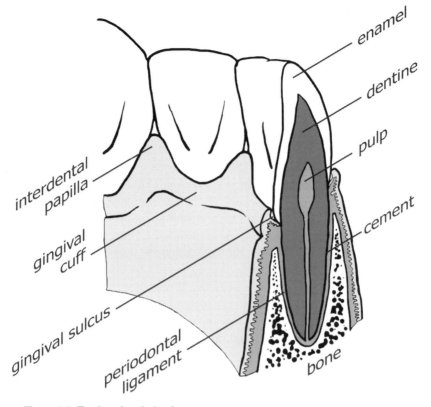

Figure 1.1 Tooth and periodontium.

All of these elements are present in even the most complicated tooth. The crown may be wider, taller or flatter. It may include extra mounds, known as *cusps*, or folds called *lophs*. Cement may cover the crown as well as the root, and there may be extra roots. The roots may become much enlarged, so that the whole tooth is composed of dentine and cement, as in the elephant's tusk. But even here, a tiny conical crown is present before the tusk becomes worn. Because of the basic uniformity in design, it is possible to apply the same terms to many types of tooth. The *coronal* end is towards the highest point of the crown. In the same way, the *apex* of the tooth is the extremity of the root and *apical* means towards this end. The *cervix* or *cervical* region is the point where crown and root meet.

Jaws, teeth and dentitions
Although in archaeology teeth are frequently found as isolated specimens, they form, along with the jaws and associated soft tissues, a complex living structure (Figure 1.1). Bone sockets (*alveolae*) are formed around the roots, and are lined with a thin layer of *alveolar bone*. Teeth are held into the sockets by a complex of fibres called the *periodontal ligament*, embedded at one end in the alveolar bone and, at

the other, in the cement coating the tooth surface. A small amount of movement is possible at this joint and is important in the function of the teeth. Some animals have exceptionally mobile periodontal ligaments for some of their teeth. One example is the long, protruding canine of some small deer (p. 134), which visibly wobbles from side to side as the animal chews.

Tooth sockets are enclosed in heavily built bony jaw structures. This robustness is useful for archaeologists, because jaws and teeth are often preferentially preserved. The lower jaw in mammals is formed by paired bones called the *mandibles*. Many genera have a joint at the front, the *mandibular symphysis*, so that some flexibility is possible, or the bones may be fused together as one unit. The upper jaw is composed of the *maxillae* and the *premaxillae*. These are the bones that make most of the face, snout, jaws and palate. The extension in which the tooth sockets are contained is called the *alveolar process* in both jaws. Attached to the surface of the bony jaw structures is a complex of muscles. Some control the lips and tongue. Others are attached to the rest of the skull and drive the chewing mechanism. Inside the mouth cavity, these soft tissues are covered by oral epithelium which has a tough, keratinised surface not unlike skin. It is gathered up around the bases of the teeth as they protrude into the mouth as gums, or *gingivae*. At this point, there is a small space, the *gingival sulcus*, running around the tooth. Between closely packed teeth, the gingivae form a small hump, the *interdental papilla*.

Different types of teeth are arranged into rows and the complete set is called a *dentition* (Figure 1.2). In plan, the tooth rows form an arch or loop in each jaw called the dental arcade. Each tooth series is symmetrical on either side of the arcade, so that a tooth from the left side is almost a mirror image of the equivalent tooth on the right. As with the rest of the body, therefore, it is possible to divide the dentition into matching left and right halves, separated by an imaginary surface called the *median sagittal plane*. This gives rise to a useful terminology. For each tooth in the dental arcade, the side which faces along the arcade towards its origin at the median sagittal plane is called *mesial*. The side which faces the opposite way is similarly called *distal*. The side which faces inside the arcade, towards the tongue, is called *lingual* in this book and the side which faces outside, towards the cheeks and lips is called *buccal* here. Other names are frequently used, so that the lingual surface may also be called *palatal* for upper teeth and, properly speaking, buccal describes the side facing the cheeks (Latin *buccae*) so that dentists often use *labial* for the side of teeth facing the lips (Latin *labia*). This is a problem for a book on comparative dental anatomy because the distinction is fairly meaningless for many mammals. One possibility is to use another term, such as *vestibular* (the vestibule is the space between teeth, cheeks and lips), or *facial*. Many people, however, just use the word buccal in a general sense, and that is what is done here. The biting or chewing surface of the teeth, which faces teeth in the other jaw, is called *occlusal*. In sharp teeth, there is not really an occlusal surface as such, so it is usual to talk about its occlusal 'edge' or 'point' instead or, in the case of incisors, its *incisal* edge.

Tooth form in mammals

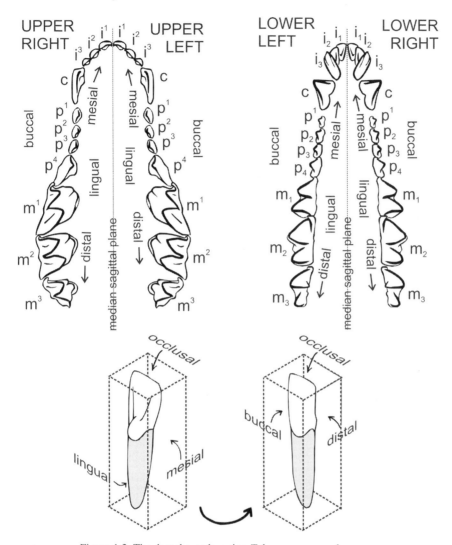

Figure 1.2 The dental arcade, using *Talpa* as an example.

The teeth in one jaw can be divided into types. At the simplest level, it is often possible to split the dentition into anterior teeth and cheek teeth. *Anterior* means at the front of the jaw, in the direction that the animal travels. *Cheek teeth* run along the sides of the jaw. It is also usually possible to divide teeth into four classes: *incisors*, *canines*, *premolars* and *molars*. Incisors and canines are grouped together as anterior teeth; premolars and molars as cheek teeth. In addition, almost all mammals have two dentitions: a *deciduous* or milk dentition when young which is replaced by a *permanent* dentition when mature. Deciduous teeth show similarities to the permanent teeth of the same genus, but tend to have lower crowns with a marked waisting at the cervix and widely spreading roots. These accommodate the crowns of the permanent teeth growing inside the jaws underneath and, as these continue to develop, the deciduous roots are resorbed. When they are lost (*exfoliated*), little of their roots remains.

The deciduous and permanent dentitions can be divided into quadrants, either side of the median sagittal plane: upper left, upper right, lower left and lower right. It needs to be remembered that left and right refer to the sides of the animal being examined, not the observer. One half of the dentition can be summarised simply as a *dental formula*. The left half of the generalised permanent dentition for eutherian (placental) mammals is shown in this book as follows:

$$i\frac{3}{3}, c\frac{1}{1}, p\frac{4}{4}, m\frac{3}{3}$$

where i denotes permanent incisors, c means canines, p premolars and m molars. In each quadrant of the generalised mammal dentition, there are three incisors, one canine, four premolars and three molars which are numbered from the most mesial tooth in each series as follows:

$$i^1, i^2, i^3, c, p^1, p^2, p^3, p^4, m^1, m^2, m^3$$

A useful shorthand device is to employ superscript numbers, as above, to denote upper teeth and subscript numbers to denote lower teeth:

$$i_1, i_2, i_3, c, p_1, p_2, p_3, p_4, m_1, m_2, m_3$$

It is important, when writing these by hand, to make the superscripts and subscripts really clear. The numbering of the generalised, unreduced dentition is retained in those genera for which evolution has reduced the numbers of teeth in different classes. In the premolar class, the reduction is usually from the mesial end of the tooth row so, where there are only two premolars in each quadrant, these may be referred to as third and fourth premolars.

The generalised eutherian deciduous dentition does not have any molars, so its dental formula is as follows:

$$di\frac{3}{3}, dc\frac{1}{1}, dp\frac{4}{4}$$

where di denotes deciduous incisors, dc deciduous canines and dp deciduous premolars. The different tooth classes are numbered in the same way as for permanent teeth. Not all mammals have a deciduous dentition. Many rodents, for example, do not have a functional one although some deciduous teeth do erupt. Seals and sea lions shed at least some of their deciduous teeth before birth, into the amniotic fluid (Kubota & Matsumoto, 1963). The situation in insectivores is complex and it is often difficult to decide which tooth belongs to which dentition. In order to reduce space, this chapter deals in detail only with permanent dentitions, except for the deciduous dentitions of people and the main domestic animals – cats, dogs, cattle/sheep/goat, camels, pigs and horses. These do, however, give an idea of how deciduous dentitions look in their wild relatives.

Chewing

Dentitions fulfil a number of specialised purposes. These might include grooming of fur, social display, offence or defence, but primarily, teeth are for feeding with. They might be used for rooting up vegetation, or stirring up the sea-bottom sediments, peeling off layers of bark or killing and holding onto prey. Other functions might be crushing, slicing or grinding of food. Most of these functions involve movement of the lower jaw against the upper. This is accomplished by the jaw muscles, but is also to do with the hinge points, the *temporomandibular joints* (TMJ for short) situated in front of the ear region. The complexity of these joints varies. For those animals that cut up their food by shearing one sharp-ridged tooth against another, jaw movement needs only to be up and down. Here the TMJ is limited in movement. The same is true of animals that merely crush food between sharp sets of cusps. So the carnivores and insectivores tend to have a simplified TMJ, with strong bony ridges limiting its movement. Indeed, in the badger *Meles* it is often not possible to separate the lower jaw from the skull. Other animals make grinding movements with the jaws. In humans, this is a complex process of combined up and down, side to side and back and forth motions. The human TMJ is correspondingly complex. Most herbivores specialise in side to side (e.g. ungulates) or back and forth (e.g. rodents) lower-jaw movements. An appropriate range of movement is promoted by the bony and ligamentous structures of the TMJ.

The muscles of the jaws also reflect specialisation. Two main muscles are involved in chewing. The *temporalis* arises from a lever-like protrusion of bone on the mandible, just in front of the TMJ. At its other end, the temporalis spreads onto the side of the cranium, or brain box. The *masseter* arises from the angle of the jaw and is attached to the side of the face. In specialised carnivores, a large temporalis is attached to a long bony lever and exerts a powerful leverage on the mandible. In ungulates, a large masseter is attached to a broadened area on the angle of the mandible, to swing the jaw from side to side in chewing. The muscle is particularly large in non-ruminant ungulates such as the horse, where tough plant food is chewed only once, and relatively smaller in ruminants that spend more time chewing the partially digested cud (below). Rodents use their masseter muscles to pull the jaw forward and their temporalis muscles to pull it back again. So, although this book considers teeth alone, it is essential to remember that they are part of a functional dentition in a living animal.

Form, function and identification

Teeth are highly variable structures. They are very closely adapted to the jobs that they have to do. For this reason, they are particularly useful for identification of mammal remains from archaeological sites. It is often possible to identify a single tooth.

Names of living genera are given according to Corbet (1978; Corbet & Hill, 1990). In several instances, this results in widely recognised genera being lumped into others. Examples are fallow deer *Dama*, which is incorporated into the genus *Cervus*, and polecat *Putorius*, which becomes part of *Mustela*. In terms of dental

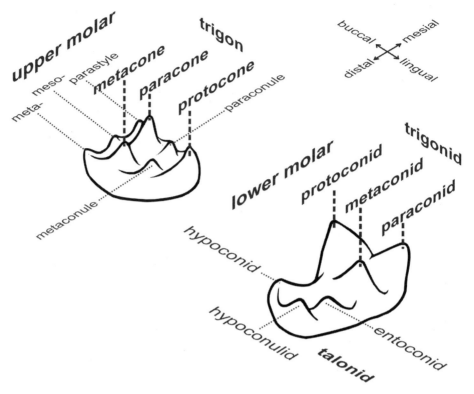

Figure 1.3 Tribosphenic molar crowns.

identification, this in fact makes sound sense, irrespective of any taxonomic merits. The grouping of orders and families below is purely a matter of convenience and is not meant to imply phylogeny. Names for fossil genera follow Kurtén & Anderson (1980), Stuart (1982) and Kurtén (1968). For fossils, there is considerable variation in the names used in different publications and museum collections, and care is needed.

Henry Fairfield Osborn and the tritubercular theory
To talk about the highly complex anatomy of teeth, it is necessary to have a vocabulary. The most widely used terms for comparative anatomy were originated by a palaeontologist, Henry Fairfield Osborn. The molars of many early mammals from the lower Tertiary have a characteristic form, with major cusps arranged into triangular elements (Figure 1.3 and see Romer, 1966). Upper molars in these fossils have a single triangle of this kind, with one major cusp on the lingual side and two on the buccal. The lower molars have a similar main triangle, but with one cusp on the buccal and the other two on the lingual side. In addition, there is a lower 'heel' attached to the distal side of the main lower molar triangle. This also has three major cusps.

Osborn's (1907) essay 'Evolution of mammalian molar teeth' has formed the basis of subsequent work on classification. He suggested that mammalian molar evolution could all be explained as elaborations of the basic triangular form. Mammals evolved from the reptiles during the Mesozoic era. The reptilian tooth (Hershkovitz, 1971) has a single main cusp, elongated mesiodistally, with small accessory cusps at each end. All three cusps are joined by crests. Osborn suggested that the single lingual cusp of the upper mammal molar was derived from the main cusp of the reptilian tooth. So he named that cusp in the mammal upper molar the *protocone* (Figure 1.3). Similarly, the single buccal cusp of the main lower molar triangle was named the *protoconid*. The other two cusps in the primitive upper molar he named *metacone* (the more distally positioned one) and *paracone* (the more mesial). This triangle of upper molar cusps is usually called the *trigon*. In the same way, Osborn named the two lingual cusps of the main lower molar triangle *metaconid* (more distal) and *paraconid* (more mesial). This more prominent triangle of the lower molars he called the *trigonid* (often shortened to trigon) and the low heel he named the *talonid*. The three cusps of the talonid were called *hypoconid, hypoconulid* and *entoconid*. Additional cusps are often present in the upper molars – the *protoconule* (most palaeontologists would prefer to call it 'paraconule') and *metaconule*. Around the cervix of the primitive molar crown is usually a prominent bulge or shelf called the *cingulum*. In some groups, additional cusps arise from the cingulum on the buccal side of upper molars. These are called '-styles'. The *metastyle* is distal to the metacone, the *parastyle* to mesial of the paracone and the *mesostyle* lies in between. A fourth cusp is present on the distolingual corner of many upper molars (Figure 1.4), the *hypocone*. This basic form of the mammal molar is termed *tribosphenic*.

It is now known that evolution followed a different path from that believed by Osborn (Butler, 1978), but the names he invented have stuck. Various nomenclatures more related to modern ideas of dental phylogeny have been developed, but the original 'Osbornian' names are by now universal. The fact is that they can be used to label cusps most effectively, and everyone knows what they mean.

Cusps may be joined by a variety of crests or ridges. In some groups, the cusps have fused together into folds '-lophs'. A major fold joining paracone, metacone and -styles is called the *ectoloph*. A similar fold joining the protocone and paracone is the *protoloph* (also called paraloph). Another, joining metacone and hypocone, is the *metaloph*. These are especially prominent in the rhinoceros (p. 126). In the lower molars the *metalophid* (protolophid) joins metaconid and protoconid, and the *hypolophid* joins entoconid with hypoconid. Elaborations of these schemes can describe any form of molar.

Where the crown is low it is termed *brachydont*. Herbivorous mammals often have higher, *hypsodont*, crowns. Where cusps are rounded in form, the tooth is called *bunodont*. Cusps and ridges arranged into sharp cutting edges are called *secodont*. Where the cusps have coalesced into folds, molars are described as *lophodont* or *selenodont* (Figure 1.4). The term lophodont is normally used where the long axis

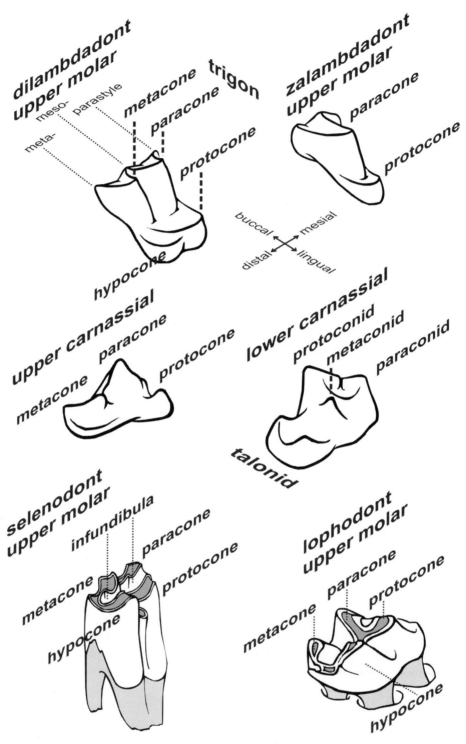

Figure 1.4 Variations on tribosphenic form. Top and bottom; dilambdadont, zalambdadont, selenodont and lophodont molars. Middle: carnivore upper carnassial (fourth premolar) and lower carnassial (first molar).

of the folds is mainly buccolingual; selenodont for crescentic folds with long axis running mesiodistally.

Other teeth can be described by similar terms. Hershkovitz (1971) suggested that all teeth could be related to the basic reptilian form. Teeth could be described as 'caninised' when the one main cusp and one, both, or neither of the small accessory cusps was developed. Canines may remain similar to the basic reptilian form, or may have only the main cusp developed. Incisors might also remain in the basic form, or have the accessory cusps and connecting ridges developed to give a spatulate form. 'Molarisation' was the word Hershkovitz used to describe addition of cusps to the reptilian pattern. The degree of molarisation increases away from the canine. Cusps of Osborn's tribosphenic arrangement, or variants of it, can therefore be labelled with increasing definition along the premolar/molar series.

Teeth with shearing edges
Most (but not all) insectivorous mammals retain the main elements of the tribosphenic pattern, as well as the primitive eutherian number of teeth. Many of the primitive mammals were probably insectivores themselves (Osborn, 1981), so similarities in form might be expected. Jaw movement is mainly up and down. Parastyle, paracone, mesostyle, metacone and metastyle are joined by a high, sharp W-shaped ridge (Figure 1.4). This type of upper molar is called *dilambdadont*. The cusps of the lower molar trigonid and talonid are joined by similar ridges, arranged as two 'V's that fit inside the upper molar ridges and shear against them. To lingual of the ridge in the upper molars, the protocone and hypocone form lower, crescentic crests. Molars of this type are found in many of the Insectivora and Microchiroptera. Another upper molar condition, *zalambdadont*, is with a single 'V' ridge joining paracone, protocone and metacone. It is found in insectivores such as tenrecs, solenodons and golden moles (below).

In the carnivores, some cheek teeth are modified into blades (Figure 1.4, middle row). In upper teeth, this is done by great enlargement of the metacone/metastyle crest and pronounced reduction of the protocone. In lower teeth, the ridges of protoconid and paraconid are similarly enlarged. The incisors of primates and some ruminants are spatulate, with a high cutting edge. This is formed by enlargement and coalescence of several cusps. The tips of the cusps can be seen as low mounds, *mamelons*, on the unworn tooth.

Bunodont molars
An alternative to cutting up food by shearing of sharp crests is crushing or grinding between sets of bunodont molars. Cusps are low, rounded, separated by grooves and fissures, and distributed over broad occlusal surfaces. Jaw movement may involve a simple up and down crushing action, or a more complex one involving sideways motion, so that cusps slide along grooves between cusps in the opposite jaw, with a grinding effect. Omnivorous animals such as Primates, bears and the suiform artiodactyls all have bunodont molars. Generalised carnivores within the Mustelidae,

Viverridae and Canidae have some bunodont teeth, or bunodont elements of teeth, used for crushing bone and other hard tissues.

Teeth with infoldings and infundibula

In this chapter, *infolding* is used to describe folds in the side of prismatic crowns. These folds extend the whole height (or most of it) of the crown and somewhat resemble pleats in a skirt. When the cusp tips have worn away, the pattern of infoldings is visible in section, exposed in the occlusal surface. Teeth whose main structure involves infolding of this kind include: the cheek teeth of the porcupine and beaver, the multiple prismatic molars of the voles and lemmings and the molars of the Lagomorpha (rabbits, hares and pikas). The long, ever-growing incisors of rodents and lagomorphs are frequently grooved by a similar infolding down their whole length. In larger animals, infoldings from the crown sides are major features of cheek teeth in the rhinoceros, ruminants and horse.

There may also be deep depressions in the crown, down from the occlusal surface (Figure 1.4, bottom left). Each is called an *infundibulum* (Latin, funnel). Infundibula may be isolated, or may communicate for part of their length with infoldings from the sides of the crown. When the cusps have worn away, they are represented by bands of dentine exposed in the occlusal surface of the tooth, lined inside and out with layers of enamel. The simplest example of this is the single infundibulum of the horse incisor. Other teeth with a single prominent infundibulum include the upper premolars of ruminants. Paired infundibula are found in the characteristic ruminant molars. These teeth are rather like two upper premolar elements, joined side by side. The infundibula separate crescentic bands of dentine and enamel to produce the selenodont form. This arrangement is found in the camels, deer, all bovids and the horse. Selenodont crowns are usually further complicated by infoldings from the side, producing prominent ridges and buttresses. These often communicate with the infundibula near the unworn occlusal surface. Gradually the infundibula become isolated as wear progresses and the tooth goes through various changes in the pattern of exposed enamel layers.

Elephant cheek teeth have infoldings of a different kind. These come in from either side, to meet in the middle, splitting the tooth into multiple lophodont folds (p. 118). The folds extend right across the breadth of the crown, forming regular plates or laminae, which are stacked one against another.

The function of all these complex folds is to present, at the occlusal plane, a series of enamel ridges. Teeth of this kind only function properly when at least partly worn, so that the enamel ridges are exposed. Opposing sets of ridges cut against one another, like a milling machine, as the lower jaw oscillates from side to side (selenodont) or back and forth (lophodont).

Intermediate forms in the rodents

A graded series of cheek teeth is found in members of the Muridae and Hesperomyinae (rats and mice), Cricetinae (hamsters), Gerbillinae (gerbils), Zapodidae (birch mice), Heteromyidae (pocket and kangaroo mice) and Dipodidae (jerboas). Some species in each of these groups have teeth which could be called bunodont, with separate cusps, albeit connected by high ridges. Other species would be more correctly called lophodont, with cusps fused together to form much taller folds. Yet other species are intermediate between the two. Although separate from this series, the multiple prisms of the microtine molars can be seen as the ultimate expression of a trend towards higher crowns.

Another form which is difficult to classify is found in the Sciuridae (squirrels) and Gliridae (dormice). Here, cusps are coalesced into ridges, but the crown is brachydont and the ridges are so low that they can scarcely be classified as lophs in the same sense as some of the Gerbillinae.

Subclasses Eutheria, Prototheria and Metatheria

Almost all of the mammals whose teeth are described in this chapter are eutherians – the placental mammals that have worldwide distribution. The egg-laying Prototheria or monotremes are now just represented by the duckbilled platypus and echidna. They are confined to Australasia and so are excluded. In addition, none of them has teeth in adult life, although the platypus *Ornithorhynchus* has very small teeth which are replaced by horn-like plates. Metatheria, or marsupials, are confined to Australasia and the Americas and only one genus is found in the Holarctic region.

Order Marsupialia

Marsupials are divided into two groups on the grounds of their anterior teeth. Polyprotodonts have four to five equal-sized peg-like upper incisors. Diprotodonts have one or three incisors, the first being larger than the others, with reduced or absent canines. Other dental peculiarities of the marsupials in general are possession of tubular enamel (but see p. 158) and five to seven postcanine (i.e. to distal of the canine – denoted pc in this chapter) teeth, which are difficult to divide into premolars and molars. Many marsupials retain primitive tribosphenic molariform cheek teeth.

Polyprotodont marsupials include:

- the extinct Tasmanian wolf (Thylaciniidae)
- marsupial 'cats' and the Tasmanian 'devil' (Dasyuridae), with their large canines and carnassial-like postcanines
- marsupial 'mice' (Dasyuridae), which are mostly insectivorous, with low cutting ridges on their cheek teeth
- the numbat (Myrmecobiidae), an anteater with teeth confined to the very front of the long jaws, and numbering up to 52, the record tooth count for a land mammal
- bandicoots (Peramelidae)
- marsupial 'moles' (Notoryctidae) with somewhat reduced dentitions
- opossums (Didelphidae).

Diprotodonts include:

- the 'shrew' or 'rat' opossums (Coenolastidae) with large first lower incisors
- possums, caucuses and koala (Phalangeridae) which have large and persistently growing lower first incisors, separated by a diastema from the rest of the dentition
- the wombat (Phascalomidae) with very rodent-like dentition
- the kangaroos and wallabies (Macropodidae) which are grazers, having a continuously erupting series of cheek teeth, rather like the arrangement in the elephants.

Family Didelphidae
Opossums: *Didelphis*[AM]

$$i\frac{5}{4}, c\frac{1}{1}, pc\frac{7}{7}$$

The Virginia opossum is the sole marsupial Holarctic resident, a nocturnal omnivore that does well alongside human habitations. It makes dens in caves, hollow trees and buildings. As with all other marsupials, there is only one deciduous tooth per jaw quadrant (Figure 1.5) and this is replaced by a tooth denoted the permanent third premolar (third postcanine tooth). As the other teeth are not replaced, it is not clear whether or not they represent deciduous or permanent dentitions in the eutherian mammal sense (above). They are usually labelled as first to third premolars and first to fourth molars. In *Didelphis* these molar-form teeth are similar to the primitive tribosphenic pattern (above), with the cusps raised into high ridges. The premolar-form teeth are backward-hooking points and the incisors are small and spatulate.

Order Insectivora
Not all insectivores actually eat insects, although most feed on a range of small invertebrates and their incisors act as forceps for catching their prey. The molars in moles (Talpidae) and shrews (Soricidae) have a classic dilambdadont form of cutting edges (above). This is expanded into hypsodonty in elephant shrews (Macroscelididae), and reduced to isolated cusps in hedgehogs (Erinaceidae). The three remaining insectivore families, not covered in this book, have the simpler zalambdadont molar form: tenrecs and otter shrews (Tenrecidae) from Madagascar and West Africa, solenodons (Solendontidae) from the West Indies and golden moles (Chrysochloridae) from southern Africa.

Family Talpidae
Moles: *Talpa*[EU,AS], *Parascalops*[AM], *Scapanulus*[ASE], *Condylura*[AM], *Scapanus*[AM], *Scalopus*[AM], *Scaptonyx*[ASE]; desmans: *Galemys*[EU], *Desmana*[ASW]; shrew moles: *Neurotrichus*[AM], *Urotrichus*[JA]

$$i\frac{2-3}{1-3}, c\frac{1}{0-1}, p\frac{2-4}{2-4}, m\frac{3}{3}$$

Most of the moles are specialist underground burrowers, including *Talpa, Scapanus, Scalopus, Scaptonyx* and *Parascalops*. The star-nosed mole *Condylura*, with

Tooth form in mammals

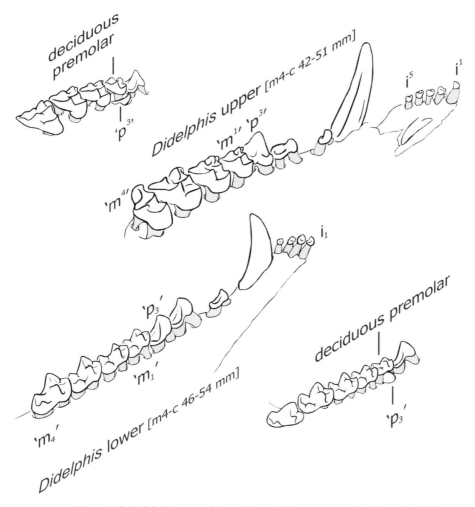

Figure 1.5 *Didelphis* upper right and lower left permanent dentitions, including 'deciduous premolar' at the point of being replaced by 'permanent premolar' (see text).

its unusual fleshy nose appendages, tunnels in damp soil, swims and includes aquatic creatures in its diet. American (*Neurotrichus*) and Japanese shrew moles (*Urotrichus*) have somewhat shrew-like bodies and shorter jaws, with forefeet that are not so strongly broadened into digging claws as in most moles, and are more active on the surface. In all genera, the moles have relatively heavy jaws and larger, stouter dentition than shrews (Soricidae). The incisors, canines (except in *Talpa*) and most of the premolar row are reduced to points or spatulate forms. In complete dentitions, genera may be distinguished by the teeth present or absent in the incisor, canine and premolar rows (Figures 1.6–1.7). In upper molars, the paracone and metacone form a high 'W' ridge and the protocone makes a lower lingual 'shelf'. The hypocone is absent, but a metaconule and paraconule are variably present, as bulges on the mesial/distal edges of the protocone or as fully developed cusps.

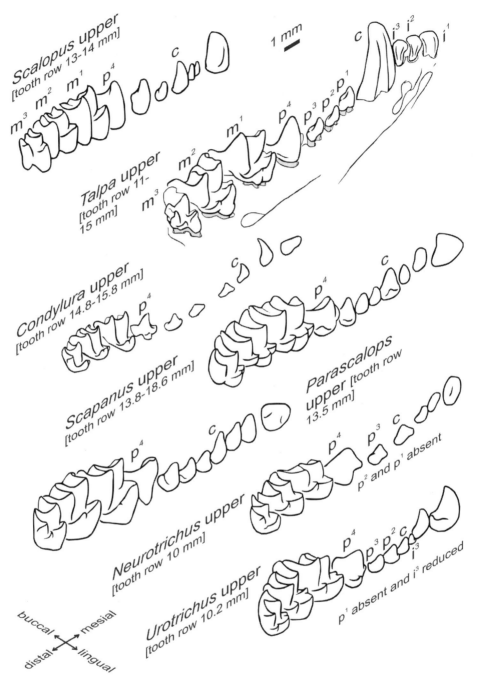

Figure 1.6 Talpidae (excluding desmans), upper right permanent dentitions.

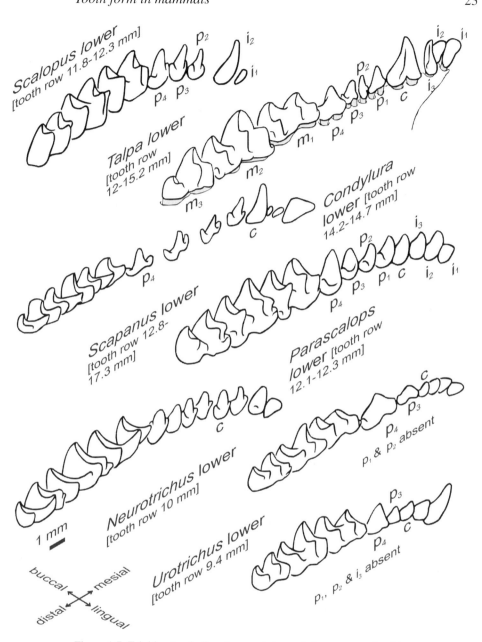

Figure 1.7 Talpidae (excluding desmans), lower left permanent dentitions.

Amongst the moles, they are largest in *Parascalops*, followed by *Urotrichus*. The metaconule is moderately developed, but paraconule less so, in *Talpa*, *Scapanus*, *Scapanulus* and *Neurotrichus*. Both are effectively missing in *Scalopus*, so that the protocone forms a single lobe. *Condylura* has distinctive dentition, with small anterior teeth/premolars well separated by diastemata, and a three-lobed lingual

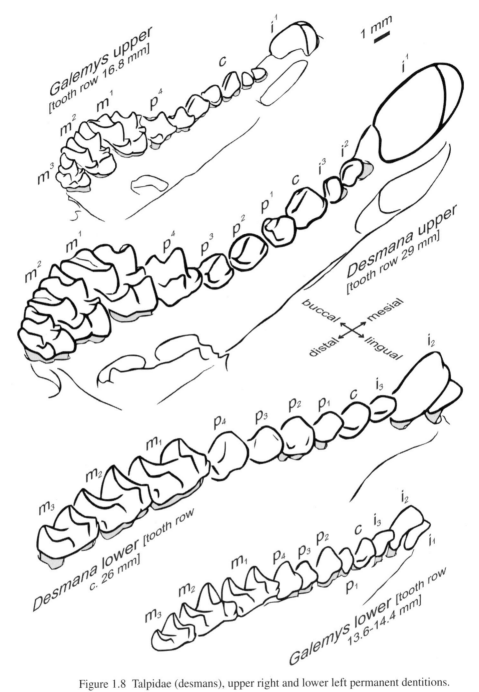

Figure 1.8 Talpidae (desmans), upper right and lower left permanent dentitions.

protocone shelf in the upper molars which is much lower than the high paracone and metacone ridges.

Desmans have webbed feet and a long flexible snout, and are specialised for life in streams and lakes. They are active swimmers that take a variety of small aquatic creatures. Both genera are larger than typical moles (*Desmana* more so than *Galemys*), with a more heavily built dentition. The distinctive feature is an exceptionally large, heavy triangular upper first incisor – almost a tusk. Otherwise, the anterior teeth and premolar row are strongly reduced. The high dilambdadont 'W' blade of the upper molars is characteristically split into two 'V' ridges (Figure 1.8). In addition, they have strongly developed hypocones and paraconules, giving them three-lobed lingual shelves.

Family Erinaceidae
Hedgehogs: *Erinaceus*[EU,AF], *Hemiechinus*[AF,ASE], *Paraechinus*[AF,ASW,ASC]

$$i\frac{2-3}{3}, c\frac{1}{1}, p\frac{3-4}{2-4}, m\frac{3}{3}$$

Hedgehogs have a wide distribution, in many environments from woodland to desert, and eat a wide range of foods – seeds, leaves, eggs, small invertebrates, amphibians, reptiles and mammals. Their generalised dentition reflects this varied diet. Molars (particularly upper), although recognisably dilambdadont, have low broad cusps whose function is clearly crushing rather than cutting. The first incisors are long and pointed but the remaining teeth are reduced, with the exception of the multicusped upper fourth premolar. The three very similar genera can to some extent be distinguished by variation in reduction of the upper third premolar and second incisor (Figure 1.9). *Erinaceus* is least reduced, and *Paraechinus* most, although there is variation within genera.

Family Soricidae
Shrews with 5 upper unicuspid teeth: *Sorex*[EU,AS,JA,AM], *Microsorex*[AM], *Blarina*[AM], *Blarinella*[ASE]; shrews with 4 upper unicuspids: *Neomys*[ASC,ASN], *Suncus*[EU,ASE], *Cryptotis*[AM]; shrews with 3 upper unicuspids: *Crocidura*[EU,AF,AS,JA], *Chimerogale*[ASC,ASE], *Soriculus*[ASE], *Notiosorex*[AM]; shrews with 2 upper unicuspids: *Diplomesodon*[ASC]

$$i\frac{3}{1}, c\frac{0-1}{0}, p\frac{1-3}{1}, m\frac{3}{3}$$

The largest family of insectivores in terms of species and smallest in size, shrews are Holarctic in distribution and forage for small invertebrates through the litter layer of marshy areas, forests, and even to some extent deserts. Their jaws and teeth are minuscule, requiring a good microscope for identification, but the first incisors are relatively enlarged as 'forceps' for catching their tiny prey. The upper first incisor is beak-like, with two backward hooking points, and the lower first

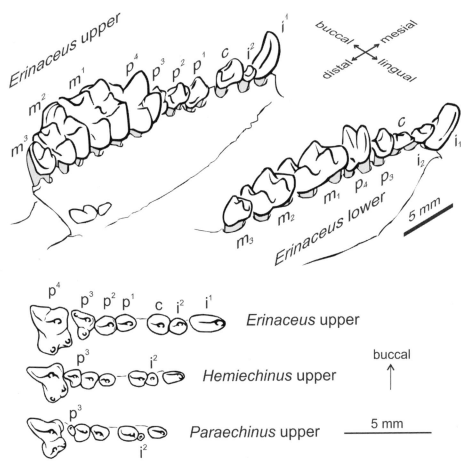

Figure 1.9 *Erinaceus*, upper right and lower left permanent dentitions. Details of variation in upper right incisors, canines and premolars for Erinaceidae.

incisor is elongated into a blade with a variable number of low cusplets along its edge (Figure 1.10). Other incisors, canines and premolars are reduced to tiny points, and are collectively called the unicuspids. Upper unicuspids vary in number between genera (see listing above and Figure 1.11). The exception is the upper fourth premolar, which is always high and strongly blade-like, whereas the upper molars have in its most prominent form the dilambdadont condition (above). Some genera may have red-orange pigmented enamel at the cusp tips (notably *Blarina, Blarinella, Sorex, Neomys* and *Microsorex*), which may or may not survive in fossil material (Stuart, 1982).

Family Macroscelididae
Elephant shrews: *Elephantulus*[AF]

$$i\frac{3}{3}, c\frac{1}{1}, p\frac{4}{4}, m\frac{2}{2-3}$$

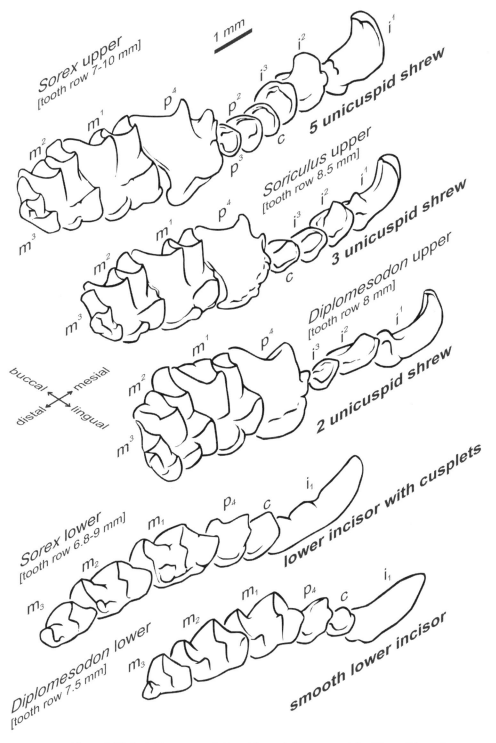

Figure 1.10 Soricidae, upper right and lower left permanent dentitions. Dimensions are lengths of tooth row from first incisor to third molar.

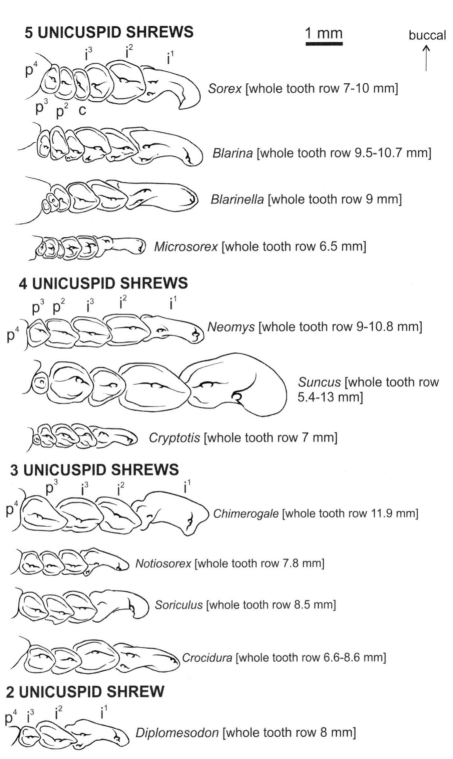

Figure 1.11 Soricidae, details of variation in upper right incisors, canines and premolars (unicuspids). Dimensions are lengths of tooth row from first incisor to third molar.

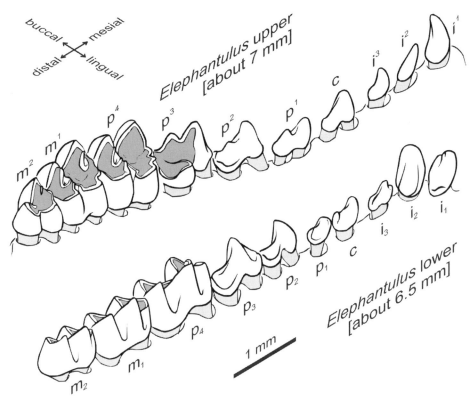

Figure 1.12 *Elephantulus*, upper right and lower left permanent dentitions. Dimensions are lengths of tooth row from first incisor to third molar.

Elephant shrews take their name from a long, mobile snout, which they use to probe for ants, termites and a variety of other food. Their molariform cheek tooth crowns are tall and wear into high enamel ridges, enclosing dentine basins (Figure 1.12). The remaining teeth are reduced to low blades and points, except for the taller point of the upper first incisor and the spatulae of the lower first and second incisors.

Order Chiroptera
The Chiroptera are divided into two suborders: the Megachiroptera (flying foxes and Old World fruit bats) and Microchiroptera (the remaining 18 families of bats).

Suborder Microchiroptera
Most microchiropteran bats hunt insects in the air using a sophisticated echolocation apparatus, but some prey on small ground-living animals, others catch fish, some drink blood, and still others eat fruit, or gather pollen and nectar from flowers. As with the shrews (above), the jaws and teeth are small, and require a microscope for identification. Many insect hunters are characterised by a short face and jaw, with reduction of upper incisors to accommodate the elaborate noses which

focus the sonar pulses emitted from their nostrils. The lower incisors have multiple cusps along their incisal edge, arranged like a comb, and presumably for grooming. Canines in both jaws tend to be tall and sharply pointed, with a pronounced cingulum at the base of the crown from which accessory cusps sometimes arise. Most of the premolars are reduced to low points and the number varies between taxa, but the upper fourth is a tall, sharp blade. The upper molars are strongly dilambdadont, with tall 'W'-shaped ridges and a low lingual shelf, and the lower molars have marked double 'V' ridges. As with the canine, all premolars and molars tend to have a pronounced and sharply defined cingulum. The chief distinguishing features between the bulk of microchiropteran families and genera are the number of premolars, form of upper fourth premolar, reduction of the upper third molar and variation in the hypocone of the other upper molars. Some of the families, however, have strongly divergent dentitions which reflect their highly individual lifestyles, notably the vampire bats (Desmodontidae) and American leaf-nosed bats (Phyllostomatidae).

Family Vespertilionidae
Common bats: *Myotis*^{EU,AF,AS,JA}, *Pipistrellus*^{EU,AF,AS,JA,AM}, *Nyctalus*^{EU,AS}, *Eptesicus*^{EU,AF,AS,AM}, *Vespertilio*^{EU,AS}, *Otonycteris*^{AF,ASW}, *Barbastella*^{EU,AF,AS}, *Plecotus*^{EU,AF,AS}, *Miniopteris*^{EU,AF}, *Scotophilus*^{ASE}, *Lasionycteris*^{AM}, *Nycticeius*^{AF,ASW,AM}, *Lasiurus*^{AM}, *Euderma*^{AM}, *Murina*^{ASW,JA}, *Antrozous*^{AM}

$$i\frac{1-2}{2-3}, c\frac{1}{1}, p\frac{1-3}{2-3}, m\frac{3}{3}$$

The vespertilionid bats include over 300 species in 37 genera, of which 16 range into the Holarctic. Practically all are insect hunters and their dentition conforms to a similar general plan. The lower incisors usually have a comb-like imbrication along their incisal edges. The fourth premolar is present in all genera – in the upper jaw it rises to a sharp point, with a flaring blade to distal, and in the lower jaw also it has a prominent point. There is variation in the number of the much smaller more distal premolars, in both the upper and lower dentitions. For the upper dentition: *Myotis* retains both the third and second premolars; *Pipistrellus*, *Nyctalus*, *Barbastella*, *Plecotus*, *Miniopteris*, *Lasionycteris*, *Euderma* and *Murina* have only the third premolar; *Antrozous*, *Vespertilio*, *Otonycteris*, *Eptesicus*, *Scotophilus* and *Nycticeius* have lost both (examples in Figure 1.13); in *Lasiurus*, the third premolar is present in some species, but absent in others; *Miniopteris* and *Murina* are distinguished by a particularly large upper third premolar. For the lower dentition (Figure 1.14): *Myotis*, *Miniopteris*, *Plecotus*, and *Lasionycteris* retain both the third and second premolars; *Pipistrellus*, *Nyctalus*, *Barbastella*, *Euderma*, *Antrozous*, *Vespertilio*, *Otonycteris*, *Eptesicus*, *Lasiurus*, *Scotophilus*, *Murina* and *Nycticeius* have only the third premolar. The remaining third lower premolar in the latter group is mostly much reduced, except in *Murina*, *Nyctalus* and *Nycticeius*.

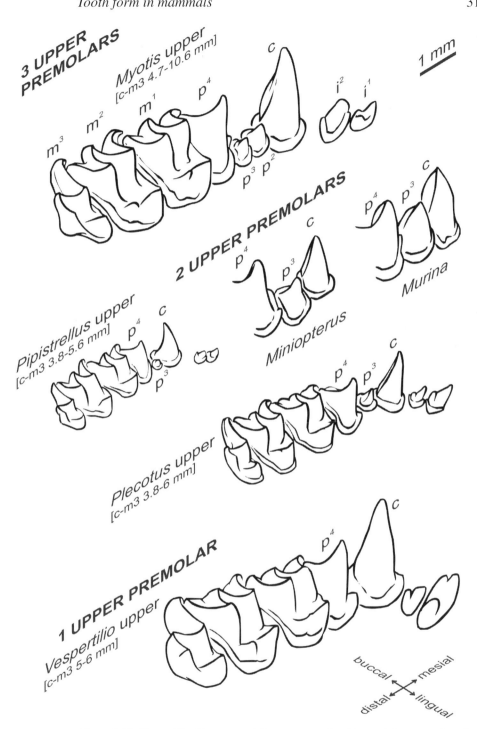

Figure 1.13 Vespertilionidae, upper right permanent dentitions, giving examples of dentitions with three, two and one premolars. Dimensions are lengths of tooth row from mesial side of canine to distal side of third molar.

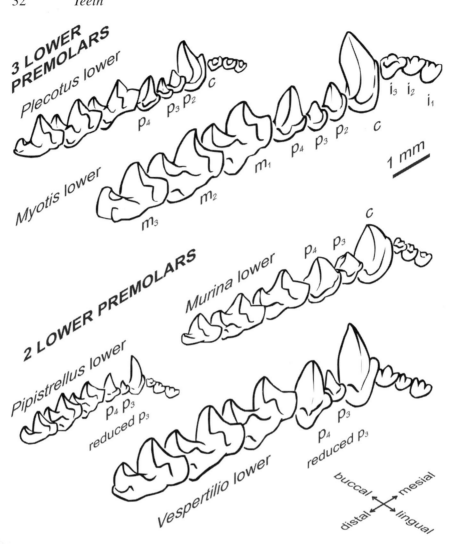

Figure 1.14 Vespertilionidae, lower permanent dentitions, giving examples of dentitions with three and two premolars. Dimensions are lengths of tooth row from mesial side of canine to distal side of third molar.

Family Molossidae
Free-tailed bats: *Tadarida*[EU,AF,ASW,AMC], mastiff bats: *Eumops*[AMC]

$$i\frac{1}{2-3}, c\frac{1}{1}, p\frac{1-2}{2}, m\frac{3}{3}$$

The single upper incisors are relatively large and pointed (Figure 1.15), whereas the lower incisors have comb-like imbrications and are much reduced, particularly in *Eumops*. Upper third premolars are also strongly reduced, or even absent in some

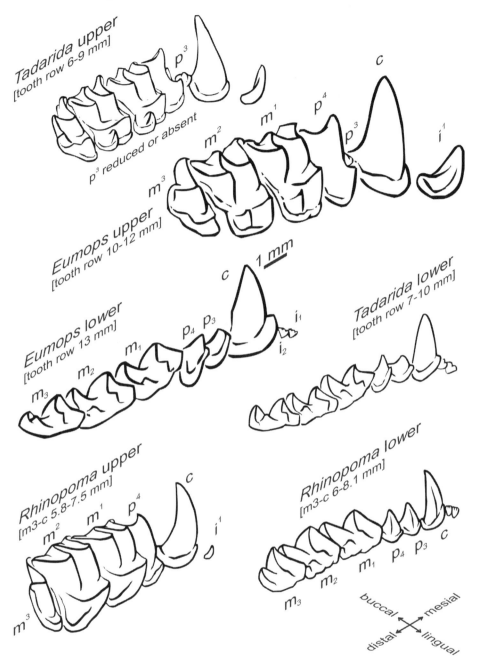

Figure 1.15 Molossidae and Rhinopomatidae, upper right and lower left permanent dentitions. Dimensions are lengths of tooth row from mesial side of canine to distal side of third molar.

specimens, but the lower third premolar is not reduced. The hypocone in *Tadarida* upper molars is a prominent shelf-like lobe, somewhat less marked in *Eumops*.

Family Rhinopomatidae
Mouse-tailed bats: *Rhinopoma*[AF,ASW,ASC]

$$i\frac{1}{2}, c\frac{1}{1}, p\frac{1}{2}, m\frac{3}{3}$$

Upper incisors are tiny pegs and lower incisors have comb-like imbrications. Molars and premolars are robustly built, with only one upper premolar and two lower (Figure 1.15). The angles of the 'W' ridge of the upper molars are rather rounded and the prominent hypocone makes a clear lower shelf.

Family Emballonuridae
Sheath-tailed bats: *Coelura*[ASW], *Taphozous*[AF,ASW]

$$i\frac{0}{3}, c\frac{1}{1}, p\frac{2}{2}, m\frac{3}{3}$$

Upper incisors are missing or reduced to tiny pegs and the small lower incisors bear comb-like imbrications (Figure 1.16). The fourth upper premolar has a high point (especially in *Taphozous*) and the third is a small button. The two lower premolars are similar in size. In upper molars, the hypocone is not clearly separated from the protocone shelf. The distal part of the upper and lower third molars is markedly reduced in *Taphozous*, but not *Coelura*.

Family Mormoopidae
Leaf-chinned bats: *Pteronotus*[AMC], *Mormoops*[AMC]

$$i\frac{0}{3}, c\frac{1}{1}, p\frac{2}{2}, m\frac{3}{3}$$

Upper incisors have two cusps and lower incisors are comb-like. These bats have a reduced molar row (Figure 1.16) relative to premolars and anterior teeth (most marked in *Mormoops*), with relatively low crowns and a clearly separated hypocone shelf in the upper molars. There are three lower premolars; the third being smallest.

Family Nycteridae
Slit-faced bats: *Nycteris*[AF,ASW]

$$i\frac{2}{3}, c\frac{1}{1}, p\frac{1}{2}, m\frac{3}{3}$$

Upper incisors are small, with two cusps, and the tiny lower incisors have fine imbrications. The single upper premolar (Figure 1.17) and three molars have tall crowns, flaring out above the cervix. The lower fourth premolar is reduced to a tiny vestige.

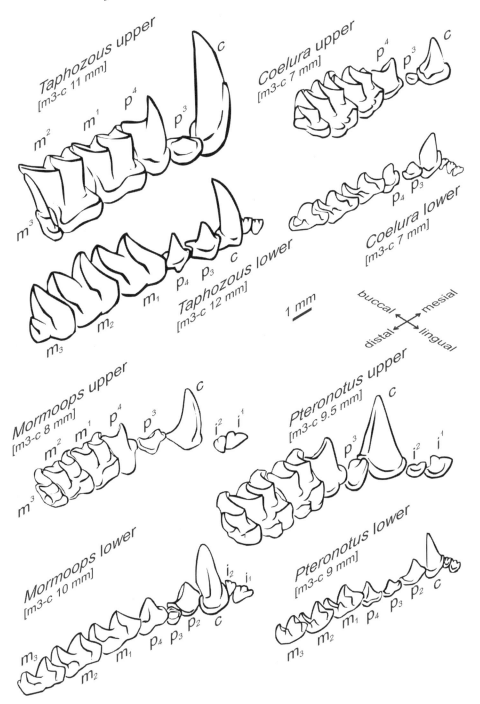

Figure 1.16 Emballonuridae and Mormoopidae, upper right and lower left permanent dentitions. Dimensions are lengths of tooth row from mesial side of canine to distal side of third molar.

36 Teeth

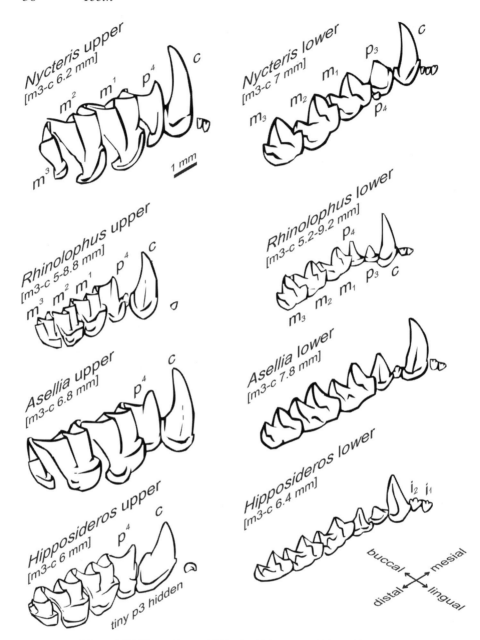

Figure 1.17 Nycteridae and Rhinolophidae, upper right and lower left permanent dentitions. Dimensions are lengths of tooth row from mesial side of canine to distal side of third molar.

Family Rhinolophidae
Horseshoe bats: *Rhinolophus*[EU,AS,JA]; trident-noseleaf bats: *Asellia*[AF,ASW], leaf-nosed bats: *Hipposideros*[ASW], *Triaenops*[ASC]

$$i\frac{0-1}{2}, c\frac{1}{1}, p\frac{1-2}{2-3}, m\frac{3}{3}$$

The upper incisors (Figure 1.17) are absent (*Asellia*) or greatly reduced and the two small lower incisors are comb-like. The upper third premolar is absent in *Asellia* and tiny in *Rhinolophus* and *Hipposideros*. In upper molars, the hypocone is reduced to a small, but clearly separate, low shelf. There are two lower premolars, with the third reduced in *Asellia*. *Triaenops* is similar to *Rhinolophus* and *Hipposideros*.

Family Phyllostomatidae
Neotropical fruit bats: *Artibeus*[AMC]; wrinkle-faced bat: *Centurio*[AMC]; white-lined bat: *Chiroderma*[AMC]; yellow-shouldered bats: *Sturnira*[AMC]; short-tailed leaf-nosed bat: *Carollia*[AMC]; big-eared bats: *Macrotus*[AMC]; long-nosed bats: *Leptonycteris*[AMC], *Choeroniscus*[AMC]; long-tongued bats: *Glossophaga*[AMC], *Hylonicteris*[AMC]

$$i\frac{2}{0-2}, c\frac{1}{1}, p\frac{2}{2-3}, m\frac{2-3}{2-3}$$

This is a very variable family of bats, including some quite exceptional genera. *Artibeus*, *Centurio*, *Chiroderma*, *Sturnia* and *Carollia* occupy a similar niche to Old World fruit bats (below) and have stout, low-crowned dentitions with strongly reduced or absent third molars (Figure 1.18). Their shelf-like lingual elements form a broad crushing area. By contrast (Figure 1.19), *Choeroniscus*, *Choeronycteris*, *Leptonycteris* and *Hylonycteris* are nectar and pollen feeders, with a long, narrow snout, long brush-tipped tongue and strongly reduced teeth (especially *Choeroniscus*). *Glossophaga* also eats nectar and pollen, but takes other parts of the flower, fruit and insects as well, and reflects this in a less strongly reduced dentition. *Macrotus* feeds on insects and fruit, and has a more generalised microchiropteran dentition (Figure 1.20).

Family Desmodontidae
Vampire bats: *Desmodus*[AMC], *Diaemus*[AMC], *Diphylla*[AMC]

$$i\frac{1-2}{2}, c\frac{1}{1}, p\frac{1}{2}, m\frac{1-2}{1-2}$$

The vampire bats are famous for feeding on blood. Tiny skin incisions are made with their very sharp lancet-like incisors and the blood is lapped up with the tongue into a tube made with the lips. The upper incisors (Figure 1.21) dominate the dentition and the postcanine teeth are strongly reduced (particularly in *Desmodus*). *Diphylla*

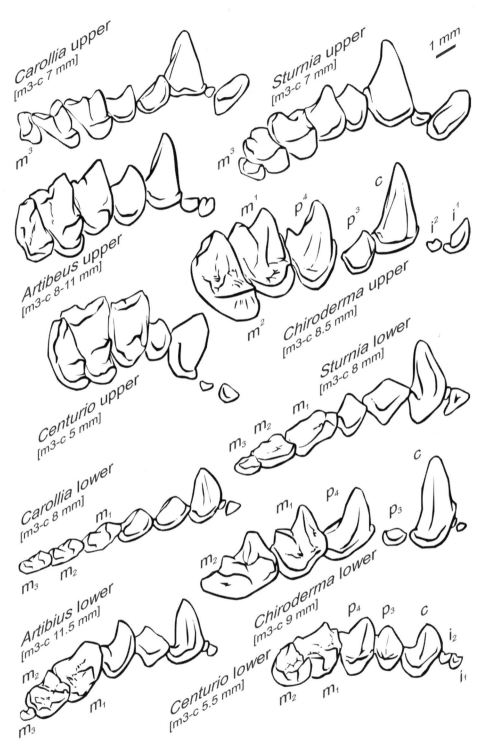

Figure 1.18 Phyllostomatidae, New World fruit bats, upper right and lower left permanent dentitions. Dimensions are lengths of tooth row from mesial side of canine to distal side of third molar.

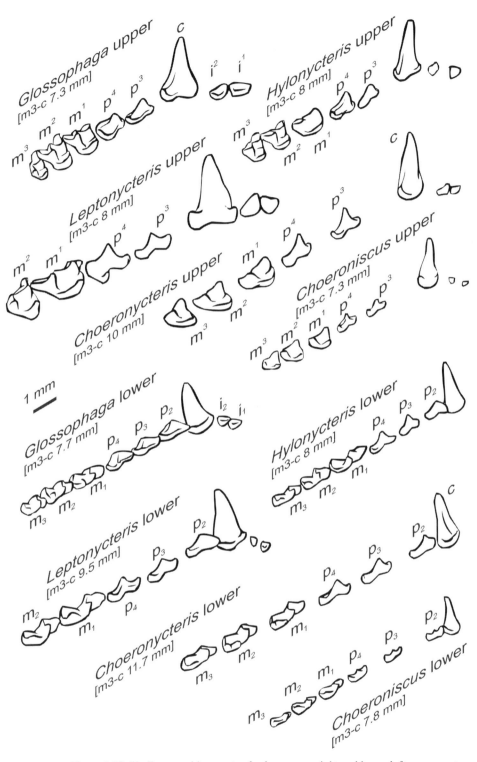

Figure 1.19 Phyllostomatidae, nectar feeders, upper right and lower left permanent dentitions. Dimensions are lengths of tooth row from mesial side of canine to distal side of third molar.

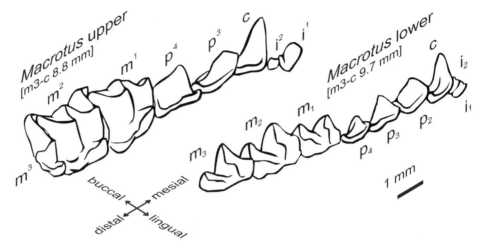

Figure 1.20 Phyllostomatidae, *Macrotus*, upper right and lower left permanent dentitions. Dimensions are lengths of tooth row from mesial side of canine to distal side of third molar.

is distinguished by broad lower incisors with a complex, strongly comb-like incisal edge.

Family Natalidae
Funnel-eared bats: *Natalus*[AMC]

$$i\frac{2}{3}, c\frac{1}{1}, p\frac{3}{3}, m\frac{3}{3}$$

The Natalidae are insectivorous bats with a generalised microchiropteran dentition (Figure 1.22). Their premolar row is long, with three similar-sized teeth in the upper and lower jaw. By contrast, the molars are relatively small and low crowned.

Suborder Megachiroptera
This suborder contains only one family, the Pteropodidae, including tropical Old World fruit bats and flying foxes. Most are frugivorous, but some have other specialisations. The molars are highly characteristic, with a single, broad mesiodistal groove running down their occlusal surfaces. A combination of these teeth, a ridged palate and the tongue is used to crush fruit and extract pulp and juice.

Family Pteropodidae
Egyptian fruit bat: *Rousettus*[AF]; Ryukyu flying fox: *Pteropus*[JA]; straw-coloured fruit bat: *Eilodon*[ASW]

$$i\frac{2}{2}, c\frac{1}{1}, p\frac{1-2}{3}, m\frac{3}{3}$$

All incisors are tiny pegs, but the tall canines have substantial crowns (Figure 1.23). Third molars are strongly reduced in *Eilodon* and *Pteropus*. The premolars are

Tooth form in mammals

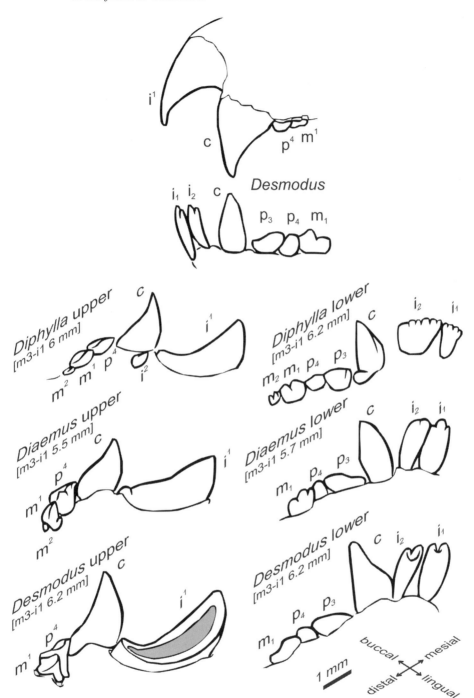

Figure 1.21 Desmodontidae. Top figure shows buccal view of upper and lower left permanent dentitions of *Desmodus*. Lower figure shows upper right and lower left permanent dentitions. Dimensions are lengths of tooth row from first incisor to most distal molar.

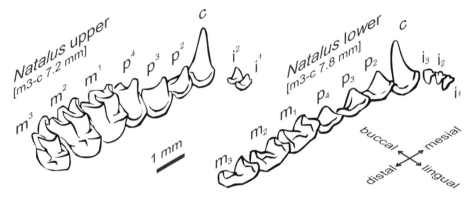

Figure 1.22 Natalidae, upper right and lower left permanent dentitions. Dimensions are lengths of tooth row from mesial side of canine to distal side of third molar.

curving points, hollowed-out distally, but the lower fourth premolar takes on a molar-like form. In *Pteropus* the upper third premolar is missing and the lower second premolar is markedly small.

Order Primates

Two suborders of Primates are recognised: Prosimii and Anthropoidea. The Prosimii are all arboreal climbers, feeding on leaves, fruits, insects and small vertebrates. Their molars (three per jaw quadrant) are usually bunodont and premolars (two or three per quadrant) are simple pegs or blades, or sometimes become 'molarised'. Upper anterior teeth are typically reduced and spaced and, in most prosimians, lower anterior teeth form a 'dental comb', used for grooming fur. This comb is most strongly developed in the thin, needle-like lower incisors and canines of lemurs (Lemuridae), which were distributed in many forms throughout the Old World during the Tertiary period, but are now confined to Madagascar. Dental combs are also well developed in the lorises, pottos and bushbabies (Lorisidae). In the indri and its relatives (Indriidae), again from Madagascar, there is a dental comb, but the number of anterior teeth involved is reduced. The smallest of all Primates, tree shrews (Tupaiidae) from the forests of South-east Asia, also have a dental comb but it is even further reduced. Tree shrews also have dilambdadont upper molars like insectivores. The tarsiers (Tarsiidae), also from South-east Asia, are insectivorous and have simple conical anterior teeth and premolars, with no dental comb. Finally, the extraordinary aye-aye of Madagascar (Daubentoniidae) occupies a niche not unlike a woodpecker, using its long, thin finger to probe tree bark for insects. It has a greatly reduced cheek tooth row, no premolars (except the diminutive upper fourth) or canines, and chisel-like, persistently growing first incisors.

The Anthropoidea consists of marmosets, tamarins, monkeys, gibbons, apes and humans. In all families, the incisors (two per jaw quadrant) are spatulate and canines vary from prominent 'daggers' to low cones. The molars are bunodont, usually with four to five cusps well developed. Premolars are also mostly bunodont, with one

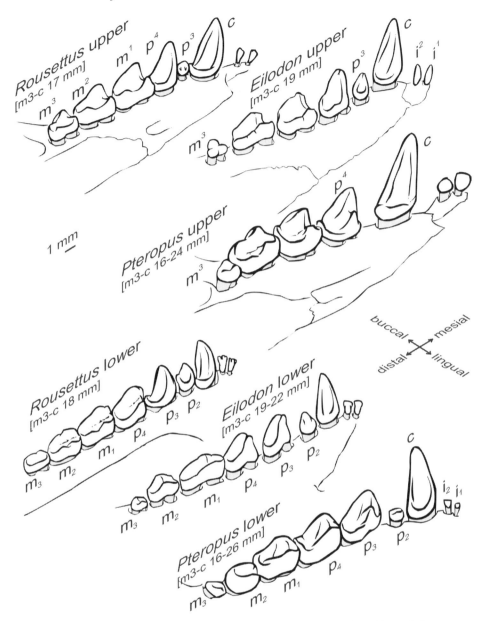

Figure 1.23 Pteropodidae, Old World fruit bats, upper right and lower left permanent dentitions. Dimensions are lengths of tooth row from mesial side of canine to distal side of third molar.

or two main cusps, except for the lower third premolar, which is often blade-like. The marmosets and tamarins (Callithricidae) of central/southern America are mostly insectivorous. Almost all have two molars (upper with four cusps, lower with three) and three premolars per jaw quadrant. New World monkeys (Cebidae) have three molars (usually four cusped) and three premolars per quadrant. They are vegetarians, but insects, eggs and small vertebrates are also taken. The remaining Primates all have three molars and two premolars only per jaw quadrant. In Old World monkeys (Cercopithecidae), the four molar cusps are arranged into two buccolingual folds (bilophodont) and the lower third premolar is usually developed into a pronounced blade which cuts against the upper canine (sectorial premolar). Many Old World monkeys are arboreal, but some live on the ground. The latter have elongated snouts, with enlarged sectorial premolars and canines. All the rest of the Primates have non-bilophodont molars, usually with four cusps in upper teeth and five in lower. The highly arboreal gibbons and siamangs (Hylobatidae) are found only in the forests of South-east Asia. Apes (Pongidae) are also arboreal, but less exclusively so, and are larger animals. Both Hylobatidae and Pongidae have a sectorial premolar and canine complex, but this is absent in people (Hominidae) where the canines are also reduced.

Detailed accounts of Primate dentitions can be found in James (1960) and Swindler (2002; 1976). Only two families are found in the present-day Holarctic region.

Family Cercopithecidae
Macaques: *Macaca*[AS]; baboons: *Papio*[AF,ASW]

$$i\frac{2}{2}, c\frac{1}{1}, p\frac{2}{2}, m\frac{3}{3}$$

The canines are prominent and show strong sexual dimorphism, with the lower third premolar formed into a stout sectorial blade which is more elongated in *Papio* than *Macaca* (Figure 1.24). Molar crowns are high for a monkey, with four or five cusps arranged into bilophodont (above) folds. The dentition of *Macaca* is smaller than that of *Papio* and the crowns are relatively narrower buccolingually. The distal cusp of the lower third premolar is more pronounced in *Papio*.

Family Hominidae
Humans: *Homo*[EU,AF,AS,AM]

$$di\frac{2}{2}, dc\frac{1}{1}, dp\frac{2}{2}, \rightarrow i\frac{2}{2} c\frac{1}{1}, p\frac{2}{2}, m\frac{3}{3}$$

Canines are much reduced, taking on the form of incisors (Figure 1.25). The cheek teeth are bunodont, with thick enamel, and are lower and broader than those of monkeys. Premolars have two main cusps and molars have three, four or five. The upper third molar is usually reduced. In dental anatomy texts (e.g. Jordan *et al.*, 1992), the two premolars are often called first and second but for compatibility with other mammals in this book they are here called third and fourth. Human deciduous teeth have lower crowns which flare out from the cervix, with widely spaced, spindly roots. Once again, what are here called the third and fourth deciduous premolars are often called first and second molars in dental anatomy texts.

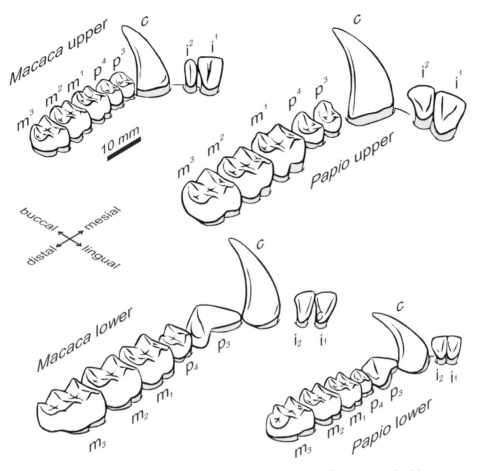

Figure 1.24 Cercopothecidae, upper right and lower left permanent dentitions.

Order Carnivora

Teeth in carnivores are adapted primarily for particular roles in hunting, catching, killing and butchering prey. The incisors are small relative to the rest of the dentition, with pointed crowns packed close together into a comb-like structure which is used for grooming the animal's protective fur. Throughout the dentition, the cingulum forms a broad cervical bulge which protects the gingivae. The canines are prominent points (Figure 1.26), in a range of sizes and sharpness, which are used for holding and dispatching prey either by strangulation or stabbing. Most of the premolars are compressed buccolingually, into backward-hooking, pointed blades which do not necessarily meet when the jaws are closed, and presumably have a function in catching and holding prey. The most characteristic teeth of the carnivores, however, are the blade-like *carnassial* teeth that shear against one another to cut through meat and skin. In the permanent dentition, the carnassials are the upper fourth premolar and lower first molar and, in the deciduous dentition, the upper third

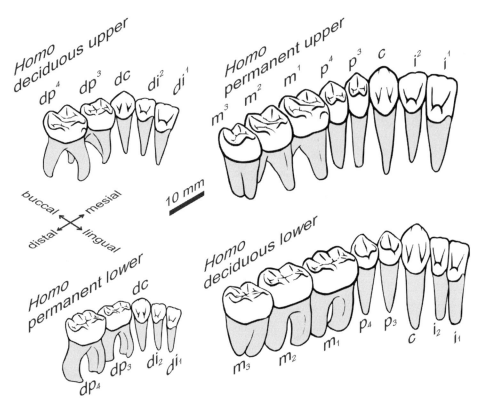

Figure 1.25 Hominidae, human upper right and lower left permanent and deciduous dentitions.

premolar and lower fourth premolar take on this role (Figure 1.27). In the upper carnassial, the main cutting edge is formed by development of the paracone and the metacone/metastyle crest (Figure 1.4), with the protocone reduced to a low mound in most genera. The parastyle is present at the mesial base of the paracone, but varies greatly in prominence. In lower carnassials, the main cutting edge is formed from high ridges of the protoconid and paraconid, whilst the other cusp of the trigonid area, the metaconid, is reduced. The talonid of this tooth remains as a low crushing area in most families, but is much reduced in cats and hyaenas. Other molars, where present, are bunodont and adapted to crushing bones or other hard foods. The upper molars characteristically have higher buccal cusps, with a lower lingual area, and show a large variation in relation to function which is very helpful for identification.

The form of the teeth follows the dietary specialisation of the animals, which varies both within and between families. More generalised carnivores and omnivores tend to have less developed carnassial blades, with more strongly developed crushing molars. Specialist hunters have more strongly blade-like carnassials and reduced crushing teeth. Thus most of the variation useful in identification is seen in

Tooth form in mammals 47

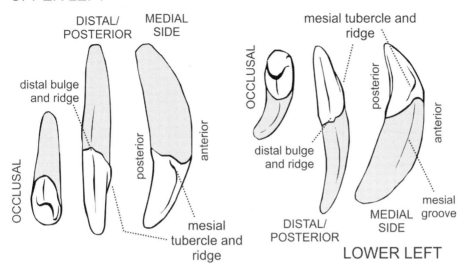

Figure 1.26 Carnivore canines. The canines curve mostly in an anterior–posterior plane. Lower canines also curve to lateral, with a more curved root, often marked with a mesial groove. Upper and lower crowns are marked on the distal side by a ridge, and bulge at the cervix, on the posterior side of the tooth. The mesial side is diagonally placed because the dental arcade turns a corner at this point. It is marked with a small tubercle and ridge at the cervix, which curves to occlusal at this point.

the cheek teeth. Many of the carnivore families originally had a worldwide distribution – except for Australasia and a number of other islands. This includes the bears (Ursidae), the badgers, weasels etc. (Mustelidae), dogs (Canidae) and cats (Felidae). The hyaenas (Hyaenidae) are nowadays confined to Africa and western Asia, but were once found in Europe too. The civets, genets and mongooses (Viverridae) are found in south-west Europe, Africa and southern Asia, whereas the racoons and their relatives (Procyonidae) are confined to the New World and a small area of eastern Asia.

Family Canidae
Dogs, wolf: *Canis*[EU,AF,AS,AM]; Cape hunting dog: *Lycaon*[AF]; foxes: *Vulpes*[EU,AF,AS,AM], *Alopex*[EU,AS,AMN], *Urocyon*[AM]; dhole: *Cuon*[AS]; racoon-dog: *Nyctereutes*[ASE,JA]

$$\text{di}\frac{3}{3}, \text{dc}\frac{1}{1}, \text{dp}\frac{3}{3} \rightarrow \text{i}\frac{3}{3}, \text{c}\frac{1}{1}, \text{p}\frac{4}{4}, \text{m}\frac{2}{3}$$

This is a family of generalised carnivores, taking a range of prey from small insects to large herbivores, either as solitary or pack hunters, and some of the smaller genera also eat plant foods. Their dentition is less reduced and specialised than in most other carnivore families (Figures 1.27–1.29). Permanent upper carnassials

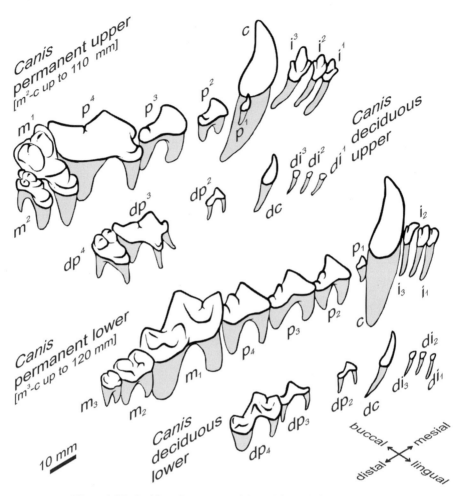

Figure 1.27 Canidae, *Canis* upper right and lower left permanent and deciduous dentitions. Dimensions are lengths of tooth row from canine to most distal molar.

have a long blade, with a protocone of moderate size, whilst lower carnassials have a prominent protoconid/paraconid blade associated with a somewhat reduced metaconid, and an extensive talonid 'heel' at the distal end. Deciduous carnassials are similar in form, with finer blades and spindly, wide-spaced roots. The permanent upper molars are relatively large for a carnivore in most of the canid genera, to form an extensive crushing area. *Canis* and *Lycaon* are the biggest members of the family, with a very similar, robustly built dentition (with perhaps a slightly greater degree of first premolar reduction in *Canis*). Domestication of wolf/dog in the archaeological record is marked by dental reduction (p. 270), and more recent dogs show a very wide range in size, including the smallest canids. *Cuon* is smaller than wolf, and distinguished by higher, markedly blade-like carnassials, prominent pointed buccal cusps in the upper first molar and reduction in the rest of the crushing molars.

Tooth form in mammals 49

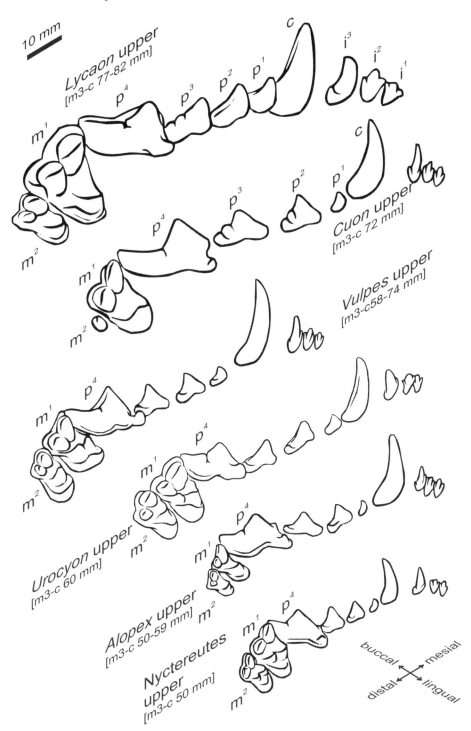

Figure 1.28 Canidae, upper right permanent dentitions. Dimensions are lengths of tooth row from canine to second molar.

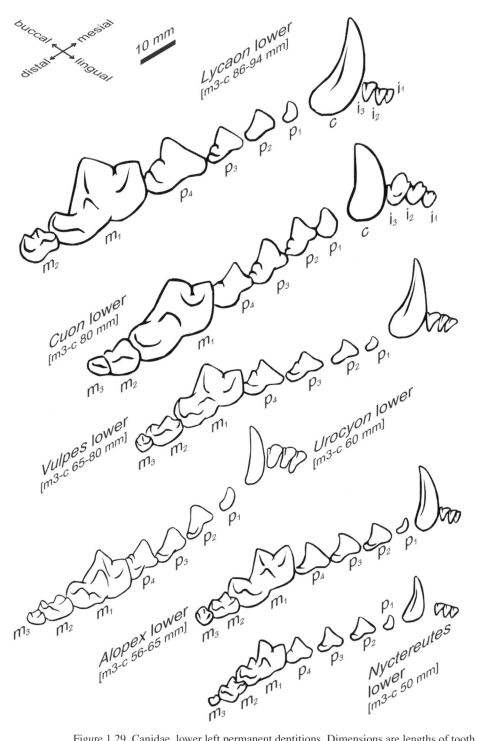

Figure 1.29 Canidae, lower left permanent dentitions. Dimensions are lengths of tooth row from canine to second molar.

Tooth form in mammals 51

The three genera of foxes all have the protocone of their upper carnassials placed somewhat more to mesial than the other genera, and show a varying degree of molar reduction (with *Vulpes* least and *Alopex* most reduced). *Nyctereutes* is one of the smallest wild canids and has a distinctive extra cusp on the buccal side of the lower carnassial.

Family Viverridae
Genets: *Genetta*[EU,AF,ASW]; mongooses: *Herpestes*[EU,AF,AS], *Ichneumia*[ASW]; palm civet: *Paguma*[AS]

$$i\frac{3}{3}, c\frac{1}{1}, p\frac{4}{4}, m\frac{2}{2}$$

The Viverridae are all small, eating a diet ranging from small animals to nuts and fruit. Some genera are more committed carnivores than others, and the carnassial/molar complex in the family varies from low, crushing teeth, to high carnassial blades with reduced crushing surfaces (Figure 1.30). Thus *Paguma* has low, broad carnassials with an expanded talonid crushing 'heel' in the lower first molar, and a broad upper first molar. *Ichneumia* has a higher blade on its carnassials, but these incorporate a broad protocone and talonid, and the molars are expanded. *Herpestes* and *Genetta* have much more markedly carnivorous dentitions, with a reduction in the protocone and talonid of the carnassials, and a strong reduction in the upper molars and lower second molar. Viverrid carnassials typically have a prominent parastyle (in the upper) and metaconid (in the lower).

Family Mustelidae
Polecats, weasels, stoats, mink: *Mustela*[EU,AF,AS,AMN], *Vormela*[AS], *Poecilictus*[AF]; martens: *Martes*[EU,AS,AMN]; skunks: *Spilogale*[AM], *Mephitis*[AM], *Conepatus*[AM]; glutton/wolverine: *Gulo*[AF,AS,AMN]; honey badger/ratel: *Mellivora*[AF,ASW]; badgers: *Meles*[EU,AS], *Arctonyx*[ASE], *Taxidea*[AM]; river otters: *Lutra*[EU,AS,AM]; sea otters: *Enhydra*[PAC]

$$i\frac{3}{3}, c\frac{1}{1}, p\frac{2-4}{3-4}, m\frac{1}{2}$$

This is a large and highly variable family (Figures 1.31–1.34). Diet in the more generalised forms consists of small vertebrates, birds and eggs, a wide range of invertebrates and a proportion of berries or similar plant foods. The smallest mustelids are fierce and agile hunters of small, tunnel-dwelling rodents, whilst *Gulo*, the largest member of the family, is powerful enough to bring down a large deer on its own (although it relies largely on carrion). The badgers and skunks, by contrast, are omnivorous and forage for a wide range of foodstuffs. Several mustelids live in and beside water, including the mink and, particularly, the otters. River otters live on riverbanks or seashores and hunt fish, crustaceans and amphibians, whereas sea otters specialise in shellfish and visit the land only occasionally – they have unique

52 Teeth

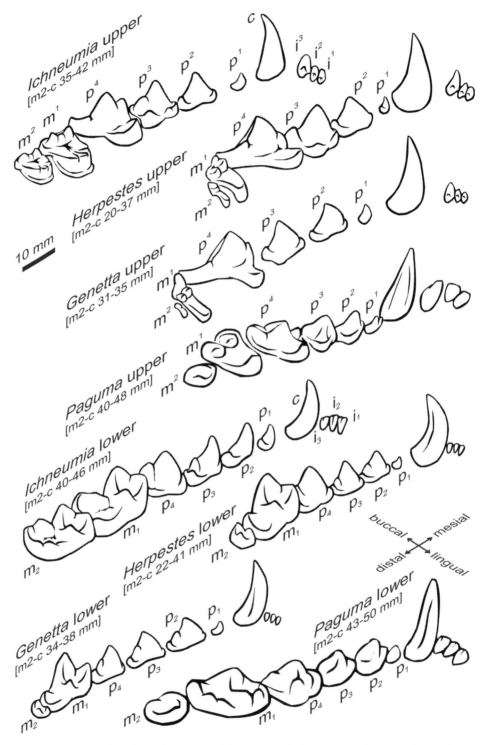

Figure 1.30 Viverridae, upper right and lower left permanent dentitions. Dimensions are lengths of tooth row from canine to most distal molar.

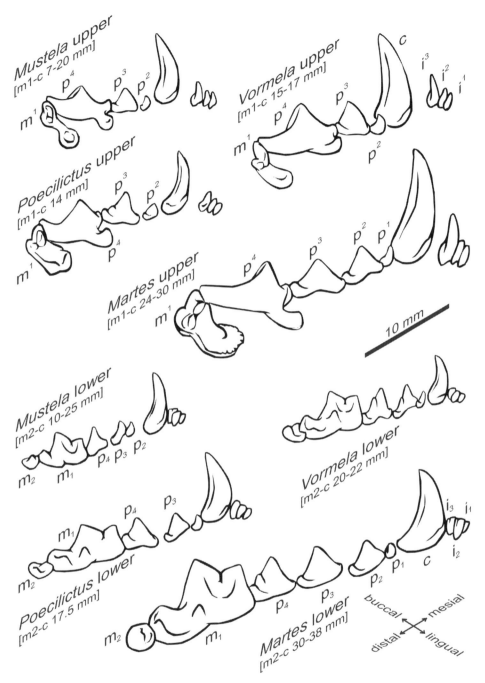

Figure 1.31 Mustelidae, small-sized (weasels, stoats, polecats and martens). Upper right and lower left permanent dentitions. Dimensions are lengths of tooth row from canine to most distal molar.

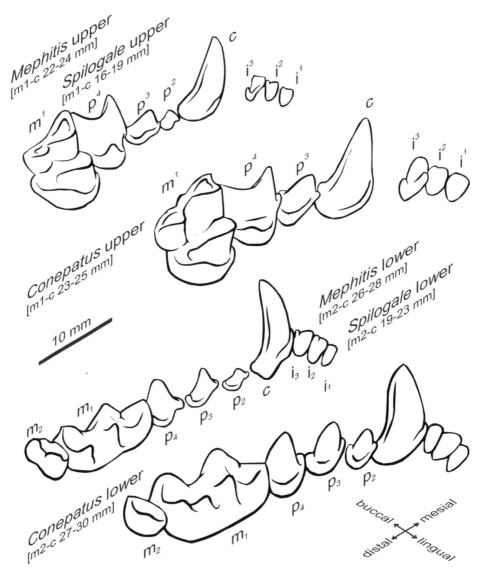

Figure 1.32 Mustelidae, skunks. Upper right and lower left permanent dentitions. Dimensions are lengths of tooth row from canine to most distal molar.

cheek teeth with broad, cushion-like cusps. The river otter *Lutra* has a distinctive upper carnassial with a marked blade and broad, shelf-like protocone, and a distinctive upper first molar with a rectangular occlusal outline. The more carnivorous small mustelids (*Mustela*, *Vormela*, *Poecilictus* and *Martes*) have high blade-like carnassials, with moderately developed protocone and parastyle in the upper teeth, and reduced metaconid and talonid in the lower teeth. Upper first molars and lower second molars in these forms are reduced. *Martes* has relatively large upper first

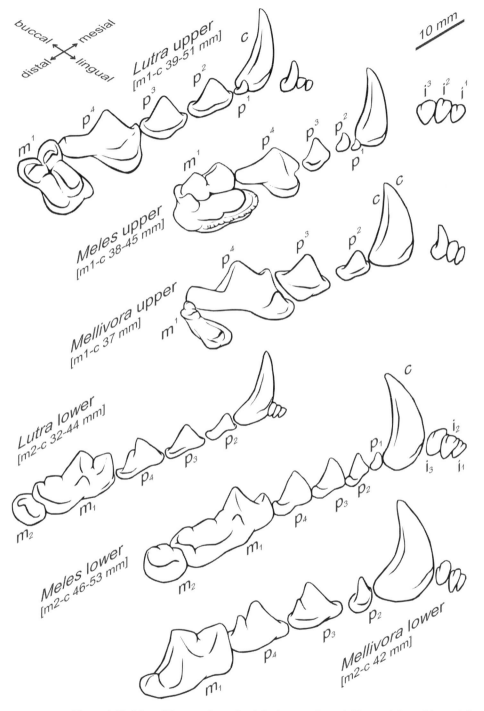

Figure 1.33 Mustelidae, medium-sized (badgers and otter). Upper right and lower left permanent dentitions. Dimensions are lengths of tooth row from canine to most distal molar.

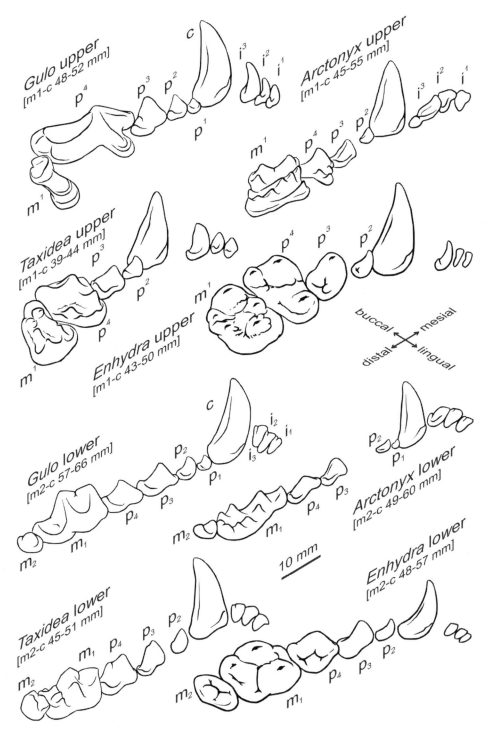

Figure 1.34 Mustelidae, large-sized (wolverine, badgers and sea otter). Upper right and lower left permanent dentitions. Dimensions are lengths of tooth row from canine to most distal molar.

molars, large second premolars and retains small first premolars, whilst *Mustela* has a reduced lower third premolar and lower carnassial talonid 'heel'. *Mellivora* and *Gulo* dentitions are larger and more heavily built versions of the small mustelid carnivorous form. Badgers show a broadening of carnassial crowns, accompanied by a lowering of their blades, expansion of the protocone and talonid, and decoration by additional cusps. The upper first molars in badgers are very greatly expanded and elaborate teeth. Skunks also have an expanded protocone and talonid in their carnassials, but retain a tall blade, and their broad upper first molars have a distinctive form in which the buccal cusps are high and blade-like (although they wear down rapidly). *Mephitis* and *Spilogale* are quite similar in form, but not size, whereas *Conepatus* shows a broadening of the crushing areas of the dentition and a reduction in the upper premolar row.

Family Hyaenidae
Hyaenas: *Hyaena*[AF,AS], *Crocuta*[AF, [EU], [AS]]

$$i\frac{3}{3}, c\frac{1}{1}, p\frac{4}{3}, m\frac{0}{1}$$

Hyaenas are specialist hunters of large herbivores, and are often described as having the most powerful jaws of any mammal. Their upper carnassials are exceptionally robust, with a pronounced parastyle forming part of the blade (Figure 1.35). The metacone part of the blade is relatively more elongated in *Crocuta* than it is in *Hyaena*. The lower carnassials have a tall and stoutly built protoconid/paraconid blade, with a greatly reduced metaconid and talonid (this reduction is more pronounced in *Crocuta* than it is in *Hyaena*). The upper third, and lower third and fourth premolars are raised into high, robust cones which are used to crush bones – the 'bone hammers'. The upper molars and lower second molar are absent.

Family Felidae
Cats: *Felis*[EU,AF,AS,AM]; lions, leopards, tigers: *Panthera*[AF,AS,[EU],[AM]]; cheetah: *Acinonyx*[AF,AS]; sabre-tooths: *Smilodon*[[AM]], *Homotherium*[[AM]]

$$di\frac{3}{3}, dc\frac{1}{1}, dp\frac{3}{2} \rightarrow i\frac{3}{3}, c\frac{1}{1}, p\frac{1-2}{2}, m\frac{1}{1}$$

Like the Hyaenidae, the Felidae are highly specialist carnivores, but the family contains many more species in a much bigger range of sizes. Anatomically, there are few differences except in size or coat and, although the dentition is instantly recognisable as felid, it is often difficult to make any finer distinctions. The permanent upper carnassial has a high blade, part of which is formed by a prominent parastyle, and a reduced protocone (Figure 1.36). The lower carnassial is almost entirely composed of the protoconid/paraconid blade, with no trace of the metaconid and little sign of the talonid. In the permanent upper dentition, the first premolar is absent and the second is greatly reduced (and sometimes missing). In larger felids

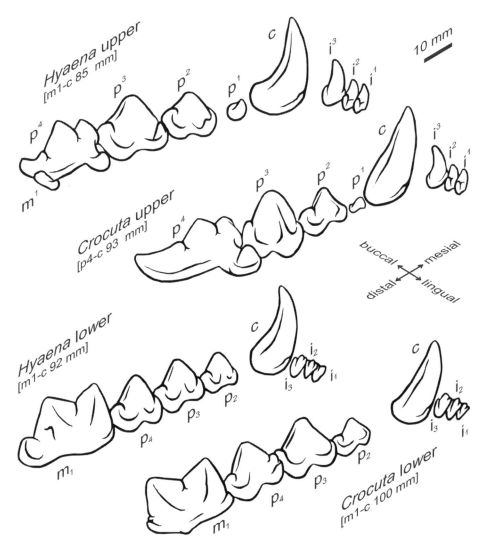

Figure 1.35 Hyaenidae, upper right and lower left permanent dentitions. Dimensions are lengths of tooth row from canine to most distal premolar or molar.

such as the lion, there are diastemata around the small upper second premolar, but these gaps close up in smaller felids or in the shorter-faced forms such as the snow leopard and cheetah. The first and second premolars are absent in the lower jaw, leaving a pronounced diastema which is occupied by the upper canine when the jaw is closed (and therefore the length of the diastema matches the upper canine size). The single upper molar (sometimes absent) is reduced to a small bar-like crown tucked in beside the upper carnassial. The deciduous carnassials are high, sharp blades, and the lower carnassial is a very distinctive form. This overall form of dentition is practically indistinguishable between *Panthera* and *Felis* and, although

Tooth form in mammals

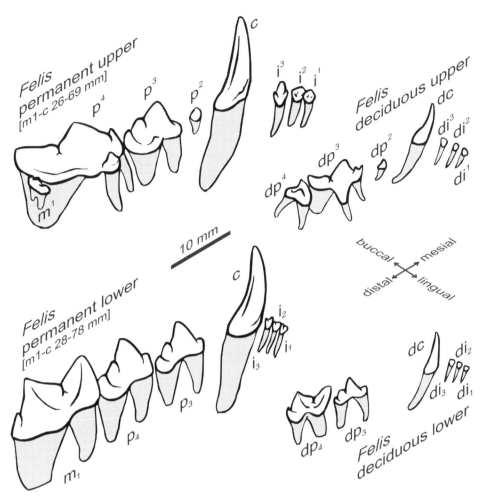

Figure 1.36 Felidae, *Felis* upper right and lower left permanent and deciduous dentitions. Dimensions are lengths of tooth row from canine to most distal premolar or molar.

Panthera contains the largest felid species (the Pleistocene lions of Eurasia and North America were amongst the largest Carnivora of all time), there is a great deal of overlap in size between smaller species. The extinct sabre-tooths, however, were clearly different, both in anterior and cheek teeth (Merriam & Stock, 1932; Meade, 1961). Their upper canines were greatly enlarged, with strong ridges running down their length (marked with serrations), whereas the lower canines were so reduced that they looked almost like a continuation of the incisor row (Figure 1.37). The upper second premolar was lacking in both *Homotherium* and *Smilodon*, the upper third premolar reduced (particularly in *Homotherium*), and the parastyle of the upper carnassial was elaborated by an additional cusp. The lower third premolar

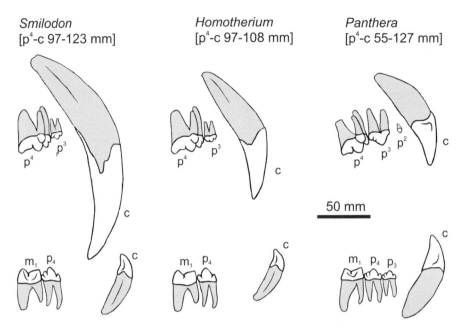

Figure 1.37 Felidae, big cats. Upper and lower permanent left canines, premolars and molars seen from lingual side. Drawn at a much smaller scale than Figure 1.36.

was missing. The function of the upper canine sabres has been the subject of much discussion (Turner & Antón, 1997).

Family Ursidae
Bears: *Ursus*[EU,AS,AM], *Selenarctos*[AS], *Tremarctos*[AM], *Arctodus*[AM]; polar bear: *Thalarctos*[ARC]; giant panda: *Ailuropoda*[ASE]

$$i\frac{3}{3}, c\frac{1}{1}, p\frac{3-4}{3-4}, m\frac{2}{3}$$

Bears have varied dietary habits. *Thalarctos* feeds mainly on seals and fish, but also takes deer, is notorious for raiding human larders, and eats tundra plants during the summer. *Ursus* and *Selenarctos* are more markedly omnivorous, with proportions of meat and plants in the diet that vary between seasons and locations, whilst the giant panda *Ailuropoda* largely feeds on bamboo, but occasionally catches and eats small creatures. The dentition of bears is robust (Figures 1.38–1.39), with reduced premolars and expanded, bunodont molars, adapted to their omnivorous diet. The upper fourth premolar is only just recognisable as the carnassial, and the lower first molar has such an expanded talonid that the carnassial blade is relatively insignificant. The cusps of the molars are gathered up to a rim around the edge of the occlusal surface, the centre of which is lower, and marked by rough tubercles. *Selenarctos*, *Tremarctos* and *Arctodus* all have four closely spaced premolars in both upper and lower cheek tooth rows, whereas *Ursus* and *Thalarctos* have three. The latter genera also have narrower lower first molars and smaller remaining molars,

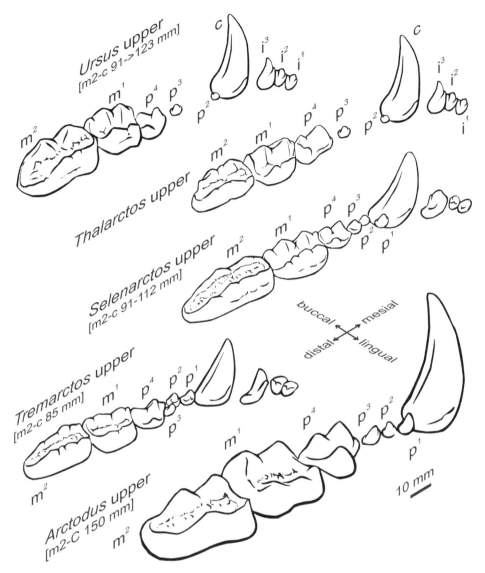

Figure 1.38 Ursidae, upper right permanent dentitions. Dimensions are lengths of tooth row from canine to second molar.

and the cusps of *Thalarctos* are particularly high and trenchant for a bear. The much larger extinct short-faced bear *Arctodus* also had relatively tall carnassial cusps, but markedly broad remaining molars. *Arctodus* and *Tremarctos* both have a distinctive extra buccal cusp at the junction between the trigonid and talonid. *Ailuropoda* is usually classified in the Ursidae, but its dentition diverges markedly from the other members of the family, particularly in the presence of a prominent parastyle in the upper carnassial and other premolars, and in the pronounced cingulum of their molars (Figure 1.40). These features are similar to those found in some of

62 Teeth

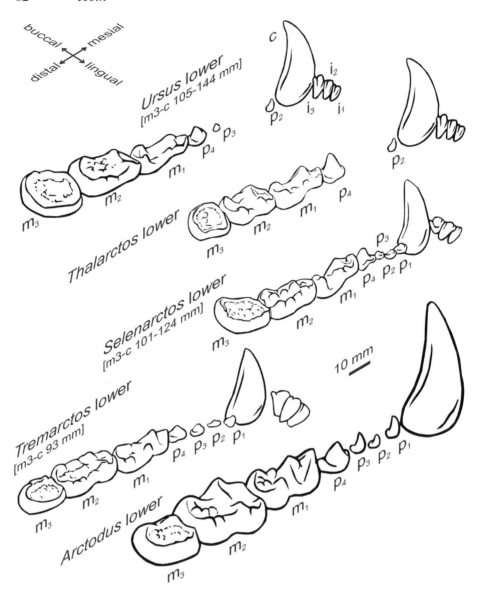

Figure 1.39 Ursidae, lower left permanent dentitions. Dimensions are lengths of tooth row from canine to third molar.

the Procyonidae, notably *Procyon* itself, and there is continuing discussion about the correct classification of *Ailuropoda* (Nowak & Paradiso, 1983).

Family Procyonidae
Racoons: *Procyon*[AM]; ringtail: *Bassariscus*[AM]; coatimundi: *Nasua*[AM]; kinkajou: *Potos*[AMC]

$$i\frac{3}{3}, c\frac{1}{1}, p\frac{3-4}{3-4}, m\frac{2-3}{2-3}$$

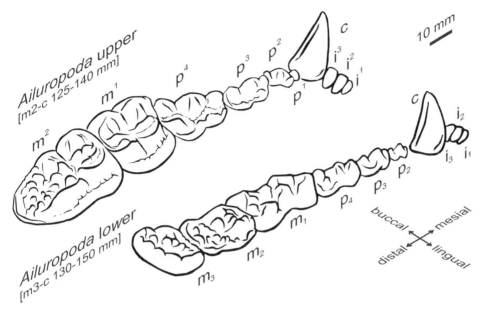

Figure 1.40 Ursidae. *Ailuropoda* upper right and lower left permanent dentitions. Dimensions are lengths of tooth row from canine to most distal molar.

Most procyonids are omnivorous, so the cheek teeth are low crowned and adapted for crushing rather than slicing. The blade of the carnassials is not strongly developed, and the fourth premolars (upper and lower) may take on the form of a molar, whilst the lower carnassial is not clearly distinguished from the rest of the lower molars (Figure 1.41). There is generally a pronounced parastyle on the upper carnassial. *Procyon* and *Nasua* follow this general description, whilst *Bassariscus* is the only procyonid in which the dentition approaches that of other carnivore families, with high carnassial blades. *Potos* is a tropical frugivore whose range only just fits into this book and has a very distinctive dentition with basin-like molar crowns that are scarcely recognisable as carnivore cheek teeth. They are much more like those of the fruit bats (below).

Order Pinnipedia
The pinnipeds include seals, sea lions, fur seals and walrus. Their diet consists of fish, squid, crustacea and molluscs. In most families, all the teeth are distal curving points, for catching and holding slippery fish. Cheek teeth (Figure 1.42) tend all to be similar and it is difficult to distinguish between premolars and molars, so they are just called *postcanines* (pc for short). The deciduous teeth are strongly reduced and mostly lost *in utero* (Kubota & Matsumoto, 1963; Kubota, 1963). There are three families: sea lions (Otariidae), found on Pacific coasts and around the oceans of the southern hemisphere; seals (Phocidae), worldwide in distribution; and walrus (Odobenidae), from the margins of the Arctic Ocean.

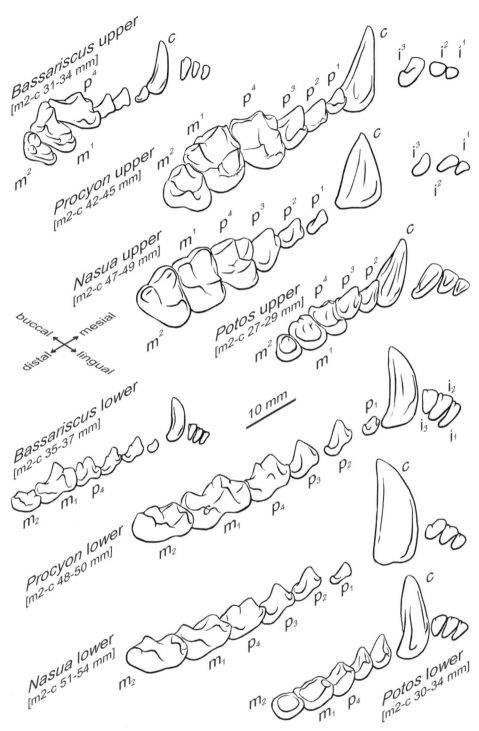

Figure 1.41 Procyonidae, upper right and lower left permanent dentitions. Dimensions are lengths of tooth row from canine to second molar.

Tooth form in mammals 65

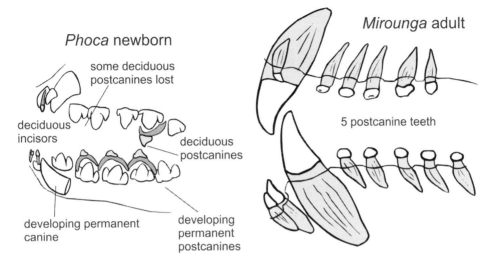

Figure 1.42 Pinnipedia, lingual view of upper and lower right dentitions. Permanent and deciduous dentitions in newborn *Phoca*. Permanent dentition of fully grown adult *Mirounga*.

Family Otariidae
Sea lions: *Zalophus*[PAC], *Eumetopias*[PAC]; fur seal: *Callorhinus*[PAC]

$$i\frac{3}{3}, c\frac{1}{1}, pc\frac{5-6}{5}$$

The upper third incisor is larger than the other two (Figure 1.43), and takes on a somewhat canine-like form. This size difference is more marked in the sea lions *Zalophus* and *Eumetopias* than in fur seals, in which the third incisor is also compressed buccal to lingual. The canines are large and stout, and show sexual dimorphism. Postcanine teeth are mostly simple points, narrower in the buccolingual diameter than the mesiodistal, with a lingual cingulum. *Callorhinus* has smaller, simpler points than the others, closely packed together. *Zalophus* has little cusplets arising from the cingulum at the mesial and distal ends of the crown in both jaws, as does *Eumetopias* in the lower dentition. Older individuals of *Eumetopias* characteristically have a large diastema between the fourth and fifth upper postcanines (this develops with age).

Family Phocidae
Subfamily Phocinae
Common, ringed, Baikal, Caspian, harp, ribbon and largha seals: *Phoca*[ATL,PAC,ARC]; grey seal: *Halichoerus*[ATL]; bearded seal: *Erignathus*[ARC]; hooded seal: *Cystophora*[ARC]; elephant seal: *Mirounga*[PAC]

$$i\frac{2-3}{2-1}, c\frac{1}{1}, pc\frac{5}{5}$$

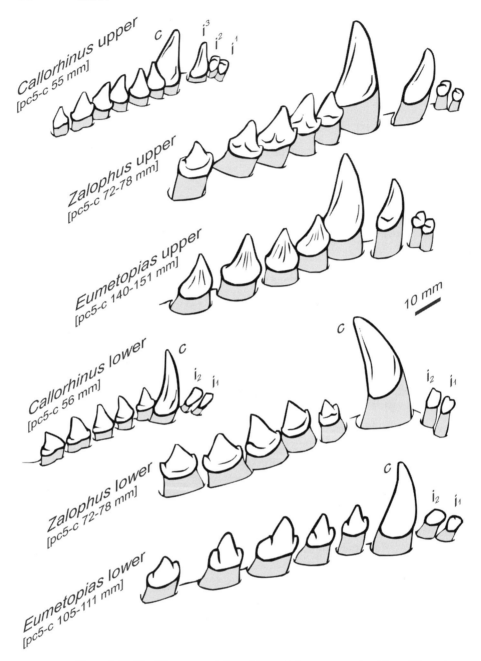

Figure 1.43 Otariidae, upper right and lower left permanent dentitions. Dimensions are lengths of tooth row from canine to most distal postcanine.

Tooth form in mammals

The incisors are backward-hooking points, often with a shelf-like cingulum on the lingual side (Figure 1.44). There are three upper and two small lower incisors in all phocids except *Cystophora*, which has two upper and one lower. As in the otariids, the upper third incisor is larger than the others, but the difference is much less marked. The canines are prominent (least so in *Erignathus*), but massive and strongly sexually dimorphic in *Mirounga*. The five postcanines in each quadrant are based on single points, with variable additional cusplets – one mesial and one or two distal. *Halichoerus* has a prominent, shelf-like cingulum on the lingual side of each postcanine, which are otherwise simple but have prominent ridges to mesial and distal which together form a recess on the lingual side. The postcanines in *Cystophora* are compressed buccal–lingual into a low blade with a rugose enamel surface, surrounded by a bulging cingulum, and small cusplets at their mesial and distal ends. *Erignathus* postcanines have low points, decorated with moderately prominent cusplets, and are not strongly held into their sockets by their roots, so they are frequently lost. *Phoca* is a rather variable genus, ranging in size from the smallest Caspian seal *P. caspis* up the largha seal *P. largha*, and with the postcanine cusplets ranging from moderate to prominent. They are least well developed in the ribbon seal *P. fasciata*, and strongest developed in the Baikal seal *P. sibirica*. *Mirounga* has simple, low pointed postcanine teeth without accessory cusplets. All phocids have deciduous dentitions, but they are tiny teeth which are exfoliated either before birth or very soon afterwards.

Subfamily Monachinae
Monk seals: *Monachus*[MED]

$$i\frac{2}{2}, c\frac{1}{1}, pc\frac{5}{5}$$

The dentition (Figure 1.44) of these seals is similar to that of the Phocinae, but the large postcanine teeth have one main point, a single distal accessory cusp, a tiny mesial cusplet and a prominent cingulum. The cingulum forms a pocket on the lingual side of the teeth.

Family Odobenidae
Walrus: *Odobenus*[PAC,ATL,ARC]

$$i\frac{1-2}{0}, c\frac{1}{1}, pc\frac{3-5}{3-4}$$

The walrus has a complex and variable succession of teeth (King, 1983), but the formula above shows the teeth that are likely to be present in a mature animal. Molluscs form the diet of the walrus and the very large permanent upper canines (up to 1 m in height) are used to turn over the bottom sediments. They erupt at four months of age with a thin cap of enamel which quickly wears away, and the tooth is formed of an ever-growing tusk (Figure 1.45) of dentine with a coating of cement. Tusks are present in both sexes, but they show strong dimorphism. The cheek teeth,

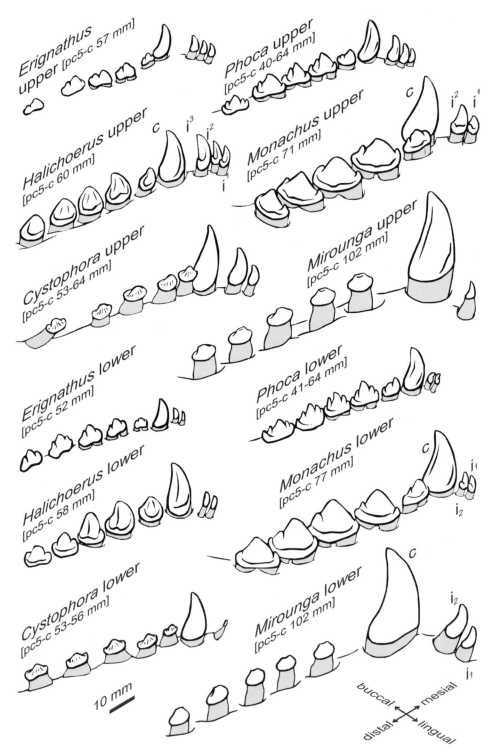

Figure 1.44 Phocidae, upper right and lower left permanent dentitions. Dimensions are lengths of tooth row from canine to most distal postcanine.

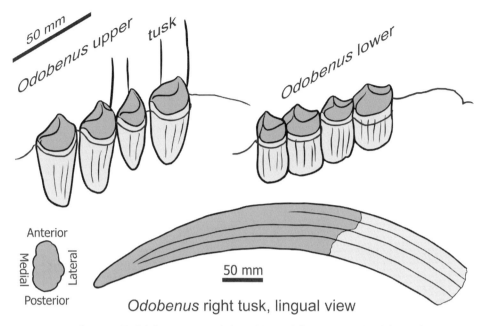

Figure 1.45 Odobenus, upper right and lower left permanent dentitions. Right tusk, seen in lingual view and section, at a smaller scale than the main drawing. Dark grey shading is dentine and paler grey cement.

which are a mixture of incisors, lower canines and postcanines, are also persistently growing, although this stops late in life. They too have a cap of enamel when first erupted, but this wears away and they appear as low, flat pegs of dentine in the jaw, with a thick layer of cement exposed around their sides.

Order Cetacea
Suborder Odontoceti
The Odontoceti are the toothed whales. Teeth in the other suborder of whales, the Mysticeti, remain only as tiny vestiges buried in the jaws, which instead bear many baleen plates. Odontoceti eat fish, squid, other small cephalopods, and shrimps. Their dentition usually consists of rows of simple, conical teeth, all alike (homodont). There may be very many in each jaw. Frequently, the enamel over the crown is very thin, but the cement may make a thick covering over the whole tooth. Teeth are often considerably worn in older individuals, so the enamel cap may be lost and the teeth reduced to simple pegs. Dental formulae are not appropriate for describing whale dentitions. Six families are commonly found in the seas around the Holarctic region. A seventh family, the Platanistidae or river dolphins, is confined to the rivers of South America, India and China.

Family Ziphiidae
Goose-beaked whale: *Ziphius*[ATL,PAC]; bottlenosed whale: *Hyperoodon*[ATL]; giant bottlenosed whale: *Berardius*[PAC]; beaked whales: *Mesoplodon*[ATL,PAC]

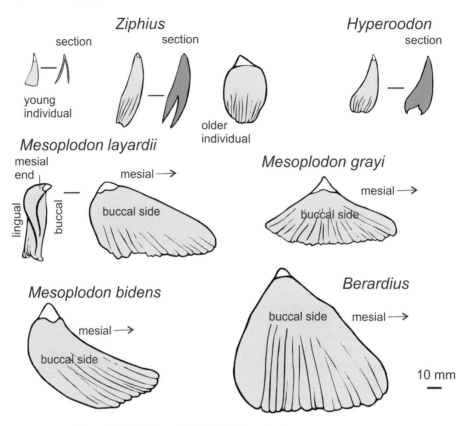

Figure 1.46 Odontoceti, Ziphiidae, lower teeth.

These whales are specialist cuttlefish feeders (Slijper, 1962) and this slow and easily caught prey does not require long rows of teeth. Only vestiges of teeth remain in the jaws, except for a single pair of large, stout teeth, almost tusks (Figure 1.46), which develop in the lower jaw. These erupt only in adults, usually only in males, and appear to be used for fighting. Cement gradually builds up to coat these teeth very thickly in older individuals, especially males. In *Ziphius* and *Hyperoodon* the teeth erupt at the anterior tip of the jaw. They start as thin-walled cones, but gradually take on an egg shape with the addition of cement (more rounded in the case of *Ziphius*). In *Mesoplodon* and *Berardius*, the teeth form gently curved plates which erupt part way along the jaw. The form of these plates varies between species and they grow largest in the extraordinary strap-toothed whale *M. layardii* from the South Pacific where they actually wrap around the upper jaw and closely limit the animals' gape.

Family Physeridae
Sperm whales: *Physeter*[ATL,PAC]; pigmy sperm whale: *Kogia*[ATL,PAC]

Like the Ziphiidae, these are cuttlefish and squid feeders. Their upper jaws have only vestigial teeth, buried deep in the periodontium, but their narrow lower jaws are armed with formidable rows of teeth. *Physeter* has 16–30 large, conical teeth (Harrison & King, 1980) on each side. They grow persistently from sharp points in young individuals, to large, tusk-like blunt worn points in older whales (Figure 1.47). Oval in transverse section, they may be up to 15 cm in diameter and up to 20 cm from root apex to occlusal tip. *Kogia* has 9–15 much smaller, narrower, hooked and pointed teeth on each side of the lower jaw.

Family Monodontidae
Narwhal: *Monodon*[ARC]; beluga or white whale: *Delphinapterus*[ARC]

Belugas have 8–10 teeth on either side of each jaw. They are rather slender (Figure 1.47) and may have accessory cusps, unlike other whales. Narwhals have four teeth, at the anterior tip of the upper jaw. Most of these are very small and do not erupt, but one of the mesial teeth, usually the left, forms a large tusk in males. This is straight and gently tapering, up to 300 cm long, and marked by a left-spiralling groove. A pulp chamber persists throughout its length, rendering the tusk somewhat fragile. In males, the mesial tooth from the opposite side of the jaw usually does not grow more than 29 cm in length and fails to erupt, but sometimes it may grow into a tusk as well. In females, both teeth are under 23 cm long.

Superfamily Delphinoidea
This is a large and varied group, mainly fish hunters, but taking a wide range of prey. It is divided into three families.

Family Stenidae
Rough-toothed dolphin: *Steno*[ATL,PAC]

There are 24 small conical teeth in each half of each jaw (Figure 1.47). Their enamel surfaces are characteristically rugose, or furrowed.

Family Phocaenidae
Common porpoise: *Phocaena*[ATL,PAC]; finless porpoise: *Neophocaena*[PAC]; dall porpoise: *Phocaenoides*[PAC]

There are up to 27 teeth in each jaw quadrant. They are small teeth (Figure 1.47) and, unlike any other odontocete, have spatulate crowns, compressed buccolingually. *Phocaenoides* has markedly reduced teeth.

Family Delphinidae
Killer whale: *Orcinus*[ATL,PAC]; false killer whale: *Pseudorca*[ATL,PAC]; pigmy killer whale: *Feresa*[ATL,PAC]; pilot whale: *Globicephala*[ATL,PAC]; common dolphin: *Delphinus*[ATL,PAC]; spinner dolphins: *Stenella*[ATL,PAC]; white-beaked/white-sided dolphins: *Lagenorhynchus*[ATL,PAC]; Fraser's dolphin: *Lagenodelphis*[ATL,PAC]; bottlenosed dolphin: *Tursiops*[ATL,PAC]; Risso's dolphin: *Grampus*[ATL,PAC]; right whale dolphin: *Lissodelphis*[PAC]

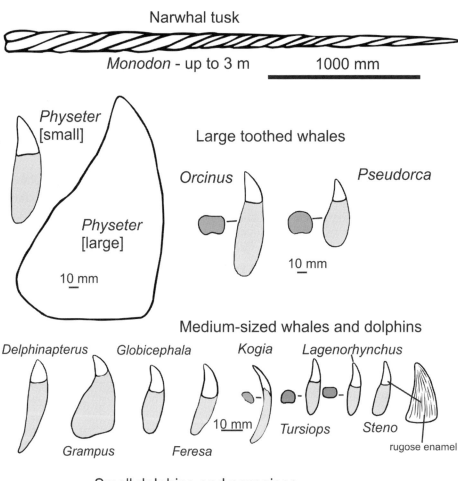

Figure 1.47 Odontoceti. Narwhal tusk, large, medium-sized and small whales, dolphins and porpoises. The largest teeth are at the top, smallest to the bottom, with a gradually increasing scale down the figure.

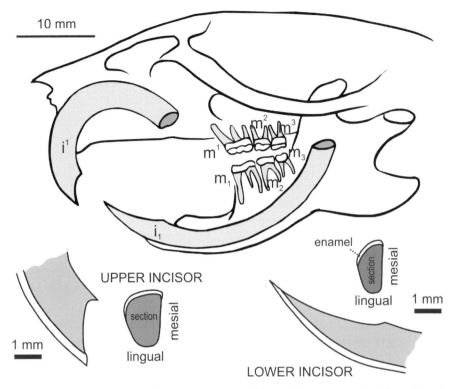

Figure 1.48 Muridae, *Rattus* upper and lower left permanent dentitions, from the buccal side.

The killer whales *Orcinus* and *Pseudorca* have a small number of relatively large teeth (Figure 1.47). The 10 or so teeth on each side of each jaw for *Orcinus* are the largest, and these killer whales are formidable predators which will even attack animals as large as the blue whale. *Pseudorca* has the same number of teeth, but they are smaller and have a different transverse section.

The teeth of other delphinids are much smaller. *Delphinus*, *Stenella*, *Lissodelphis* and *Lagenodelphis* may have 50 or more tiny sharp-pointed teeth in each jaw quadrant. The teeth are larger in *Lagenorhynchus* (22–45 teeth per quadrant), *Tursiops* (20–26 teeth per quadrant), *Feresa* (10–13 teeth per quadrant) and *Globicephala* (7–11 teeth per quadrant). *Grampus* has no teeth at all in the upper jaw, but 3–7 stout teeth on each side of the lower jaw.

Order Rodentia

All rodents have a similar dental arrangement with, in each half of each jaw, single incisors separated from three to five cheek teeth by a wide *diastema* or gap (Figure 1.48). Special folds of the upper lip fit into the diastema, protecting the rest of the mouth when the incisors are being used for gnawing. This is one of the main functions of the rodent dentition. The incisors are ever growing, with enamel only on the buccal surface. Continuous use and mutual sharpening ensures that these teeth are kept at a constant height in the jaw. Most of the actual work in gnawing is done by the lower incisors. The upper incisors are pressed into the object being gnawed, to hold the head steady, and leave characteristic pairs of rounded marks (Bang & Dahlstrom, 1972). The lower jaw moves backwards and forwards, gouging in the chisel-like lower incisors, to make double furrows which are the other characteristic of rodent gnawing. Rodent jaw articulation and musculature are specially arranged for this forward and backward movement. The cheek teeth have cusps, ridges or lophs aligned buccolingually so that, as the lower jaw moves, the enamel ridges cut against one another.

The diet of rodents is variable. All eat quantities of plant foods, ranging from soft bulbs, tubers and berries to nuts, seeds and bark. Many also eat insects, small vertebrates or carrion. The more omnivorous rodents tend to be those with low, brachydont molars, capped with separate cusps or low ridges. The more highly modified cheek tooth forms, with lophs or side infoldings, seem to be associated with a more specialised plant diet. The families of rodents below are arranged into broad groups of cheek tooth form. This is not meant in any way to imply phylogeny, but is merely a matter of convenience.

In many textbooks, the rodents are divided up into three suborders on the basis of the morphology of the masseter muscles, skull form and the enamel microstructure (Chapter 2):

Suborder Sciuromorpha includes the squirrels (Sciuridae) and scaly-tailed squirrels (Anomaluridae), with brachydont rooted cheek teeth, having buccolingual ridges across the occlusal surfaces. It also includes the beaver (Castoridae), with its hypsodont, rooted cheek teeth and complex occlusal surface, formed by a narrow, deep infolds from the sides. Four families have very simplified, tall cylindrical crowns with broad infolds from the sides – the mountain beaver (Aplodontidae), pocket gophers (Geomyidae), kangaroo rats (Heteromyidae) and the Cape jumping hare (Pedetidae).

Suborder Myomorpha have cheek teeth that vary from brachydont and cusped, to hypsodont/lophodont, to multiple prismatic crowns of persistent growth. Some families encompass all three types: mice, rats, (Muridae), hamsters, gerbils and voles (Cricetidae), jerboas (Dipodidae), jumping mice (Zapodidae). Other myomorph families have hypsodont crowns in all genera – mole rats (Spalacidae) and bamboo rats (Rhizomyidae). Yet other families all have low, brachydont crowns. The dzhalmans (Seleviniidae) have a very simplified occlusal surface. Dormice (Gliridae) and spiny dormice (Platacanthomyidae) have low buccolingual ridges running across the occlusal surfaces.

Suborder Hystricomorpha is a highly varied group. Old World hystricomorphs include the porcupines (Hystricidae), cane rats (Thryonomidae) and rock rats

(Petromyidae). Cheek teeth of these families are hypsodont, but rooted, having an occlusal pattern formed of deep folds from the lingual and buccal sides of the crown. Other Old World hystricomorphs have very simplified teeth with tall, cylindrical crowns bearing broad folds in the sides – the mole rats (Bathygeridae) and gundis (Ctenodactylidae). Romer (1966) distinguished the New World hystricomorphs as a separate group, the Caviomorpha. He split this group into four 'family groups'. First were the octodonts, spiny rats, rat chinchillas, tuco-tucos and coypus (Octodontidae, Echimyidae, Abrocomidae, Ctenomyidae and Capromyidae). These have brachydont to hypsodont teeth, rooted or persistently growing. Patterns are formed in the occlusal surface by infolds from the crown sides. Romer's second family group comprises the chinchillas and viscachas, pacarana, agoutis and pacas (Chinchillidae, Dinomyidae and Dasyproctidae). Crowns in this group are mostly made up of high plates, not joined by cement. The third family group includes guinea pigs and capybara (Caviidae and Hydrochoeridae), with high, persistently growing crowns, made up of vertical plates joined by cement. The fourth family group comprises the New World porcupines (Erethezontidae), which have rather lower, rooted crowns and an occlusal pattern formed by less prominent infolds from the sides.

Family Muridae
Old World mice and rats: *Micromys*[EU,AS], *Apodemus*[EU,AF,AS], *Mus*[EU,AF,AS,AM], *Praomys*[AF], *Acomys*[AF,ASW], *Arvicanthis*[AF], *Lemniscomys*[AF], *Rattus*[EU,AF,AS,AM], *Nesokia*[AF,AS]

$$i\frac{1}{1}, c\frac{0}{0}, p\frac{0}{0}, m\frac{3}{3}$$

This large family now ranges throughout the world, as several species of rats and mice have developed a commensal relationship with humans, although its original distribution was confined to Africa, Europe and all but the most northerly parts of Asia. Murid molar crowns are based on a cusp-cum-loph form in which the main cusps are arranged in buccal to lingual rows, joined by ridges (Figures 1.49–1.50). The separate cusps are only clearly distinguishable in the unworn state. Wear converts this arrangement into a series of incipient lophs or, in *Rattus*, *Arvicanthis*, *Lemniscomys* and especially *Nesokia*, something approaching a true loph. In the upper molars, each row consists of three cusps, the central one being larger than others whereas, in the lower molars, each row has two similarly sized cusps. There are also characteristic additional cusps: a distal one (flattened mesiodistally) in the first and second lower molars, a mesiolingual one on the upper second and third molars, and a variety of others. Of the smaller mice, *Mus* and *Praomys* are distinguished by the strongly asymmetrical mesial loph of their upper first molars, and reduced upper and lower third molars. *Acomys* also has a moderately asymmetrical mesial loph. The wood mouse *Apodemus* and harvest mouse *Micromys* both have symmetrical mesial lophs, but are clearly

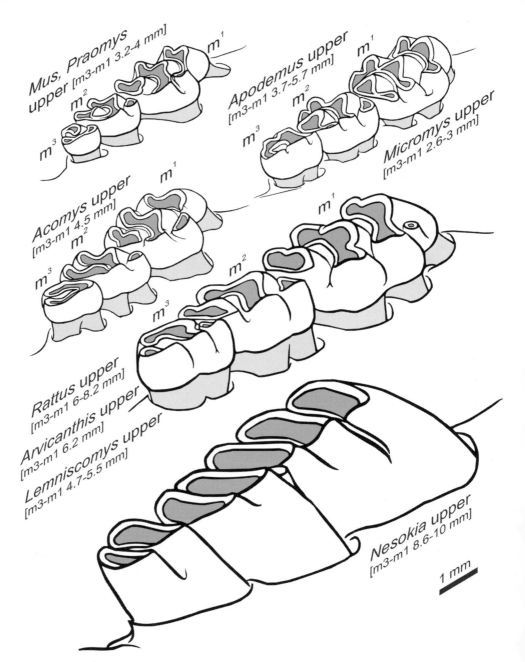

Figure 1.49 Muridae, upper right permanent cheek teeth. Dimensions are lengths of tooth row from mesial side of first molar to distal side of third molar.

distinguishable by size – *Micromys* is much the smallest mouse. The rats *Rattus*, *Arvicanthis* and *Lemniscomys* are all larger, with a rather similar form in which the cusps are somewhat more strongly gathered into lophs than in the smaller mice. *Nesokia* is distinguished by its large, fully hypsodont molars, with strongly developed lophs.

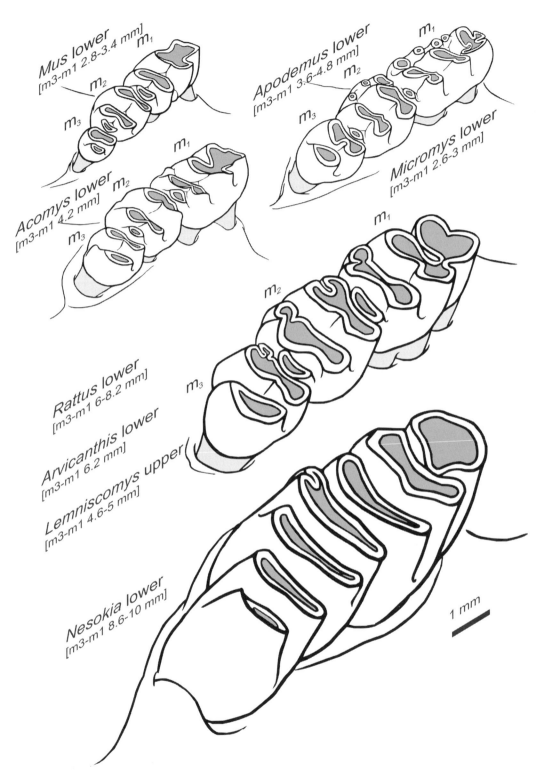

Figure 1.50 Muridae, lower left permanent cheek teeth. Dimensions are lengths of tooth row from mesial side of first molar to distal side of third molar.

Family Cricetidae

$$i\frac{1}{1}, c\frac{0}{0}, p\frac{0}{0}, m\frac{3}{3}$$

Corbet (1978) separated off the other mouse-like rodents from the Muridae, and placed most of them as a series of subfamilies within the Cricetidae. The reasons for doing this are beyond the scope of this book, but the division is convenient dentally. Although there are very great differences between subfamilies in hypsodonty and complexity, the basic form of molar crown is rather similar, with two parallel, mesiodistally running rows of cusps.

Subfamily Cricetinae
Hamsters: *Cricetus*[EU,AS], *Cricetulus*[EU,AS], *Mesocricetus*[EU,AS], *Calomyscus*[ASC], *Phodopus*[ASE]

The hamsters are amongst the lowest crowned of the cricetids. In both upper and lower jaws, the first molars have six main cusps arranged into two rows (Figure 1.51), and the second and third molars have four arranged into a rectangle. When worn, the ridges connecting the cusps form a zigzag (Figure 1.52). *Calomyscus* is distinguished by its small size, the reduced mesial cusps of its upper and lower first molars, and its reduced third molars. *Cricetulus* and *Phodopus* are also small, but have molars similar in form to the larger *Cricetus*. *Mesocricetus* is also large, but has molar crowns which are relatively narrow buccolingually, with slightly more prominent cusps.

Subfamily Hesperomyinae
New World mice and rats: *Oryzomys*[AM], *Reithrodontomys*[AM], *Peromyscus*[AM], *Ochrotomys*[AM], *Baiomys*[AMC], *Onychomys*[AM], *Sigmodon*[AM], *Neotomodon*[AMC], *Neotoma*[AM]

The molars of the lower-crowned Hesperomyinae are superficially similar to the Old World cricetines (Figures 1.53–1.54). They differ, however, in the way in which their cusp rows are staggered relative to one another, and by the form of the connecting ridges. Much the smallest of the Holarctic genera described here is *Baiomys*, the pygmy mouse. The American harvest mouse *Reithrodontomys* is not much larger and is rather similar in molar form, but takes its Latin name from its distinctive strongly grooved incisors. *Peromyscus*, *Ochrotomys* and *Onychomys* are larger still. Like *Baiomys* and *Reithrodontomys*, *Onychomys* has rather reduced third molars, but *Peromyscus* and *Ochrotomys* are also distinguished by additional folds between the main cusps, on the buccal (for upper molars) or lingual (for lower molars) sides. The largest of the low-crowned forms is the rice rat *Oryzomys*, and this also has the additional folds.

In three genera the molars are hypsodont, although they still develop separate roots. Of these, the highest crowned is *Neotoma*, the woodrat. *Sigmodon* and *Neotomodon* have less tall crowns, but can be distinguished from each other by

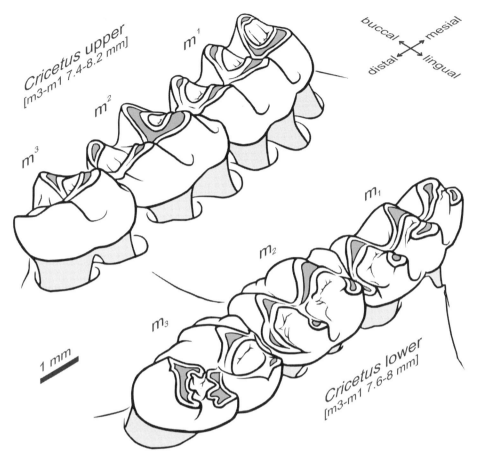

Figure 1.51 Cricetidae. *Cricetus*, isometric views of upper right and lower left permanent cheek teeth. Dimensions are lengths of tooth row from mesial side of first molar to distal side of third molar.

the details of their worn occlusal surfaces, and by the reduction of the third molars in *Neotomodon*.

Subfamily Gerbillinae
Gerbils and jirds: *Gerbillus*[AF,ASW], *Dipodillus*[AF], *Pachyuromys*[AF], *Tatera*[AF,AS], *Meriones*[AF,AS], *Sekeetamys*[AF,ASW], *Psammomys*[AF,AS], *Brachiones*[ASE], *Rhombomys*[ASC,ASE]

Gerbils are all relatively tall crowned compared with cricetines, but they vary in their level of hypsodonty (Figures 1.55–1.56). *Gerbillus* and *Dipodillus* are the lowest crowned, although this is seen mostly in the first molars and the second and third are reduced to simple lophs. Next in crown height come *Pachyuromys* and *Tatera*, in which the characteristic gerbil form of three lophs in the first molar,

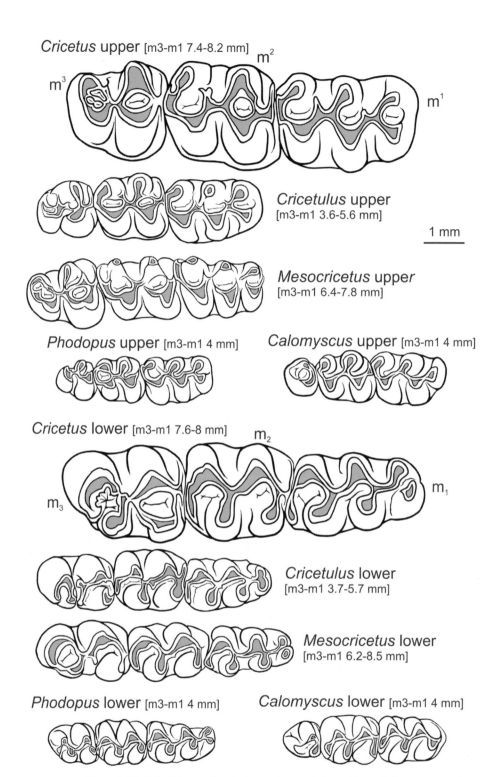

Figure 1.52 Cricetidae, occlusal views of upper right and lower left permanent cheek teeth. Dimensions are lengths of tooth row from mesial side of first molar to distal side of third molar.

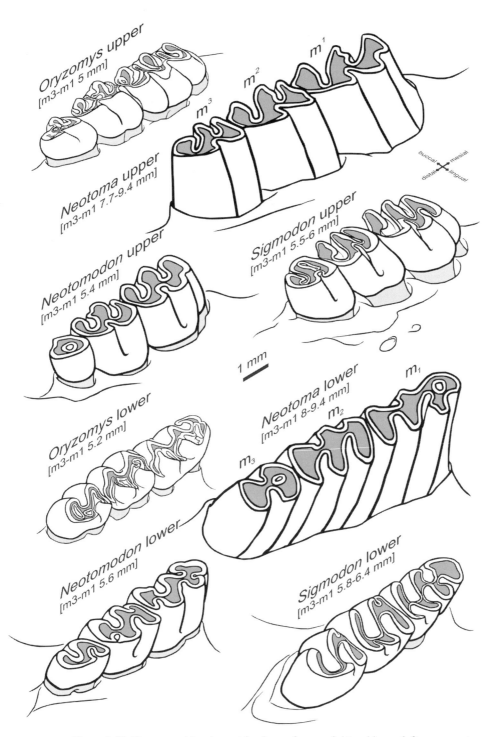

Figure 1.53 Hesperomyidae, isometric views of upper right and lower left permanent cheek teeth. Dimensions are lengths of tooth row from mesial side of first molar to distal side of third molar.

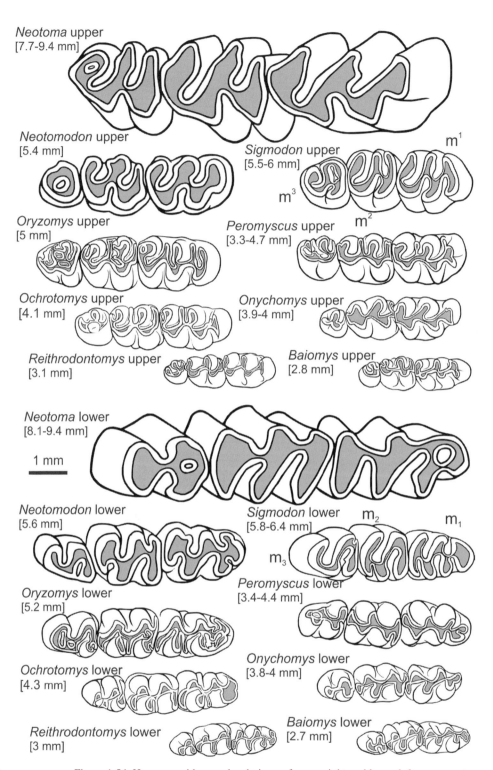

Figure 1.54 Hesperomyidae, occlusal views of upper right and lower left permanent cheek teeth. Dimensions are lengths of tooth row from mesial side of first molar to distal side of third molar.

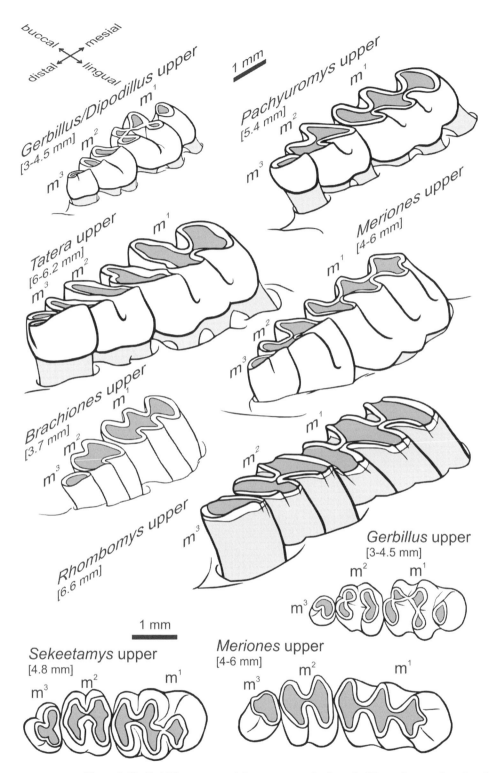

Figure 1.55 Gerbillinae, upper right permanent cheek teeth. Dimensions are lengths of tooth row from mesial side of first molar to distal side of third molar.

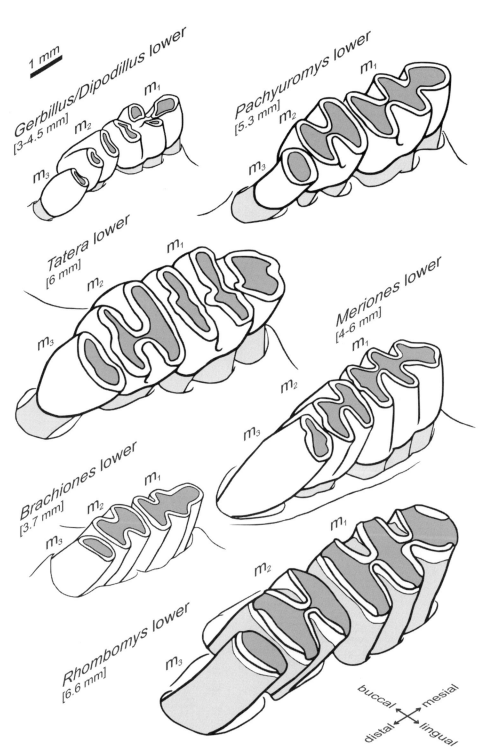

Figure 1.56 Gerbillinae, lower left permanent cheek teeth. Dimensions are lengths of tooth row from mesial side of first molar to distal side of third molar.

two in the second and one in the third is clearly established even though the teeth are prominently rooted. Higher crowned still are *Meriones* and *Sekeetamys* (with the latter distinguished by an infolding in the upper third molars). Highest of all are *Brachiones* and *Rhombomys*. The great gerbil *Rhombomys* is distinguished in three ways – it is larger than all the others, has non-rooted persistently growing molars, and the enamel jacket which surrounds the molars of the other genera is breached in a strip down the sides (rather like the voles below).

Subfamily Microtinae
Small voles with unrooted molars: $Microtus^{EU,AS,AM}$, $Pitymys^{EU,ASW,AM}$, $Hyperacrius^{ASC}$, $Alticola^{ASC,ASE}$, $Eothenomys^{ASE}$, $Lagurus^{ASC,ASE,AM}$; lemmings (with unrooted molars): $Dicrostonyx^{EU,AS,AM}$, $Lemmus^{EU,AS,AM}$, $Myopus^{EU,AS}$, $Synaptomys^{AM}$; large voles with unrooted molars: $Arvicola^{EU,AS,AM}$, $Neofiber^{AM}$; small and large voles with molars which become rooted: $Clethrionomys^{EU,AS,AM}$, $Prometheomys^{ASW}$, $Ellobius^{AS}$, $Dinaromys^{EU}$, $Phenacomys^{AM}$; $Ondatra^{AM,(EU,AS)}$

The Microtinae have very distinctive molars which consist of multiple prismatic elements (Figure 1.57), exposed by wear along a very long (mesiodistally) occlusal surface. The form of the intricately folded occlusal outlines varies between the genera, and this is the key to identification (Figures 1.58–1.61). Other generally distinctive features include the strips of enamel which are missing down the sides of the crowns. These occur in different places in different genera. All microtines have persistently growing teeth and, in most of them, roots never develop. A small number of genera (see list above) develop roots late in the life of the teeth (Figure 1.57). The prismatic elements represent the twin cusp rows of the lower-crowned cricetid forms described above, staggered relative to one another and separated by sharp infolds ('re-entrants') from the buccal and lingual sides of the crown. In most of the genera, these infolds are approximately the same depth on both sides but in the lemmings *Lemmus*, *Myopus* and *Synaptomys* the folds are strongly asymmetrical. The remaining member of the lemming group, *Dicrostonyx*, has symmetrical folds, but is exceptional in the strongly compressed nature of its prismatic elements. Most of the microtines are fairly similar in size, but *Arvicola*, *Neofiber*, *Prometheomys*, *Ellobius* and *Dinaromys* are larger, and the muskrat *Ondatra* is largest of all. *Pitymys* is distinguished by a characteristic infold in the upper second molar (Figure 1.59), although this is variably developed.

Subfamily Myospalacinae
Mole-rats, zokors: $Myospalax^{ASE}$

Zokors are solidly built rodents, with heavy claws for burrowing. Their molars are vole-like (and therefore included in Figure 1.58), persistently growing and made of multiple prismatic elements, but they differ in the form of the occlusal surface, and in the larger strips of enamel that are missing down the crown sides.

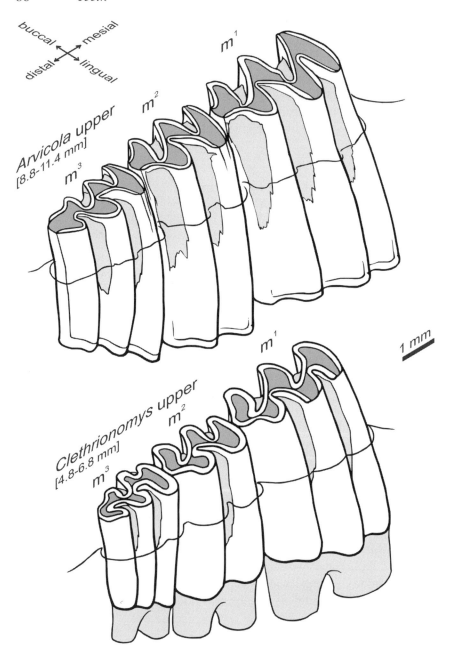

Figure 1.57 Microtinae, upper left permanent cheek teeth. Isometric views showing *Arvicola* as an ever-growing unrooted form, and *Clethrionomys* which becomes rooted as the crown stops growing. Dark grey shading is dentine and paler grey is cement.

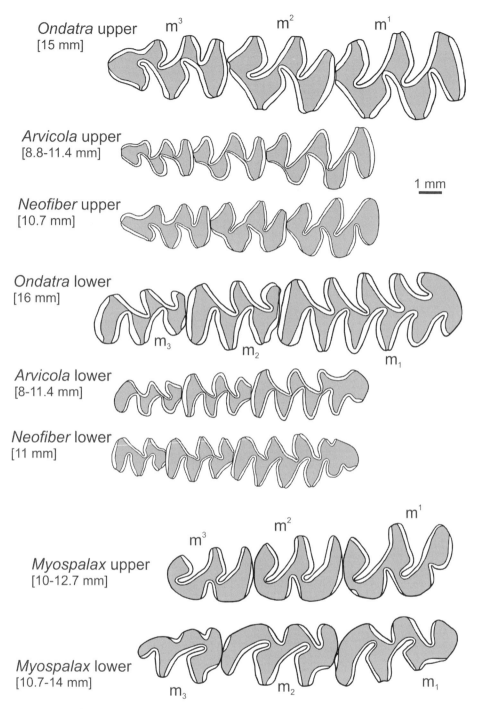

Figure 1.58 Large Microtinae, occlusal views of upper right and lower left permanent cheek teeth. *Myospalax* is not a microtine, but is included because it has similar teeth. Dimensions are lengths of tooth row from mesial side of first molar to distal side of third molar. N.B. drawn at a scale smaller than Figures 1.59–1.61.

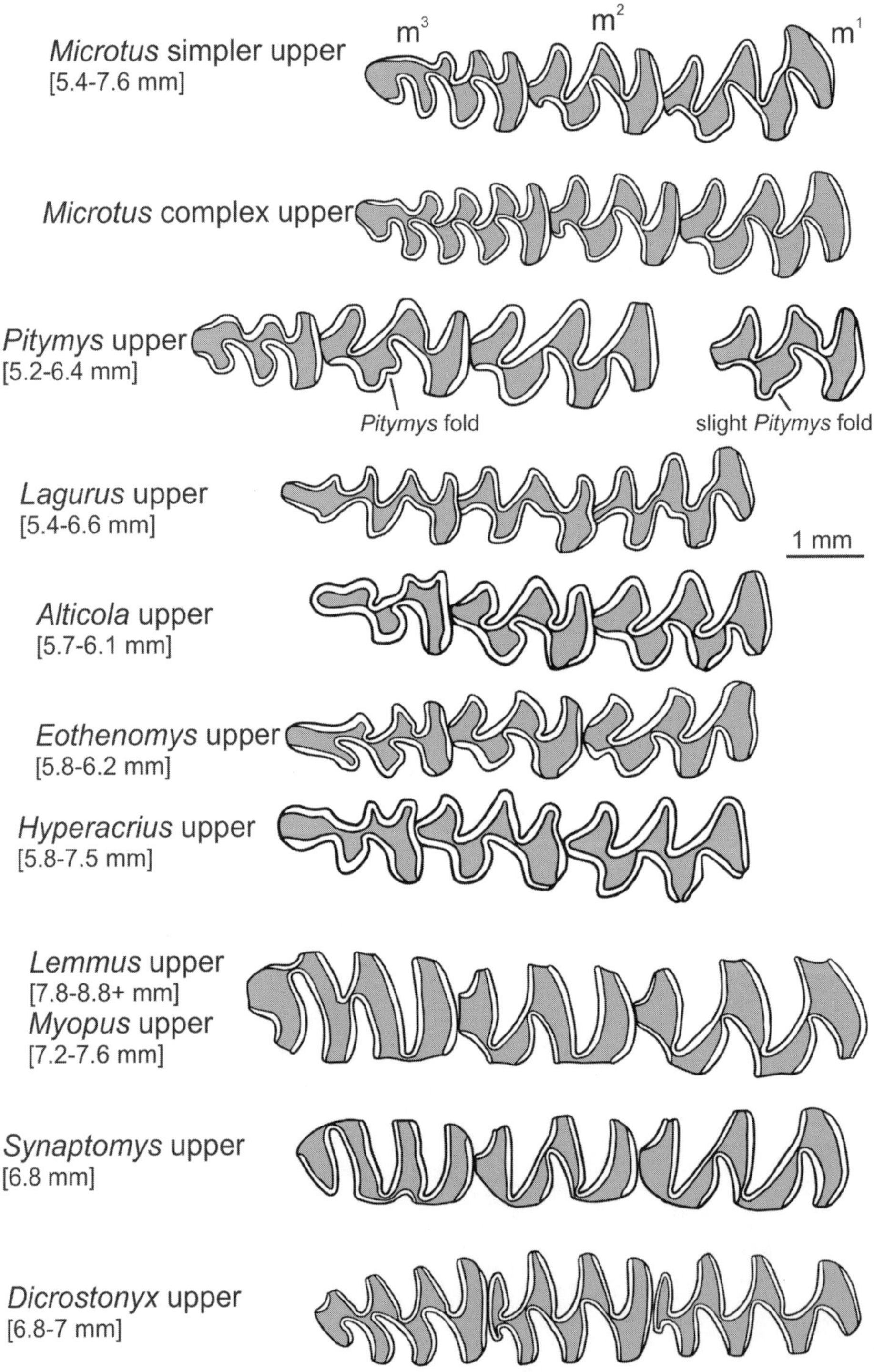

Figure 1.59 Medium-sized Microtinae, occlusal views of upper right permanent cheek teeth. Dimensions are lengths of tooth row from mesial side of first molar to distal side of third molar. N.B. drawn at a scale larger than Figure 1.59 and smaller than Figure 1.61.

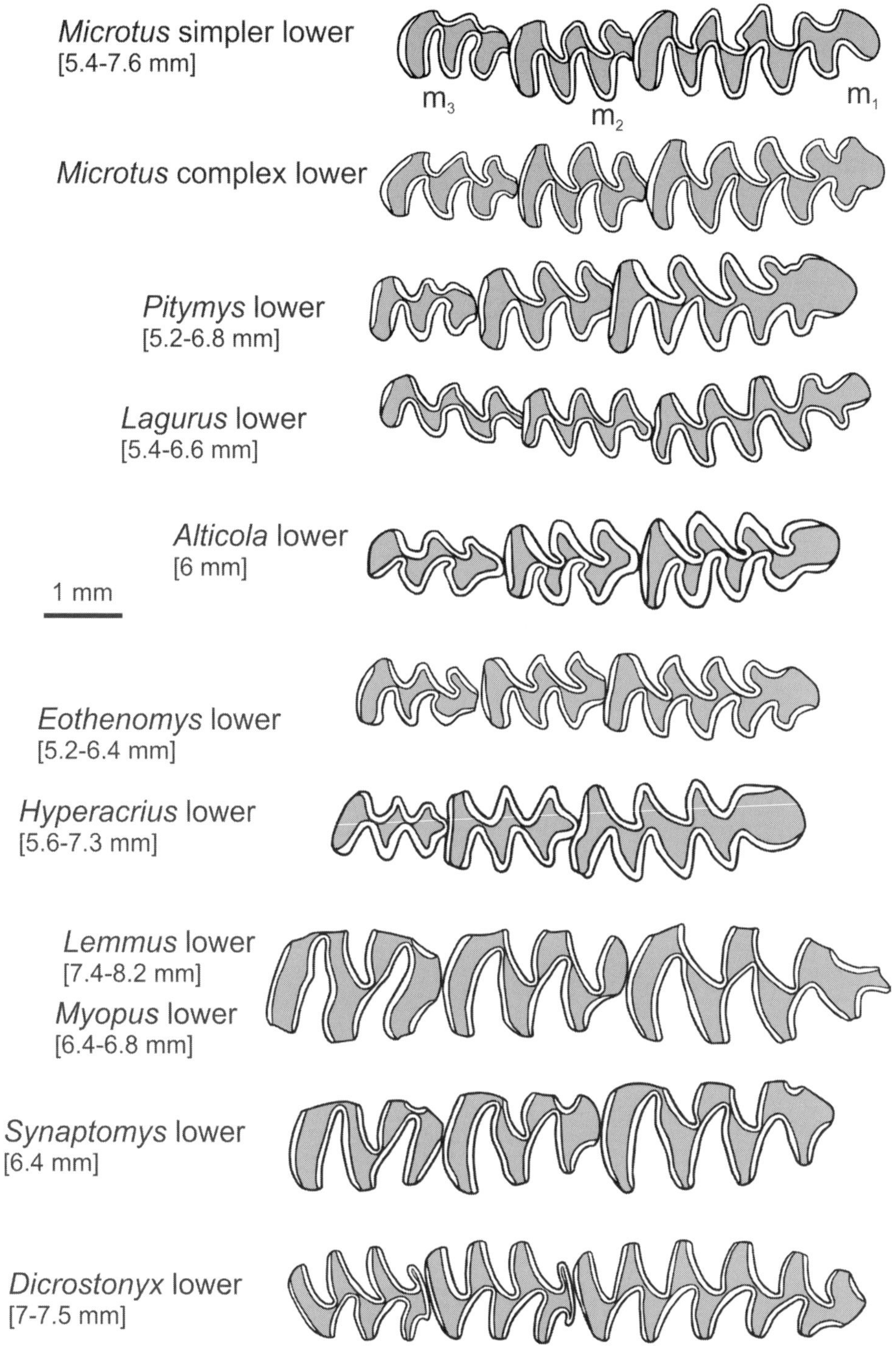

Microtus simpler lower
[5.4-7.6 mm]

m_3 m_2 m_1

Microtus complex lower

Pitymys lower
[5.2-6.8 mm]

Lagurus lower
[5.4-6.6 mm]

Alticola lower
[6 mm]

1 mm

Eothenomys lower
[5.2-6.4 mm]

Hyperacrius lower
[5.6-7.3 mm]

Lemmus lower
[7.4-8.2 mm]

Myopus lower
[6.4-6.8 mm]

Synaptomys lower
[6.4 mm]

Dicrostonyx lower
[7-7.5 mm]

Figure 1.60 Medium-sized Microtinae, occlusal views of lower left permanent cheek teeth. Dimensions are lengths of tooth row from mesial side of first molar to distal side of third molar. N.B. drawn at a scale larger than Figure 1.58 and smaller than Figure 1.61.

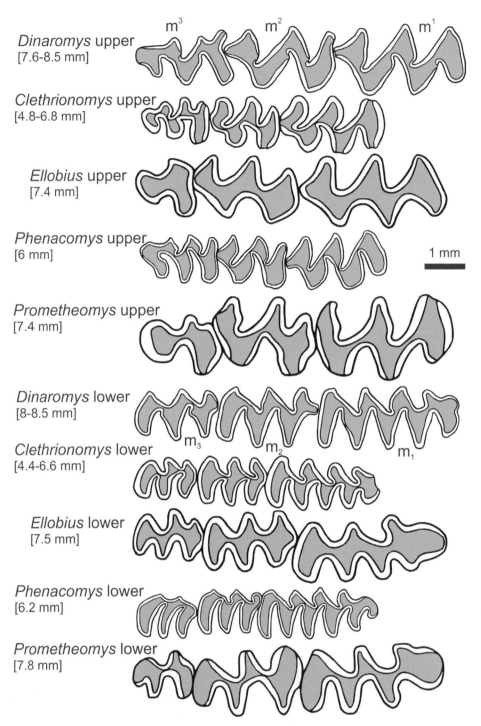

Figure 1.61 Small Microtinae, occlusal views of upper right and lower left permanent cheek teeth. Dimensions are lengths of tooth row from mesial side of first molar to distal side of third molar. N.B. drawn at a scale larger than Figure 1.60.

Family Zapodidae
Birch mice: *Sicista*[EU,AS]; jumping mice: *Eozapus*[ASE], *Zapus*[AM], *Napaeozapus*[AM]

$$i\frac{1}{1}, c\frac{0}{0}, p\frac{1}{0}, m\frac{3}{3}$$

These small nocturnal rodents scamper and jump through the woods, scrub and steppe of eastern Europe, Asia and North America. They eat berries, seeds and small invertebrates. *Eozapus*, *Zapus* and *Napaeozapus* have specially elongated hind legs for jumping, and possess more hypsodont cheek teeth than *Sicista*. The pattern of ridges and folds in these tiny teeth is highly distinctive, but it requires a good microscope to make them out (Figure 1.62). *Napaeozapus* and *Zapus* are notable as the owners of some of the most complex hypsodont folds of the mammal world, in teeth only 1–2 mm long. *Eozapus* has somewhat lower crowned teeth, in which a simplified set of folds is present, and *Sicista* has distinctly brachydont crowns with low ridges.

Family Dipodidae
Jerboas: *Salpingotus*[ASC], *Dipus*[ASC,ASE], *Stylodipus*[AS], *Jaculus*[AF,ASW,ASC], *Euchoreutes*[ASE], *Allactaga*[AS], *Allactagulus*[AS], *Pygeretmus*[ASC], *Paradipus*[ASC], *Cardiocranius*[ASC,ASE]

$$i\frac{1}{1}, c\frac{0}{0}, p\frac{0-1}{0}, m\frac{3}{3}$$

Jerboas are small jumping rodents, with massively elongated hind feet and long tails, found in steppe and desert conditions throughout North Africa, West, Central and East Asia. They dig burrows for protection from predators and climate, but emerge at night to forage for seeds, succulent plant foods and insects. Their cheek teeth crowns are based on two rows of cusps, gathered up to a variable extent into a series of transverse and oblique lophs (Figures 1.63–1.64). The lowest crowns are in the tiny teeth of *Salpingotus*, whose upper fourth premolar and third molars are reduced to tiny globular bodies. *Cardiocranius* is described as similar to *Salpingotus* (Nowak & Paradiso, 1983) but was not seen by the author. *Dipus*, *Stylodipus* and *Jaculus* have taller-crowned cheek teeth in which the lophs are simplified, with a reduction in the most mesial fold, to give a somewhat 'Z'-shaped occlusal outline in worn first and second molars. Of these three, *Dipus* has the lowest crowns and is the only one to retain an upper fourth premolar. *Allactaga*, *Allactagulus*, *Euchoreutes* and *Pygeretmus* have even taller crowns, decorated with a complex set of lophs. Of these, both *Euchoreutes* and *Allactaga* retain upper fourth premolars, whilst *Euchoreutes* is distinguished by its greatly reduced third molars, and *Allactagulus* and *Pygeretmus* do not have upper fourth premolars. The little-known Central Asiatic species *Paradipus ctenodactylus* is the highest crowned of all, with prominent infundibula which become isolated as 'islands' in worn occlusal surfaces, and broad areas of missing enamel on the buccal and lingual crown sides.

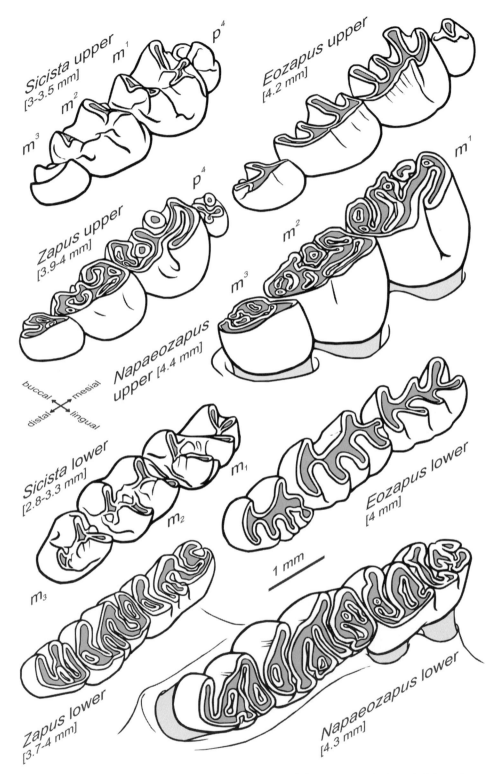

Figure 1.62 Zapodidae, upper right and lower left permanent cheek teeth. Dimensions are lengths of tooth row from mesial side of first molar to distal side of third molar.

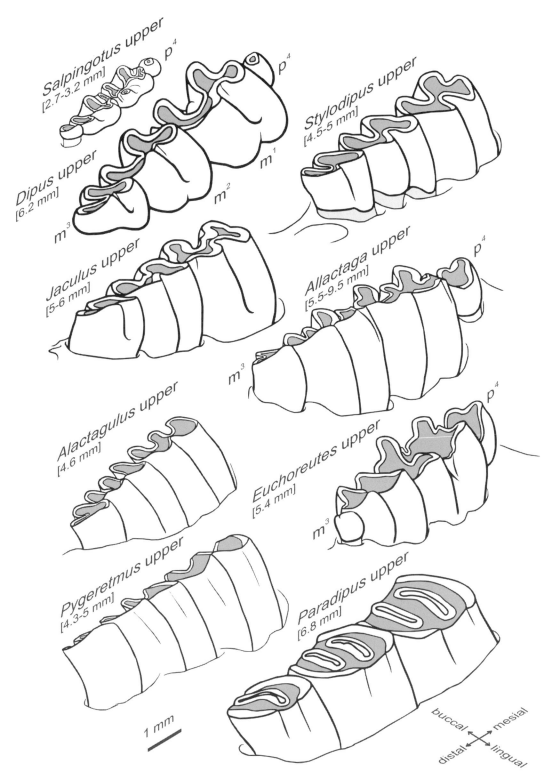

Figure 1.63 Dipodidae, upper right permanent cheek teeth. Dimensions are lengths of tooth row from mesial side of first molar to distal side of third molar.

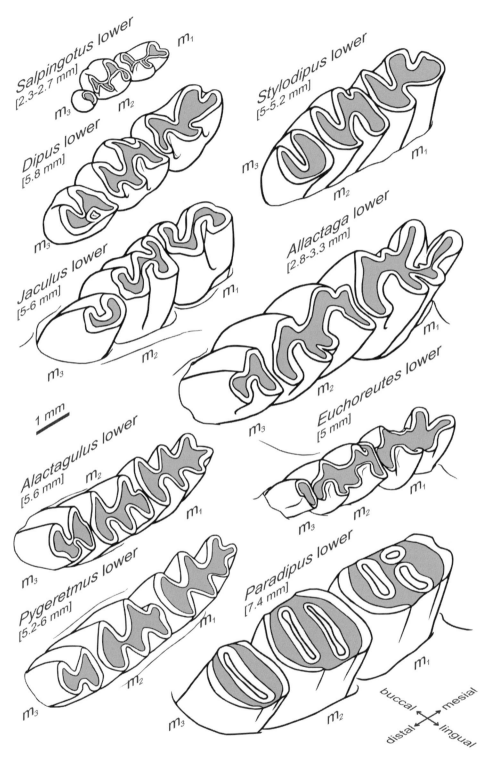

Figure 1.64 Dipodidae, lower left permanent cheek teeth. Dimensions are lengths of tooth row from mesial side of first molar to distal side of third molar.

Family Heteromyidae
Pocket mice: *Perognathus*[AM], *Liomys*[AM]; kangaroo mice: *Microdipodops*[AM]; kangaroo rats: *Dipodomys*[AM]

$$i\frac{1}{1}, c\frac{0}{0}, p\frac{1}{1}, m\frac{3}{3}$$

Most heteromyids are small, rather mouse-like creatures, with the exceptions of the kangaroo rats and mice *Dipodomys* and *Microdipodops*, which have greatly elongated hind legs for jumping, rather like those of the Dipodidae or Zapodidae. All forms share a common feature with the gophers in the possession of large cheek pouches, but they are nocturnal foragers above ground, who shelter in their burrows only during the day or in bad weather. Heteromyids are found in a variety of habitats through central and south-west North America, Central America and the most north-eastern parts of South America, and they feed largely on vegetation with the addition of some invertebrates. The cheek teeth (Figure 1.65) are gathered up into two transverse (buccal–lingual) lophs, and the crowns vary in hypsodonty from the lower-crowned *Perognathus*, to the higher crowned but still rooted *Liomys* and *Microdipodops*, to the very high crowned and persistently growing teeth of *Dipodomys*. In molars there is a small infundibulum in the deepest part of the space between lophs, which for upper molars is isolated by wear first on the lingual and then on the buccal side (reversed in the lower molars), to create an 'island' in the middle of the occlusal surface. This island is lost with further wear in the highest crowned forms. A similar pattern is seen in the upper and lower fourth premolars of *Liomys* but in the other forms the lophs of these teeth are joined first by wear more or less at the centre line of the crown. The upper incisors bear a prominent labial groove in some species, but are smooth in others.

Family Ctenodactylidae
Gundis: *Ctenodactylus*[AF], *Massoutheria*[AF]

$$i\frac{1}{1}, c\frac{0}{0}, p\frac{0}{0}, m\frac{3}{3}$$

Gundis live amongst the rocks in desert regions, feeding on a range of plant foods, and have one of the most simplified mammal dentitions. The persistently growing molars (Figure 1.66) are reduced to curving enamel jacketed pillars which wear to a characteristic hourglass form in the occlusal surface. The infolds from the side are more pronounced in *Massoutheria* than in *Ctenodactylus*.

Family Geomyidae
Pocket gophers: *Thomomys*[AM], *Geomys*[AM], *Pappogeomys*[AMC]

$$i\frac{1}{1}, c\frac{0}{0}, p\frac{1}{1}, m\frac{3}{3}$$

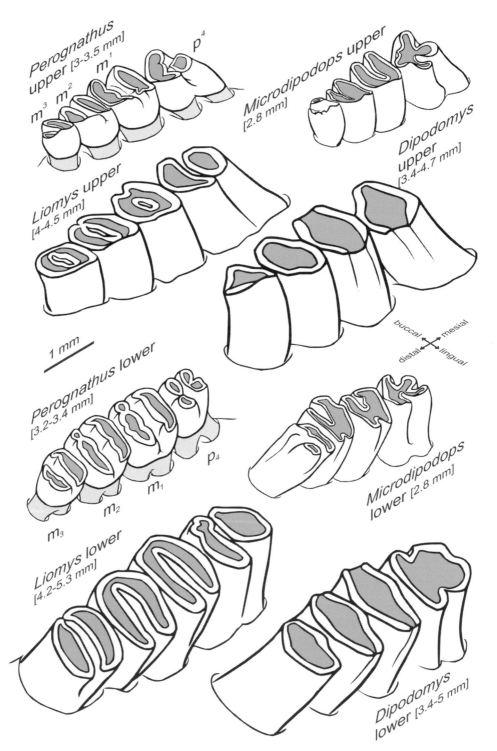

Figure 1.65 Heteromyidae, upper right and lower left permanent cheek teeth. Dimensions are lengths of tooth row from mesial side of first molar to distal side of third molar.

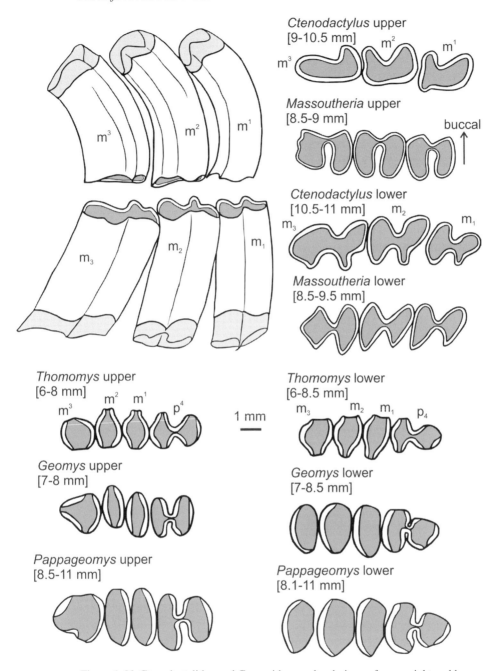

Figure 1.66 Ctenodactylidae and Geomyidae, occlusal views of upper right and lower left permanent cheek teeth. Upper and lower left permanent dentitions of *Ctenodactylus* are also shown in buccal view. Dimensions are lengths of tooth row from mesial side of first molar to distal side of third molar.

Pocket gophers are specialist diggers and burrowers, confined in range to North and Central America. They are stocky in build, with relatively robust jaws, and large incisors that can be sealed away from the rest of the mouth by special lip folds, and so used for digging. The upper incisors have labial grooves in *Geomys* and *Pappogeomys*, but are mostly smooth in *Thomomys*. Either side of the mouth, gophers have large fur-lined pockets in which they carry food. These pockets can be turned inside out for emptying and cleaning, and then pulled back in again by special muscles. Gophers mostly eat the underground parts of plants exposed in their burrows, and rarely emerge from their solitary existence. Their cheek teeth (Figure 1.66) are simple, ever-growing and non-rooted, with a thin surface coat of cement over the enamel jacket, which is broken in places to expose lines of dentine running up the sides under the cement. Upper and lower fourth premolars have a 'figure-of-eight' outline in occlusal view, but all the other cheek teeth are a simple oval in outline (slightly more complex in the third molars). *Thomomys* diverges from the other two genera in the 'teardrop' shape of the occlusal outline, whilst *Geomys* and *Pappogeomys* differ in the distribution of enamel around the molar sides.

Family Spalacidae
Mole rats: *Spalax*[EU,ASW,ASC]

$$i\frac{1}{1}, c\frac{0}{0}, p\frac{0}{0}, m\frac{3}{3}$$

Mole rats are solidly built, blind burrowing animals with specialised powerful jaws and incisors which they use for digging. They eat plant tubers and bulbs, and their cheek teeth (Figure 1.67) are strongly hypsodont, with tall folds wearing rapidly to a distinctive S-shaped occlusal pattern and eventually to a round surface with enamel 'islands'.

Family Erethezontidae
North American porcupine: *Erethizon*[AM]

$$i\frac{1}{1}, c\frac{0}{0}, p\frac{1}{1}, m\frac{3}{3}$$

Members of this family live throughout much of North America, Central America and the northern part of South America. The North American porcupine *Erethizon* forages on a wide range of plant foods, particularly the bark and cambium of trees. Its jaws are similarly powerful, but its cheek teeth (Figures 1.68–1.69) are relatively low crowned, with prominent roots and occlusal surfaces in which the infoldings persist with increasing wear.

Family Hystricidae
Old World porcupines: *Hystrix*[EU,AF]

$$i\frac{1}{1}, c\frac{0}{0}, p\frac{1}{1}, m\frac{3}{3}$$

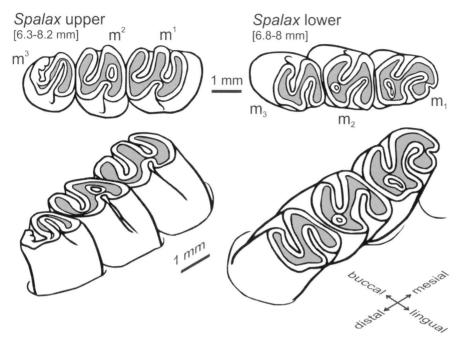

Figure 1.67 Spalacidae, upper right and lower left permanent cheek teeth. Dimensions are lengths of tooth row from mesial side of first molar to distal side of third molar.

Old World porcupines live in a variety of environments, foraging at night on many different plant foods, occasionally invertebrate animals and rarely carrion. Bones are collected along with other objects in porcupine burrows or dens, and are gnawed possibly to keep the large incisors trimmed. They are large rodents, with powerful jaws and strongly hypsodont, but still rooted, cheek teeth (Figure 1.69). In little-worn upper teeth, there are two or three buccal infoldings and one lingual infolding. Further down the tall crowns, these folds become infundibula which make isolated 'islands' in the worn occlusal surface, and create a pattern which is characteristic of *Hystrix*. The arrangement for lower cheek teeth is similar, but reversed.

Family Castoridae
Beavers: *Castor*[EU,AS,AM], *Castoroides*[AM]

$$i\frac{1}{1}, c\frac{0}{0}, p\frac{1}{1}, m\frac{3}{3}$$

At one time found throughout northern Europe, Asia and America, beavers feed on bark, twigs, leaves and roots of trees, and they cut timber with their powerful incisors for dams and lodges. The cheek teeth (Figure 1.69) of both jaws are strongly hypsodont, parallel-sided and curving along their height, but still rooted, with a square occlusal outline. There are three buccal infoldings and one lingual infolding in each upper cheek tooth (three lingual and one buccal infoldings in the lower cheek

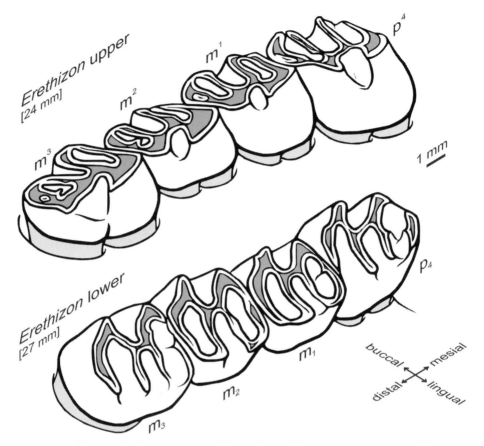

Figure 1.68 *Erethizon*, upper right and lower left permanent cheek teeth. Dimensions are lengths of tooth row from mesial side of fourth premolar to distal side of third molar.

teeth), persisting down the full height of the crown, and producing a characteristic occlusal pattern of 'V'-shaped lamellae. A giant fossil form *Castoroides* was present in North America (Kurtén & Anderson, 1980).

Family Aplodontidae
Mountain beaver/sewellel: *Aplodontia*[AM]

$$i\frac{1}{1}, c\frac{0}{0}, p\frac{2}{1}, m\frac{3}{3}$$

The mountain beaver (which in fact bears little relation to the true beaver *Castor*) shares features of skull morphology with the earliest known rodents in the fossil record, and is regarded as the most primitive living rodent (Kurtén & Anderson, 1980). It is nocturnal, solitary, confined in range to a small area in north-west North America, lives in a burrow system and eats a wide range of plant foods. The cheek

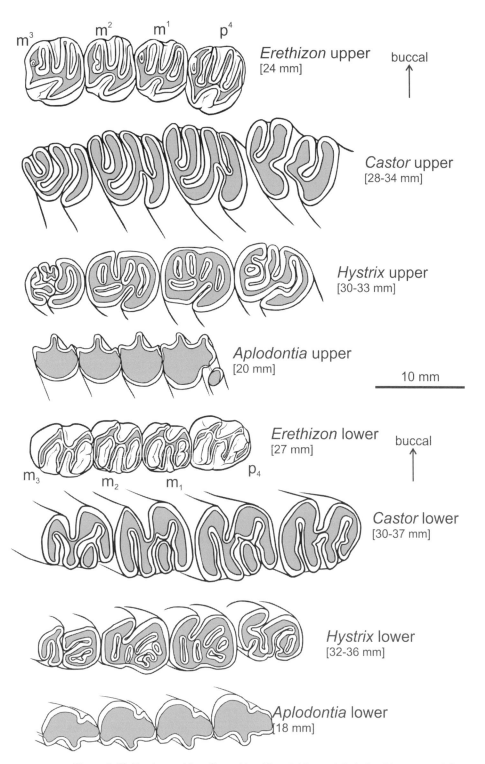

Figure 1.69 Erethezontidae, Castoridae, Hystricidae and Aplodontidae, upper right and lower left permanent cheek teeth. Dimensions are lengths of tooth row from mesial side of fourth premolar to distal side of third molar.

teeth (Figure 1.69) are persistently growing and open rooted, with a 'D'-shaped occlusal outline marked by a distinctive ridge down the buccal (in upper teeth) or lingual (in lower teeth) sides of the crown. Both premolars and molars have a similar form, with the exception of the greatly reduced upper third premolar.

Family Sciuridae
Tree squirrels: *Sciurus*[EU,AS,JA,AM], *Tamiasciurus*[AM], *Callosciurus*[ASE]; chipmunks: *Tamias*[AM], *Eutamias*[ASE,JA,AM]; rock squirrels: *Sciurotamias*[ASE]; ground squirrels: *Ammospermophilus*[AM], *Xerus*[AF], *Atlantoxerus*[AF], *Spermophilus*[EU,ASC,ASE,AM]; marmots: *Marmota*[EU,ASC,ASE,AM]; prairie dogs: *Cynomys*[AM]; long-clawed ground squirrel: *Spermophilopsis*[ASC]; flying squirrels: *Glaucomys*[AM], *Pteromys*[EU,AS,JA], *Petaurista*[JA], *Aeretes*[ASE], *Trogopterus*[ASE]

$$i\frac{1}{1}, c\frac{0}{0}, p\frac{1-2}{1}, m\frac{3}{3}$$

Members of this family live in many habitats, from woodland to steppe and desert, across most of the Holarctic region. Some are arboreal climbers and leapers, others glide from branch to branch, and others live in burrows on the ground. Their food ranges from seeds to nuts, stems, leaves, roots and bulbs, insects, eggs and small vertebrates. Upper cheek teeth (Figures 1.70–1.71) are mostly brachydont and characterised by four transverse ridges curving up from low cusps on the lingual side to high buccal cusps. Lower cheek teeth are basin-like, with a high rim to mesial and buccal. Fourth premolars are similar to the molars, but third premolars are usually reduced and often absent. The least dentally specialised Sciuridae are the intermediate forms of ground squirrels, including *Ammospermophilus*, *Xerus* and *Atlantoxerus*. On one side of these are the tree squirrels and chipmunks, based on the model of *Sciurus*, with their upper first and second molars rather square in occlusal outline and bearing four well-developed transverse ridges, separated at their buccal ends by prominent styles. On the other side are the increasingly high-crowned ground squirrels, where the upper molars are more triangular in occlusal outline, with the central two transverse ridges higher and the most distal of these ridges reduced. *Spermophilus*, the genus which includes most species of ground squirrels, can be divided into two groups dentally on the basis of the height of these ridges. *Marmota* looks like a larger version of the high-crowned form (Figure 1.71). The crowns are higher still in *Cynomys*, and highest of all in the extraordinary teeth of *Spermophilopsis* which remain rooted in spite of their height. Flying squirrels (Figures 1.72–1.73) form their own series of increasing cheek tooth crown height and complexity, from the generalised forms of *Glaucomys* and *Pteromys* (nevertheless distinguished by the prominently separated cusp on the third transverse ridge of their upper molars), to the higher and more convoluted ridges of *Petaurista*, even more in *Aeretes*, and ultimately to the astonishing complexity of *Trogopterus*, which has some of the most intricately folded tooth crowns of the mammal world.

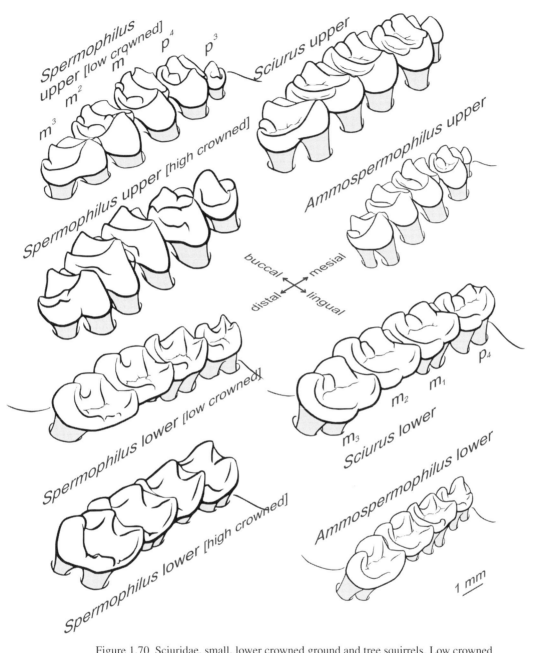

Figure 1.70 Sciuridae, small, lower crowned ground and tree squirrels. Low crowned *Sciurus*-like squirrels: *Sciurus* (8.5–11.5 mm), *Tamias* (6.3–6.5 mm), *Eutamias* (5.5–5.9 mm), *Sciurotamias* (8.8–9.8 mm), *Callosciurus* (10.5 mm). Intermediate crowned *Ammospermophilus*-like: *Ammospermophilus* (7 mm), *Atlantoxerus* and *Xerus* (9–11 mm). Higher crowned: *Spermophilus* (10.2–14.3 mm) in which some species are somewhat higher crowned than others. Dimensions are lengths of tooth row from mesial side of fourth premolar to distal side of third molar.

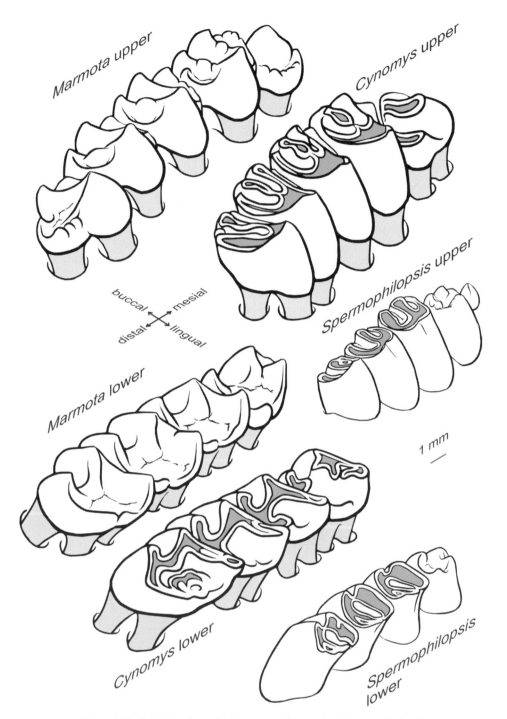

Figure 1.71 Sciuridae, large, higher crowned ground and tree squirrels. *Marmota* (18.3–22.7 mm). Hypsodont: *Cynomys* (16.5–17 mm), *Spermophilopsis* (11.5 mm). N.B. drawn at a smaller scale than Figures 1.70 and 1.72. Dimensions are lengths of tooth row from mesial side of fourth premolar to distal side of third molar.

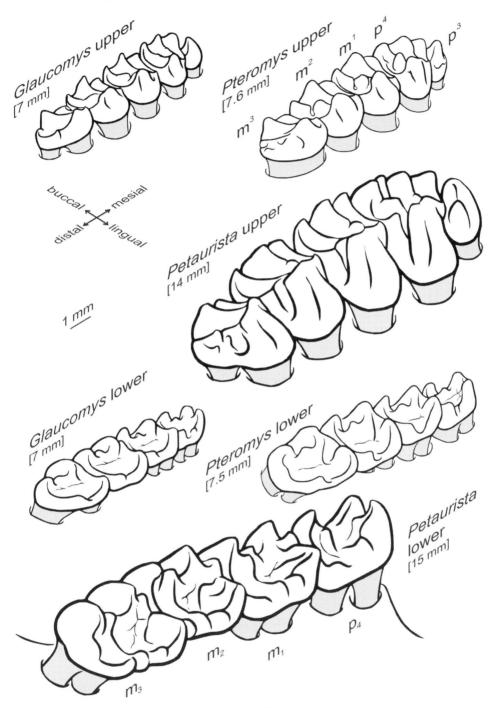

Figure 1.72 Sciuridae, flying squirrels. Upper right and lower left permanent cheek teeth. Dimensions are lengths of tooth row from mesial side of fourth premolar to distal side of third molar.

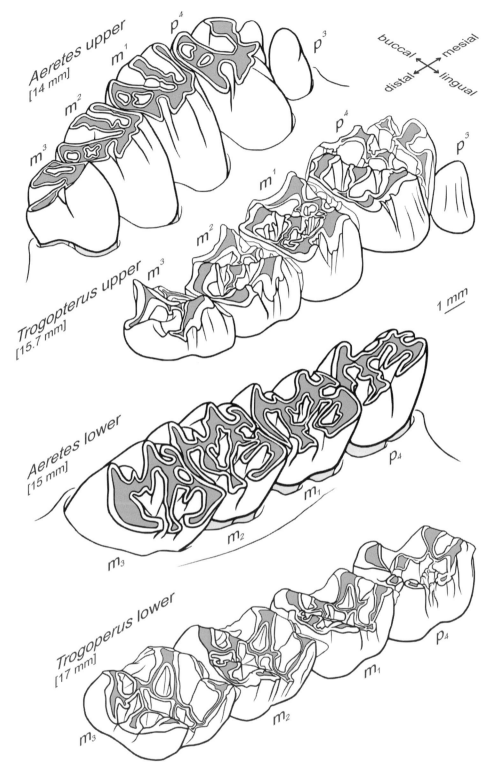

Figure 1.73 Sciuridae, high-crowned flying squirrels. Upper right and lower left permanent cheek teeth. Dimensions are lengths of tooth row from mesial side of fourth premolar to distal side of third molar.

Family Gliridae
Dormice: *Glis*[EU,ASW], *Muscardinus*[EU,ASW], *Glirulus*[JA], *Myomimus*[EU,ASW], *Dryomys*[EU,AS], *Eliomys*[EU,AF,ASW]

$$i\frac{1}{1}, c\frac{0}{0}, p\frac{1}{1}, m\frac{3}{3}$$

Dormice are confined to the Old World, living in woods and bushy vegetation, and eating nuts, seeds, fruits, insects, small birds and eggs. Their cheek teeth (Figures 1.74–1.75) are highly distinctive – brachydont, low and table-like in most genera, with multiple ridges running buccolingually across the occlusal surface. This form is most strikingly developed in *Muscardinus*, followed by *Glis*, and *Glirulus* has a similar arrangement in miniature, but with a strongly bulging cingulum. In the other living genera the occlusal ridges curve up to higher buccal (in the upper teeth) or lingual (lower teeth) cusps, most prominently in *Eliomys*, followed by *Dryomys*. *Myomimus* occupies an intermediate position dentally. Giant forms (Kurtén, 1968; Davis, 1987) were present during the Pleistocene period on the Mediterranean islands of Malta and Sicily (*Leithia*), and Mallorca (*Hypnomys*).

Family Seleviniidae
Desert dormouse: *Selevinia*[ASC]

$$i\frac{1}{1}, c\frac{0}{0}, p\frac{0}{0}, m\frac{3}{3}$$

This little animal is rare – it can be found only in the desert near Lake Balkhash in Kazakhstan, and it is very difficult to find museum specimens. It is the sole member of its family; the only family described in this book for which it proved impossible to see a specimen. For this reason, it is not illustrated, but the strongly grooved incisors are relatively large compared with the tiny cheek teeth. These are brachydont and rooted, with a pattern of low cusps on the occlusal surface.

Family Hydrochaeridae
Capybara: *Hydrochaeris*[AM]

$$i\frac{1}{1}, c\frac{0}{0}, p\frac{1}{1}, m\frac{3}{3}$$

The Capybara is today confined to South America, but was once present in the southern United States. It is the largest living rodent and has stout, continuously growing incisors, broad in transverse section and with a shallow buccal groove. The magnificent cheek tooth row (Figure 1.76) consists of high crowned and strongly lophodont crowns, with fine enamel plates quite unlike those of any other rodent in this book. In between the plates are thick layers of cement which hold them together rather loosely, so that museum specimens have a tendency to break up.

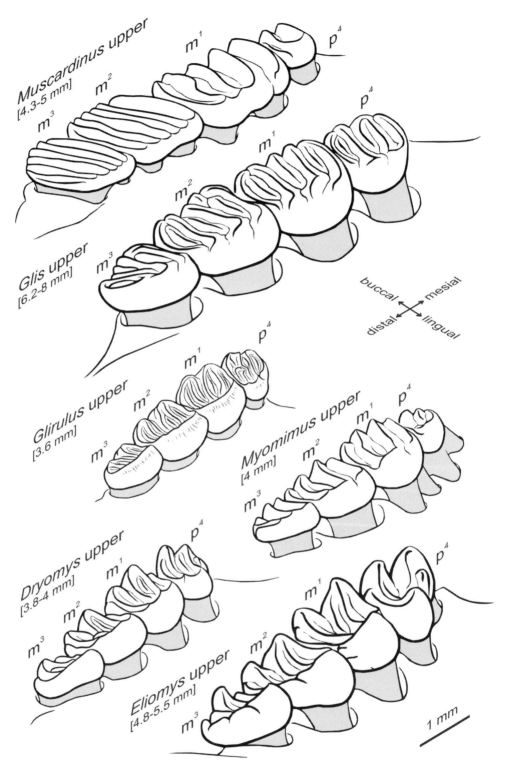

Figure 1.74 Gliridae, upper right permanent cheek teeth. Dimensions are lengths of tooth row from mesial side of fourth premolar to distal side of third molar.

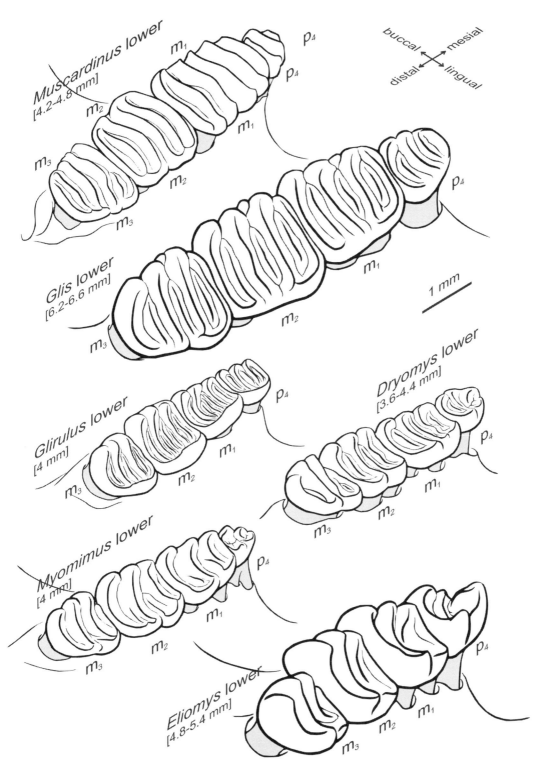

Figure 1.75 Gliridae, lower left permanent cheek teeth. Dimensions are lengths of tooth row from mesial side of fourth premolar to distal side of third molar.

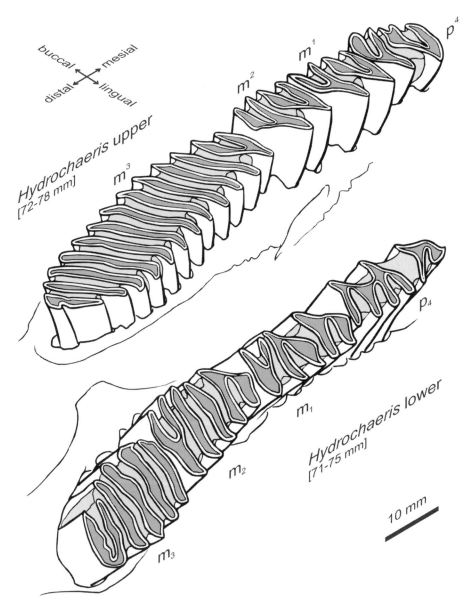

Figure 1.76 *Hydrochaeris*, upper right and lower left permanent cheek teeth. Dimensions are lengths of tooth row from mesial side of fourth premolar to distal side of third molar.

Order Lagomorpha

This group of small mammals varies rather little dentally. The upper lip and nostrils together are enclosed in a small chamber, the rhinarium, which can be opened and closed. Like the rodents, lagomorphs have no canines and the cheek teeth are separated from continuously growing incisors by a broad diastema. Unlike rodents, there is a small second pair of upper incisors (Figure 1.77) tucked in behind the large first incisors, and all incisor teeth have a complete jacket of enamel. The premolars and molars are ever-growing, prismatic and divided into two elements by a deep infolding. All the Lagomorpha are herbivorous, feeding on grass and shoots. One special feature of their digestive process is coprophagy, or reingestion of faecal pellets.

Family Leporidae
Rabbits: *Oryctolagus*[EU,AF,AS]; hares and jack rabbits: *Lepus*[EU,AS,AM]; Ryukyu rabbit: *Pentalagus*[JA]; cottontails: *Sylvilagus*[AM]

$$i\frac{2}{1}, c\frac{0}{0}, p\frac{3}{2}, m\frac{3}{3}$$

Upper incisors are deeply grooved down their buccal sides and, as both upper and lower incisors are actively used for gnawing, this leaves distinctive fourfold marks in wood that has been gnawed to get the bark (Bang & Dahlstrom, 1972). All cheek teeth (Figure 1.78) are similar in form, except for the upper second premolars, lower third premolars and heavily reduced third molars. *Oryctolagus*, *Lepus* and *Sylvilagus* all overlap in size and, when variation within species is considered, show no real distinguishing features. *Pentalagus*, however, is set apart by the elaborately 'plicated' or wrinkled enamel lining the infoldings.

Family Ochotonidae
Pikas: *Ochotona*[AS,AM]

$$i\frac{2}{1}, c\frac{0}{0}, p\frac{3}{2}, m\frac{2}{3}$$

The teeth of pikas (Figure 1.77) are smaller than those of the other leporids, and their cheek tooth row is reduced (Figure 1.78). This includes reductions in the upper second and third premolars, lower third premolar and the loss of the upper third molar. In addition, a deeper lingual infolding in lower cheek teeth makes a more symmetrical occlusal outline. An extinct ochotonid, *Prolagus*, was confined to Corsica and Sardinia during the Pleistocene (Kurtén, 1968; Davis, 1987).

Order Edentata and Order Pholidota
Living edentates comprise the sloths, armadillos and anteaters. The living sloths (Bradypodiae) are confined to South America, and have four to five simple, persistently erupting, cylindrical pegs in each quadrant of the dentition. The most mesial tooth in the row is somewhat modified into a sharp triangular peg. Tree sloths have not been found as fossils, but many extinct families of ground sloths have been found in both North and South America. Only the later families are included here. Living armadillos (Dasypodidae), with their segmented and mobile carapace, are found mostly in South America, but also spread into Central and southern North America. They tend to have larger numbers of teeth than the sloths. Probably related

112 Teeth

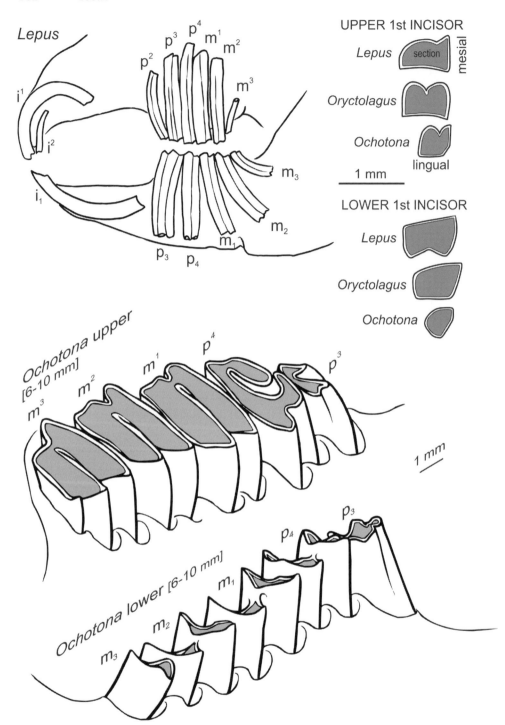

Figure 1.77 Lagomorpha. Buccal view of left *Lepus* dentition. Isometric view of upper right and lower left cheek teeth of *Ochotona*. Dimensions are lengths of tooth row from most mesial premolar to third molar.

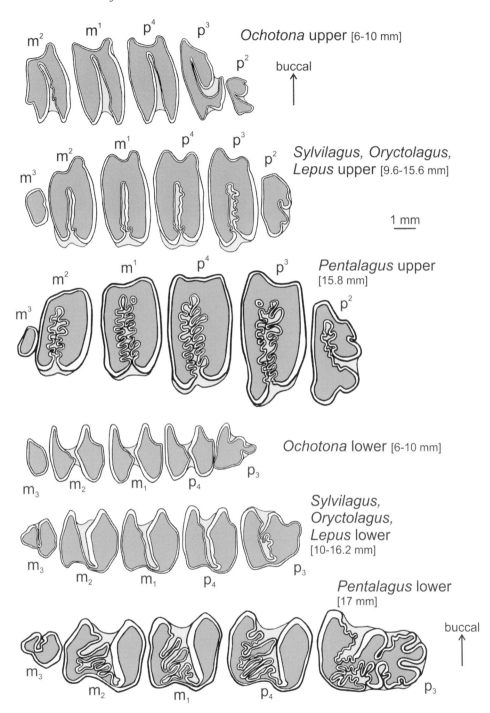

Figure 1.78 Lagomorpha, upper right and lower left cheek teeth. Dimensions are lengths of tooth row from most mesial premolar to third molar. Dentine is shaded dark grey and cement paler grey.

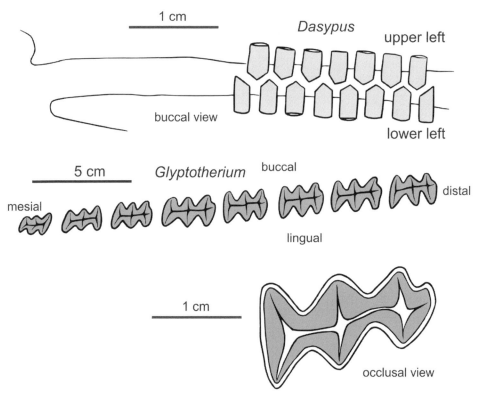

Figure 1.79 *Dasypus* buccal view of upper and lower dentitions. *Glyptotherium* occlusal views of lower left dentition.

to the armadillos were the extinct glyptodons (Glyptodontidae), which had a fused, solid carapace of bony plates. The anteaters (Myrmecophagidae) of South America have no teeth at all in their narrow, elongated jaws. Suborder Pholidota – pangolins and scaly anteaters – come from the tropics of Africa and Asia, and also have elongated jaws with no teeth.

Family Dasypodidae
Nine-banded armadillo: *Dasypus*[AMC]

$$\text{undifferentiated} \frac{7-8}{7-8}$$

Armadillos have their body covered with horny plates. They have a long, narrow snout, with a very simplified row of teeth in the posterior part of the jaws (Figure 1.79). There are seven to eight teeth in each quadrant of the dentition, all alike, and taking the form of small cylinders which are worn to a high peak in the middle of the occlusal surface. They are predominantly dentine, with a thin coating of cement.

Family Glyptodontidae
Glyptodonts: *Glyptotherium*[AMC]

$$\text{undifferentiated} \frac{8}{8}$$

The glyptodonts (Kurtén & Anderson, 1980) were an extraordinary family of extinct edentates, with a great bony carapace, and heavy armoured tail. They were a particular feature of South America, but were also present in the southern United States during the Late Pleistocene. Their name comes from the Greek *glyptos* (carved), from the sculpted appearance of their characteristically fluted teeth (Figure 1.79). Eight prismatic, continuously growing teeth were present in each quadrant of the jaw. They consist of three dentine lobes, coated in enamel and cement, with a triple infundibulum inside, lined with thin enamel, with the sides very closely pressed together. This arrangement is unique in the mammal world.

Family Megalonychidae
Jefferson's ground sloth: *Megalonyx jeffersoni*[AM]

$$\text{undifferentiated} \frac{5}{4}$$

Jefferson's ground sloth was roughly cattle-sized, about medium size for a ground sloth, and found throughout North America in the Late Pleistocene. Altogether, there are five teeth (Figure 1.80) on each side of the upper jaw and four teeth on each side of the lower jaw. Most of the persistently growing teeth have a rounded-square occlusal outline, with shallow grooves down their sides. All are worn on the occlusal surface to a concave centre and raised rim. The centre is formed from the softer dentine and at its centre is the slightly darker secondary dentine which lines the pulp chamber. From the author's own observations, the rim seems to be formed from a layer of enamel and outside this is a thick layer of cement which is also worn slightly lower than the highest part of the rim. The most mesial tooth ('caniform') in each jaw is larger than the others, and separated from the main cheek tooth row by a large diastema.

Family Megatheriidae
Rusconi's ground sloth: *Eremotherium rusconii*[AMC]; Shasta ground sloth: *Nothrotheriops shastensis*[AMC]

$$\text{undifferentiated} \frac{4-5}{3-4}$$

Eremotherium was much the largest of the ground sloths described here, weighing perhaps 3 tons, with pronounced sexual dimorphism in body size (Kurtén & Anderson, 1980). It had five teeth (Figure 1.80) on each side of the upper jaw and four on each side of the lower jaw. *Nothroptheriops* was the smallest ground sloth in

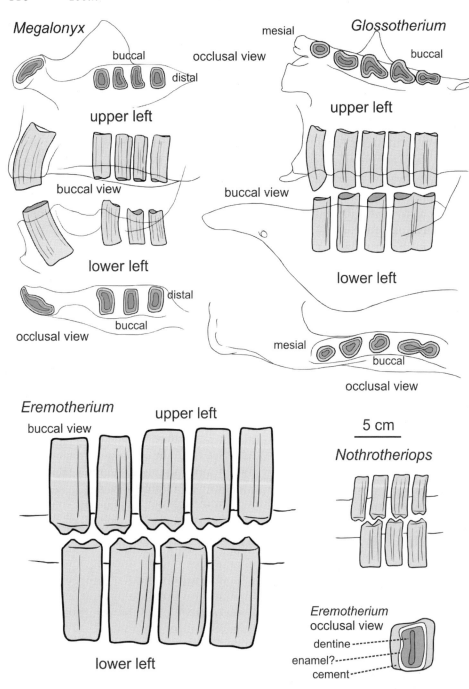

Figure 1.80 Ground sloths. *Glossotherium* and *Megalonyx*, upper and lower left dentitions, buccal and occlusal views. *Eremotherium* and *Nothrotheriops*, upper and lower left dentitions, buccal views. Detail of *Eremotherium* single tooth, occlusal view.

North America, and is best known from the caves of the south-western United States. It had a reduced dentition of four teeth on each side of the upper jaw and three on each side of the lower jaw. The teeth scale with body size and are thus smaller than those of *Eremotherium*. In both these sloths, there was no diastema and no 'caniform' mesial tooth, and all the persistently growing prismatic teeth were worn into double buccal–lingual ridges on the occlusal surface. From the author's observations, this seems to have been due to the positioning of an enamel layer within the teeth, with a particularly thick cement coating to mesial and distal, and a less resistant core of 'marbled' dentine filling the pulp chamber in between (as in walrus tusks). As a result, isolated teeth are highly distinctive.

Family Mylodontidae
Harlan's ground sloth: *Glossotherium harlani*[AM]

$$\text{undifferentiated}\frac{5}{4}$$

Harlan's ground sloth was somewhat larger than *Megalonyx*, but not so big as *Eremotherium* (above). The most distal cheek tooth (Figure 1.80) was distinguished by a figure-of-eight occlusal outline, with pronounced grooves on the buccal and lingual sides. In the lower jaw, it was also larger than the others. The most mesial tooth usually protruded above the others and was worn to a chisel-like edge, probably because not all the tooth came into contact with others in the opposing jaw. The mandible projected as an extraordinary spoon-like, toothless extension.

Order Tubulidentata

This order has just one species, the aardvark *Orycteropus afer*, which lives in Africa south of the Sahara. It has no anterior teeth, but has four to seven postcanines. Those at the front develop first, are worn and then replaced by teeth from further back in the jaw. There is no enamel, but the teeth are covered with cement. The dentine is exceptional in having multiple tubular pulp chambers, dividing the tissue into vertical hexagonal units.

Order Proboscidea

Today, only one family of proboscideans (the Elephantidae) survives but three other families are preserved in the Quaternary fossil record: the mastodonts (Mammutidae), stegodonts (Stegodontidae) and gomphotheres (Gomphotheriidae). The main feature of proboscideans, in addition to an evolutionary trend to increase in size, is the presence of a trunk, tusks, and large cheek teeth with ridged grinding surfaces. In the most recent genera, these cheek teeth erupted in sequence, one after another, to provide a continuous series of new teeth throughout life.

Family Mammutidae
American mastodon: *Mammut americanum*[AM]

$$\text{di}\frac{0}{0}, \text{dc}\frac{0}{0}, \text{dp}\frac{3}{3} \rightarrow \text{i}\frac{1}{0-1}, \text{c}\frac{0}{0}, \text{p}\frac{0}{0}, \text{m}\frac{3}{3}$$

Mastodons were common in the Americas, but became extinct in the Late Pleistocene. The name 'mastodon' (Greek μαστός, breast or udder) comes from the resemblance of this extinct animal's cheek teeth to the row of teats in an animal such as a sow. They consist of pairs of large conical cusps (Figure 1.81), raised up into lophs – two for the second (very reduced) and third deciduous premolars, three for the fourth deciduous, first and second permanent molars, and four to five for the third permanent molar. Each loph is supported by a massive root. These teeth erupted in series, from the second deciduous premolars onwards. At most ages, there were two cheek teeth in each jaw quadrant but at some stages of development there were three (Haynes, 1991). Upper cheek teeth tend to be broader buccal–lingual, and have a more clearly defined cingulum, than lower cheek teeth. The large, strongly curved upper tusks represent heavily modified incisors. In some individuals, lower tusks the size and shape of a fat cigar erupted from the tip of the lower jaw.

Family Gomphotheriidae
Gomphothere: *Cuvieronius*[AMC]

Cuvieronius was the last survivor of a substantial family of primitive elephants, in which all the cheek teeth erupted at one time, rather than in a continuous sequence as in living elephants, and there were upper and lower tusks. *Cuvieronius* retained the primitive cheek teeth, but lost the lower tusks. The upper tusks had a strip of enamel along their length, and were twisted into a spiral. It was present during the Late Pleistocene in central America and is a rare fossil find in Florida.

Family Elephantidae
Elephants: *Loxodonta*[AF], *Elephas*[AS]; mammoths: *Mammuthus*[EU],[AS],[AM]

$$\text{di}\frac{1}{0}, \text{dc}\frac{0}{0}, \text{dp}\frac{3}{3} \rightarrow \text{i}\frac{0-1}{0}, \text{c}\frac{0}{0}, \text{p}\frac{0}{0}, \text{m}\frac{3}{3}$$

The modern African and Indian elephants are the surviving representatives of genera present through the Pleistocene, during which mammoths were very common throughout the Holarctic region (they became extinct in the Late Pleistocene). In the teeth of elephants and mammoths, the cheek tooth lophs seen in mastodons are increased in height and number, to make a series of vertically arranged enamel plates (Figure 1.81). Each plate has a dentine core and the whole tooth is thickly invested with cement. As with mastodons, three deciduous premolars and three permanent molars per dentition quadrant erupt in sequence through life (p. 243). The elephants and mammoth, however, have/had only one to two teeth per quadrant in the mouth at a time. Wear starts at the mesial end of each tooth and then

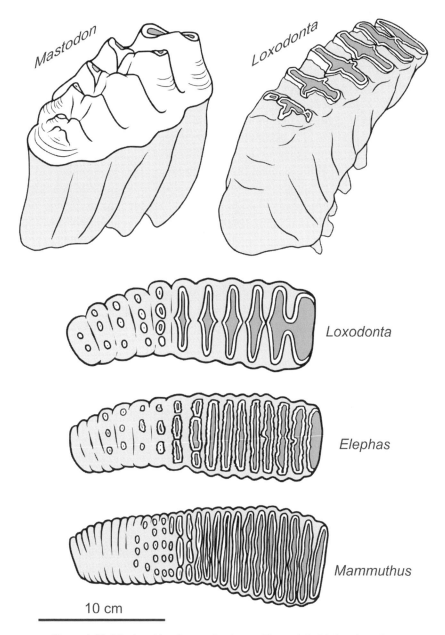

Figure 1.81 Elephantidae. Isometric views of lower left third molars. Occlusal views of lower left third molars.

progresses down the wearable tissue, exposing a set of hard enamel ridges that act like a millstone. As this goes on, the roots are resorbed and the tooth is finally lost. When this happens, the next tooth rapidly comes into wear behind (at some stages it has already started to wear). Elephants must gather huge quantities of leaves, small branches and bark, and grind them to a pulp in their efficient dental milling

120   Teeth

machine. The eruption and wear of the third molars marks the limit on each animal's life. Age can therefore be estimated from the teeth functioning in the mouth, and the different teeth in the series can be identified from the number of plates (which increases along the series, see Chapter 3) and the proportions of their occlusal area (Haynes, 1991). Second deciduous premolars have a highly reduced version of this arrangement and are distinctive teeth. *Loxodonta* differs from the other elephantids in the lozenge-shaped outline of its cheek tooth lophs, exposed in the worn occlusal surface. Both male and female elephants and mammoths have upper tusks which grow continuously throughout life – up to 16 cm per year (Miles & Grigson, 1990). The pulp chamber is represented by a deep conical depression in the base of the tusk, deeper in mature males than in females. As in mastodons, the tusks are modified incisors. At first eruption, their tips are coated with enamel and cement, which are worn away in a few years. The permanent tusks are preceded by small deciduous tusks, which are shed around one year of age. Several Mediterranean islands, the Channel Islands of California and Wrangel Island by the Bering Straits, had dwarfed forms of elephant or mammoth (Kurtén, 1968; Davis, 1984, 1987; Haynes, 1991; Lister & Bahn, 2000; Poulakakis *et al.*, 2002) during the Pleistocene period.

Order Sirenia

There are two families of Sirenia. Dugongidae, or sea cows, live along the coasts of East Africa, southern Asia and Australasia. Trichechidae, the manatees, live on West African and south-east American coasts.

Family Dugongidae
Sea cows: *Dugong*[AF,AS], *Hydrodamalus*[ASE,AM]

$$i\frac{1}{0}, c\frac{0}{0}, p\frac{0-3}{0-3}, m\frac{0-3}{0-3}$$

The living sea cows *Dugong dugon* are found in the coastal waters of warm seas, grazing on algae and other marine plants. The males have large, persistently growing tusk-like upper incisors (Figure 1.82), rooted in a downward curving extension to the skull. These tusks are mainly composed of dentine, with a cement coat, but there is a strip of enamel down the mesial side (Marsh, 1980). In females, the tusks are smaller and do not erupt. There are horny mouth pads on their palate and lower lips which help to gather food. As in proboscideans (to which they are related), sirenians erupt a continuous series of cheek teeth, replacing from behind those which have been worn and lost. In *Dugong*, up to four may be functional in each jaw quadrant at any one time. They have no enamel, but consist of dentine, coated in cement. All except the third molar are roughly cylindrical and widen towards the roots, so they produce a widening circular occlusal outline as they wear. The third molar has a groove on the buccal side, to produce a bean-shaped occlusal outline. The last Steller's sea cow, *Hydrodamalus gigas*, was

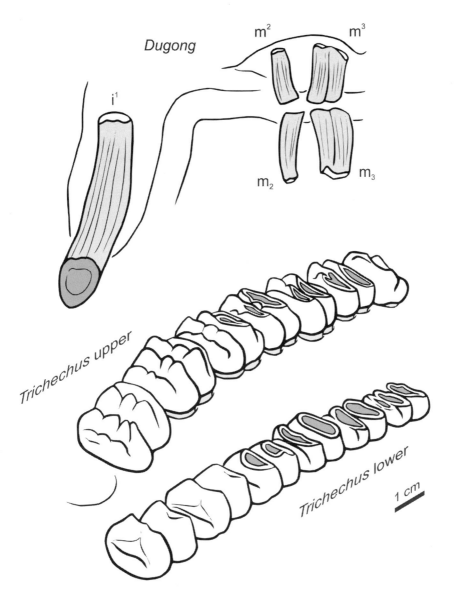

Figure 1.82 Sirenia. Buccal view of left dentition in *Dugong*. Isometric view of upper right and lower left cheek teeth of *Trichechus*.

probably slaughtered by 1768. They were giants, measuring 7.5 m or so in length (Nowak & Paradiso, 1983), with prominent horny palate plates but no teeth in their jaws.

Family Trichechidae
Manatees: *Trichechus*[AMC]

$$i\frac{2}{2}, pc\frac{6-10}{6-10}$$

The incisors in these animals are lost before maturity (Nowak & Paradiso, 1983) and, as in sea cows, horny pads at the front of the jaw are used to gather vegetation. The cheek teeth are lophodont (Figure 1.82), with enamel-coated crowns, and not dissimilar in form from those of the mastodon. Their correct identification is unclear but, as they wear and are lost, they are continuously replaced from behind by later erupting teeth. Up to seven may be present in each jaw quadrant at any one time, and the total number erupted throughout life per quadrant may be as many as ten. This represents an increase over the ancestral eutherian mammal condition (p. 12).

Order Hyracoidea
Family Procaviidae
Hyraxes: *Procavia*[AF,ASW]

$$i\frac{1}{2}, c\frac{0}{0}, p\frac{4}{4}, m\frac{3}{3}$$

These little animals eat grass, leaves and bark, with a dentition that functions like a rodent's. In fact, it is related to proboscideans, even though the cheek teeth (Figure 1.83) look most like those of a rhinoceros. They are hypsodont and the upper premolars and molars have three lophs – ectoloph, protoloph and metaloph. There is prominent, step-like cingulum. The first premolar is simple, but the teeth increase in complexity along the row to distal. Lower cheek teeth each have two U-shaped lophs – the metalophid and hypolophid. The upper incisors are relatively large, persistently growing and with enamel on all sides. Their transverse section is triangular in outline and they wear to a sharp point, rather than a chisel-like edge. The lower incisors are spatulate and together wear into a notch where the upper incisor grinds against them.

Order Perissodactyla
Perissodactyla is one of the two orders of ungulates (hoofed mammals). They are called 'odd toed' ungulates because the third digit of each limb is developed into a single hoof. All Perissodactyla are herbivores, with continuous rows of complex, hypsodont cheek teeth. Unlike the ruminants (below), they have a hindgut fermenting digestive system where the fibrous plant food is chewed only once, completely digested in a single chambered stomach, and then the cellulose is fermented to sugars in the colon and caecum (Macdonald, 1984). There are three living perissodactyl families. Tapirs (Tapiridae) live in the tropics of South America and South-east Asia. Their cheek teeth are lophodont, with two transverse plates not dissimilar from those of manatees (above). Rhinoceroses (Rhinocerotidae) in the past had a distribution throughout Europe, Africa, Asia and North America. Cheek

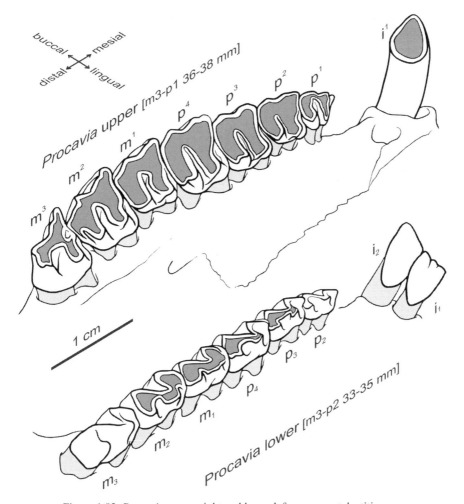

Figure 1.83 *Procavia*, upper right and lower left permanent dentitions.

teeth in this family are complicated by further lophs and ridges. Wild horse, asses and zebras (Equidae) were also widespread throughout the Holarctic, and have high crowned, selenodont cheek teeth.

Family Tapiridae
Tapirs: *Tapirus*[AMC]

$$i\frac{3}{3}, c\frac{1}{1}, p\frac{4}{4}, m\frac{3}{3}$$

Today, tapirs are confined to southern Mexico and South America but, during the Pleistocene, several extinct species of tapir were also found in southern parts of North America. Tapir incisors are spatulate, except for the third upper incisors which are tall and pointed (Figure 1.84). The small upper canines are isolated in a large diastema. The cheek teeth of living tapirs are molariform, relatively low crowned

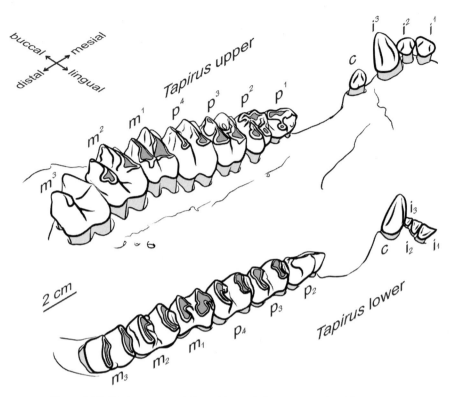

Figure 1.84 *Tapirus*, upper right and lower left permanent dentitions.

and strongly bilophodont, with a protoloph and metaloph. The ectoloph which is such a prominent feature of rhinoceros cheek teeth is only slightly developed, especially in the upper first premolar and lower second premolar. The fossil forms were somewhat larger animals and show less strong molarisation of the first and second premolars (Kurtén & Anderson, 1980).

Family Rhinocerotidae
Rhinoceroses: *Rhinoceros*[AS], *Diceros*[AF], *Ceratotherium*[AF], *Dicerorhinus*[AS], *Coelodonta*[EU,AS,AM], *Elasmotherium*[AS]

$$i\frac{0-2}{0-1}, c\frac{0}{0-1}, p\frac{3-4}{3-4}, m\frac{3}{3}$$

Rhinoceroses are divided into single- and tandem-horned varieties. Living Indian and Javan *Rhinoceros* have only one horn and the extinct giant *Elasmotherium* probably had a single very large horn arising from its forehead. Today's Sumatran rhinoceros *Dicerorhinus* and its extinct relatives, the living African black rhinoceros *Diceros* and the white rhinoceros *Ceratotherium*, and the extinct woolly rhinoceros *Coelodonta* all have two horns, arranged in tandem.

Anterior teeth are much reduced, although *Rhinoceros* and *Dicerorhinus* retain the upper first incisor and lower second incisor as large tusks (Figure 1.85). Cheek teeth form long curved rows along which complexity and size increase distally. Upper teeth are square in occlusal outline and are gathered up into three

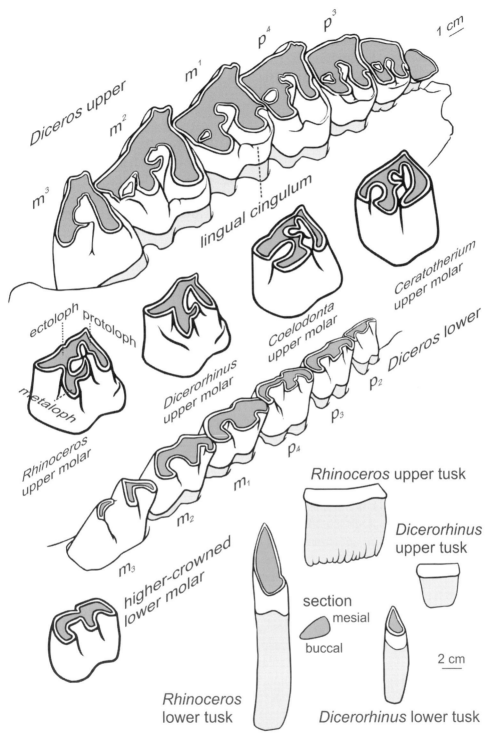

Figure 1.85 Rhinocerotidae, upper right and lower left permanent dentitions. Lingual views of the upper and lower incisor tusks.

lophs – ectoloph, protoloph and metaloph – with additional infoldings which produce a highly characteristic 'E' pattern in the worn occlusal surface. There is a prominent cingulum, above which the crown sides slope inwards to occlusal. The lower cheek teeth have paired U-shaped lophs, the metalophid and hypolophid, and are more rectangular in outline than the upper teeth. Jaw movement is from side to side, so the lophs grind diagonally across one another. *Coelodonta*, *Elasmotherium* and *Ceratotherium* are higher crowned than *Dicerorhinus*, *Diceros* and *Rhinoceros*. Their worn occlusal surfaces are flatter and their lower cheek tooth occlusal outline is different. The giant *Elasmotherium* had large, high-crowned teeth with 'plicated' or wrinkled enamel coating the lophs. *Ceratotherium* can be distinguished from *Coelodonta* by its more heavily built and chunkier crown, with copious cement coating. *Diceros* is distinguishable from *Rhinoceros* and *Dicerorhinus* by the stronger development of the lingual cingulum in upper cheek teeth. In *Dicerorhinus* this is restricted to its mesial and distal ends, and in *Rhinoceros* to mesial only.

Family Equidae
Horses, asses and zebras: *Equus*[EU,AF,AS,AM]

$$\text{di}\frac{3}{3}, \text{dc}\frac{0}{0}, \text{dp}\frac{3}{3} \rightarrow \text{i}\frac{3}{3}, \text{c}\frac{0-1}{0-1}, \text{p}\frac{3-4}{3}, \text{m}\frac{3}{3}$$

Equids have the tallest crowns (Figure 1.86) of the perissodactyls. Each tooth erupts gradually as it wears down and only becomes rooted in its later life. The crown sides act as a surface to which the periodontal ligament is bound – effectively taking the role of root – so all teeth are heavily coated with cement. Permanent incisors are conical, with no clear cervix, and have a single infundibulum which makes them somewhat trumpet-like before wear. The tall, narrow pulp chamber runs up the crown next to the infundibulum. It fills with darker secondary dentine, making the so-called 'star' on the worn occlusal surface which forms part of the traditional age estimation method for horses (Chapter 3). There is a large diastema, separating the incisors from the impressive grinding battery of the cheek teeth. In the diastema, there may be small permanent canines with a simple conical crown (often reduced or missing in females), and a rudimentary upper first premolar. All the main upper permanent cheek teeth are selenodont, with deep cement-filled infundibula that appear as islands in the worn occlusal surface. The crown is prismatic in form; square in occlusal view except for the third molar and second premolar, which are triangular. Two broad, shallow infoldings leave three narrow buttresses on the buccal side; whereas two shallow and one deep infoldings isolate two broad buttresses on the lingual side. Superficially, this is the same arrangement as for large Bovidae (below), but the occlusal pattern is more complex and the crowns higher with a much squarer appearance. Lower permanent cheek teeth have the metalophid and hypolophid developed as tall buttresses, with three small infoldings on the buccal side and two deep, two shallow infoldings on the lingual side. This

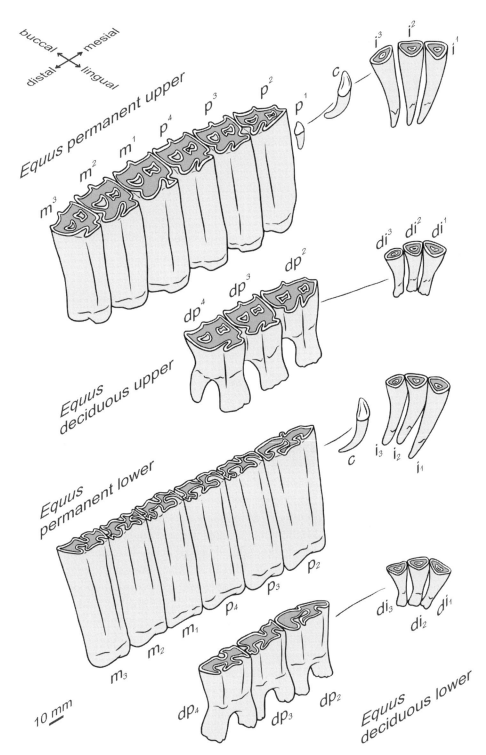

Figure 1.86 *Equus*, upper right and lower left permanent and deciduous dentitions. Dentine is shaded dark grey and cement paler grey.

gives a highly characteristic pattern on wear, with a rectangular occlusal outline except for the second premolar and third molar, which are triangular. Deciduous equid teeth are in many ways like their permanent successors, with similar occlusal dimensions in the cheek teeth, but they are lower crowned and have more prominent, wider-spread roots. It can in some cases be difficult to distinguish a deciduous cheek tooth crown, but signs of resorption on the roots can often be a deciding feature.

Order Artiodactyla
Suborder Suiformes
Suiform artiodactyls do not have a ruminant digestive system (below). Their cheek teeth are more or less bunodont and the canines are enlarged and tusk-like, with the incisors also sometimes taking on a tusk-like form.

Family Suidae
Pigs: $Sus^{EU,AF,AS,AM}$

$$di\frac{3}{3}, dc\frac{1}{1}, dp\frac{3}{3} \rightarrow i\frac{2-3}{3}, c\frac{1}{1}, p\frac{3-4}{2-4}, m\frac{3}{3}$$

The wild Suidae are confined to the Old World, but the domesticated pig *Sus scrofa* is found worldwide. Pigs are omnivorous. They root about in and on the ground with their powerful snouts for tubers, bulbs, nuts, seeds, insect larvae and carrion, and will also catch small vertebrates. There is considerable sexual dimorphism in their canines. In males, the upper canine tusks (Figures 1.87, 1.89) are large and curve upwards and outwards, with a rounded section, whereas the lower canines are slenderer, with a more sharply triangular section. These ever-growing tusks project sideways outside the mouth and are kept down in length by sharpening against one another, to produce a characteristic attrition facet. Only two sides of the section are coated with enamel. In females, both upper and lower tusks are smaller, and become rooted, with the enamel of the crown draped over the tooth like a hood in unworn teeth. The lower incisors (Figure 1.87) are not exactly tusk-like, but their tall narrow crowns and long roots give them a strongly chisel-like appearance. By contrast, the upper incisors are much smaller, curved teeth, with some superficial resemblance to female upper canines. The permanent molars and fourth deciduous premolars are based around four tall, but still bunodont cusps, with many subsidiary cusps around them – giving the unworn crown a very complex appearance. They wear down rapidly to expose the dentine. Permanent third molars and lower fourth deciduous premolars are very distinctive, with additional cusps at the distal end of their crowns. The lower permanent premolars and remaining lower deciduous premolars have their three main cusps gathered into a characteristic, blade-like line.

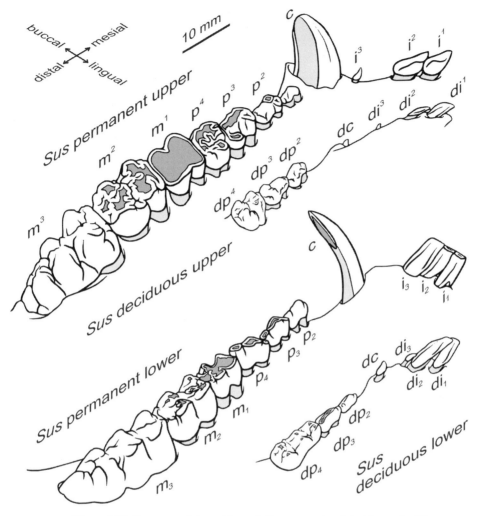

Figure 1.87 *Sus*, upper right and lower left permanent and deciduous dentitions.

Family Tayassuidae
Peccaries: *Tayassu*[AMC], *Mylohyus*[AM], *Platygonus*[AM]

$$i\frac{2}{2-3}, c\frac{1}{1}, p\frac{3}{3}, m\frac{3}{3}$$

The peccaries are the pig-like animals of the New World, mostly confined to South and Central America, but spreading into the USA. Their main points of difference with the Suidae are their somewhat narrower snouts (and therefore narrower dental arcade), their generally less robust dentition (Figure 1.88) in which the subsidiary cusps of the molars are reduced, and their sharp canine tusks (Figure 1.89), which curve in a cranial–caudal plane rather than out to lateral. Males and females are the same size, and the tusks are rooted (open-rooted in the young), with a clearly defined crown coated with enamel. Upper canines develop an attrition facet on their mesial surface, and lower canines on the distal surface. The collared peccary *Tayassu* is

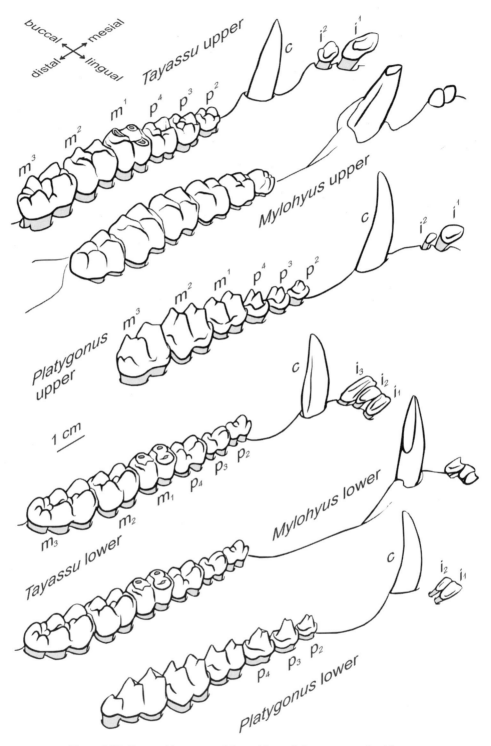

Figure 1.88 Tayassuidae, upper right and lower left permanent dentitions.

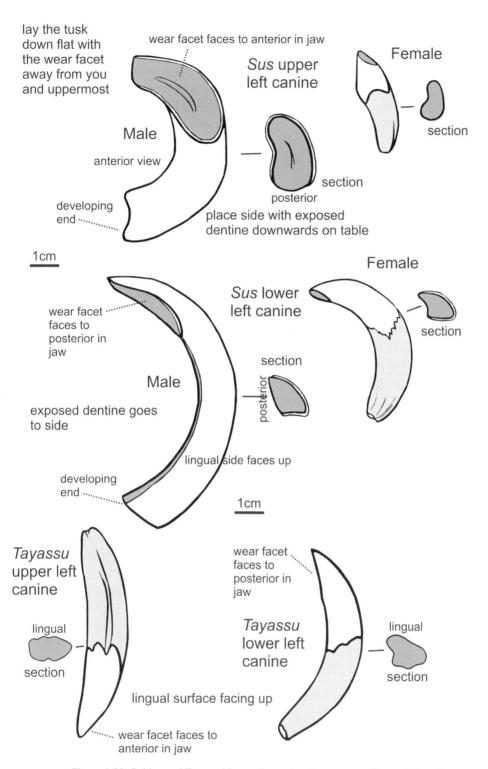

Figure 1.89 Suidae and Tayassuidae canine tusks. *Sus* upper and lower left canines, males and females. *Tayassu* upper and lower left canines, lingual views and sections. The tusks are shown as they naturally lie when placed on a table.

found today in Texas, New Mexico and Arizona, but the slightly larger (extinct) flat-headed peccary *Platygonus* was common right across the USA during the last cold stage, and the extinct long-nosed peccary *Mylohyus* is a less common find from the eastern USA. Its characteristic features are a very narrow dental arcade and a long diastema between the canine tusks and the cheek teeth.

Family Hippopotamidae
Hippopotamus: *Hippopotamus*[AF, [EU]]

$$i\frac{2-3}{1-3}, c\frac{1}{1}, p\frac{4}{4}, m\frac{3}{3}$$

The living hippopotamus is an amphibious creature, living mostly in or near water and emerging to graze at night. It eats mainly grass and a range of other plant foods. Males grow rather larger than females, and this is reflected in the size of their canine tusks, although the form of these teeth does not differ between the sexes. They are ever-growing and develop characteristic attrition facets where they meet (Figures 1.90–1.91). Lower canines are distinguished from upper by their sections in which, like pig tusks, enamel coats only two sides. Before wear, however, there is a layer of cement coating the sides of the tusks, and this may flake away in archaeological specimens. Females have section dimensions one half to two thirds the size of males. The first and second permanent incisors are also developed into tusks, but they are much less strongly curved and project forwards in the jaw without meeting, and therefore they do not have the characteristic attrition facets of the canines. They do wear, however, usually to polished points. Once again, they can be distinguished by their form in section, with the lower incisors being coated all round in a layer of enamel, whereas the upper incisors just have an enamel stripe along their lateral sides. Hippopotamus cheek teeth are often described as bunodont, although the main cusps are much taller than those of the Suidae. Premolars have a single large cusp, which rises to a rounded point, with a pronounced cingulum and low accessory cusps at the base. The molars have four high cusps, which enclose an infundibulum between each pair, and again a pronounced cingulum at their bases. Lower permanent third molars have extra cusps on their distal ends, as in pigs. Hippopotamus was widespread through Eurasia during the Pleistocene, but has been confined to Africa during the Holocene. Some Mediterranean islands had dwarfed forms of hippopotamus during the Pleistocene period (Davis, 1984, 1987; Spaan, 1996).

Suborder Ruminantia (Pecora)

$$di\frac{0}{3}, dc\frac{0}{1}, dp\frac{3}{3} \rightarrow i\frac{0}{3}, c\frac{0-1}{1}, p\frac{2-3}{2-3}, m\frac{3}{3}$$

All the members of this suborder are herbivorous, and all have very similar dentitions. The ruminant digestive system includes a chamber (the *rumen*) as part

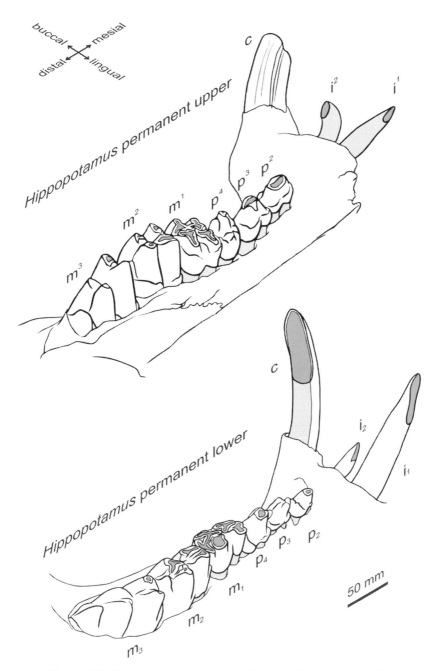

Figure 1.90 *Hippopotamus*, upper right and lower left permanent dentitions.

of a complex stomach, in which a resident flora of micro-organisms ferment the cellulose (to sugars) which forms the major part of plant foods. Partly digested food is periodically regurgitated and rechewed (chewing the cud) before finally being passed into the other chambers of the stomach, along with micro-organisms, which are also digested. Grass, leaves, twigs, moss and bark are gathered by the spatulate

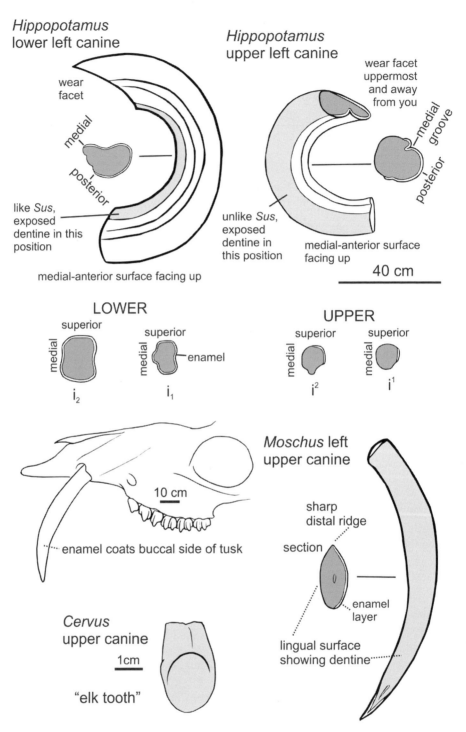

Figure 1.91 Upper part of figure shows *Hippopotamus* tusks: medial-anterior views and sections of upper and lower left canines; sections of upper and lower first and second incisors. Lower part of figure shows deer canines (see text).

lower incisors and canines, which are arranged in a fan and work against a horny pad on the toothless anterior portion of the upper jaw. The prehensile tongue can also be used to gather fodder. There is a large diastema, over which the food is passed to the grinding battery of hypsodont cheek teeth.

All the premolars and molars (Figures 1.92–1.93) have persistently growing, cement-coated crowns that act as an extension of the root trunk when the teeth are first erupted into the mouth, but eventually become rooted. The degree of hypsodonty varies between different ruminants. The permanent upper premolars have a single infundibulum, which appears in the occlusal surface as a crescentic island. On the buccal side, narrow mesial and distal buttresses flank a broad central bulge, and the lingual surface is curved, to give a 'D'-shaped occlusal outline when worn. The permanent upper molars have a double infundibulum, with three narrow buttresses separating two broad bulges on the buccal side, and two curved lingual surfaces giving a 'B'-shaped occlusal outline after wear. There is frequently an additional cusp arising between the lingual bulges, at their base, which rises to a variable height. Permanent lower molars are similar in general plan, but narrower buccal/lingual, and with their features reversed so that the twin bulges of the 'B' are on the buccal side instead. The lower third molar has a variably developed additional distal element and is the largest tooth in the dentition. Permanent lower premolars mostly do not have an infundibulum, but instead show a pattern of infoldings that gives them an occlusal outline rather like a '£' sign when worn. In some genera, an infundibulum develops in the distal part of the fourth premolar. The deciduous upper second premolar is similar to a permanent premolar, whereas the deciduous upper third and fourth premolars look like permanent molars. In the lower deciduous dentition, the second and third premolars are also like permanent premolars, but the deciduous fourth premolar is an exceptional tooth with a triple infundibulum, and a third small root in the middle of its buccal side. As with all deciduous teeth, those of the ruminants are distinguished by their smaller crown size, more spindly and wider spread roots, and root resorption before they are exfoliated.

There are four families of ruminants. The Cervidae, or deer, are mostly distinguished by bearing antlers and include some wide-ranging species: red deer (called wapiti or elk in America) is Holarctic in distribution, and moose (called elk in Europe) and reindeer (caribou) occupy the far north of Eurasia and America. Fallow and roe deer are found throughout Eurasia, but other species are much more restricted in range. Axis deer live in India and Indochina, whereas Père David's deer were confined to a single herd in the Imperial Hunting Park near Beijing, although they once probably ranged through northern China. White-tailed and mule deer are found throughout the Americas, and South America has its own special cervids including marsh deer, pampas deer, guemals and huemals, brocket deer and pudu, the world's smallest deer. The upper canine may be present as a small tooth in any deer but in some it is enlarged into a tusk that projects outside the mouth in males (Figure 1.91). These are the small, antlerless deer of Asia; musk deer, Chinese water deer, muntjacs and tufted deer.

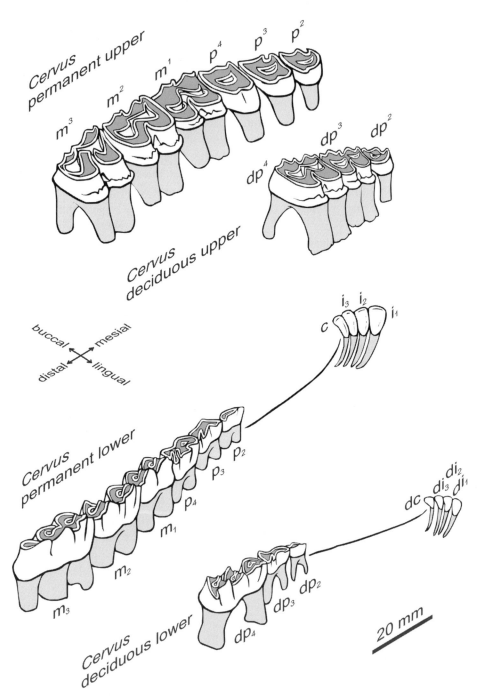

Figure 1.92 *Cervus*, upper right and lower left permanent and deciduous dentitions.

The Bovidae all have horns (with a bony core covered by a permanent horny sheath), particularly the males, and have more hypsodont teeth than other ruminants. Subdivision of the bovids is controversial, but many authors divide them into five subfamilies. The Bovinae are large, solidly built beasts including oxen, bison, buffalo, kudu, eland and bluebuck. Cephalophinae, or duikers, are all small

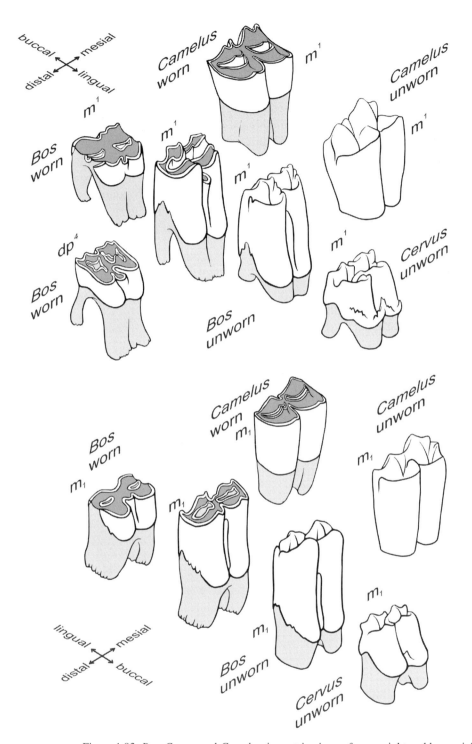

Figure 1.93 *Bos*, *Cervus* and *Camelus*, isometric views of upper right and lower right permanent first molars (*Bos* deciduous fourth premolar), in various states of wear. *Cervus* is shown only in the unworn state, because Figure 1.92 illustrates the worn state. The unworn *Camelus* molars do not have any developing roots at the state of development shown, but they are present in the worn specimens.

and lightly built. Subfamily Hippotraginae includes a variety of large, exotically horned African antelopes, such as kob, redbuck, roan and sable antelopes, oryx, addax, sassabies, hartebeest and wildebeest. Antilopinae comprises smaller African and Asian antelopes such as the klipspringer, dik-dik, blackbuck, impala, gerenuk and gazelle. The Caprinae include small Eurasian and North American forms: sheep, goats, ibex, saiga, chiru, chamois, American mountain goat and musk ox.

The family Giraffidae (giraffes and okapi) has skin-covered bony outgrowths on the skull, instead of horns, and displays a unique double-lobe crowned lower canine. The Antilocapridae includes only one living species, the pronghorn, confined to America (although there are a number of fossil forms). The horns of this creature have a bony core, with a horny sheath that is shed annually.

Family Cervidae
Deer: $Cervus^{EU,AS,AM}$, $Capreolus^{EU,AS}$, $Elaphurus^{ASE}$, $Odocoileus^{AM}$, $Navahoceros^{[AM]}$, $Sangamona^{[AM]}$; musk deer: $Moschus^{AS}$; moose/elk: $Alces^{EU,AS,AM}$; stag moose: $Cervalces^{[AM]}$; giant deer: $Megaceros^{[EU]}$; reindeer/caribou: $Rangifer^{EU,AS,AM}$

The deer generally have a highly characteristic form to their cheek teeth, with much lower crowns and more prominent roots than other ruminants. There is a strongly developed cingulum bulge (Figures 1.92–1.93), particularly on the lingual side of the upper teeth, in association with a much reduced accessory cusp between the twin curves. This form is also present on the buccal side of the lower teeth, but is slightly less prominent. Above the cingulum bulge, the crown narrows markedly, so that the worn occlusal area is smaller than the maximum diameters of the tooth. In most cervids, the ridges or buttresses on the buccal sides of the upper cheek teeth are especially strongly developed, and spread outwards above the cervix of the tooth. The ridges were somewhat reduced in the extinct American Pleistocene form *Sangamona* (Kurtén, 1979) and in the European giant deer *Megaceros*.

Dimensions of deer dentitions (Figure 1.94) range from the cow-sized giant deer and moose, to the somewhat smaller red deer/American elk *Cervus*, down to the small sheep-sized teeth of roe *Capreolus* and musk deer *Moschus*. The extinct American forms *Navahoceros* and *Sangamona* were a little larger than *Cervus*, and *Cervalces* was about the same size as a living moose. They differed little in dental form from today's deer. *Cervus*, as the term is used here, includes a number of species which others place in separate genera. This particularly includes fallow deer (often called *Dama dama*), and dental details (Lister, 1996) may be used to distinguish this from red deer (*Cervus elaphus*). The medium-sized reindeer *Rangifer* has a relatively longer and narrower premolar tooth row than other deer. Permanent upper canines are variably present, usually as a much reduced point (Figure 1.91). Both male and female *Cervus* may have these, and they show enough sexual dimorphism to allow them to be distinguished (d'Errico & Vanhaeren, 2002). They are common finds on European prehistoric sites, pierced in the thin root for a cord or sewing, and such 'elk teeth' have long been prized in North America as ivory

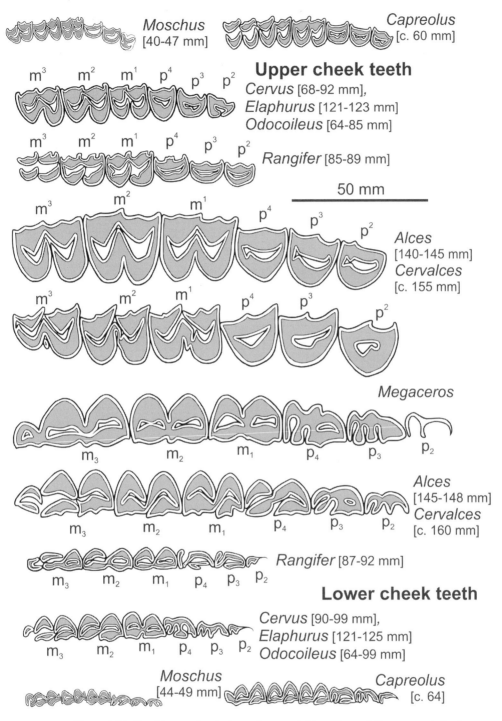

Figure 1.94 Cervidae, occlusal views of upper right and lower left permanent cheek teeth. N.B. Scale is larger than Figure 1.95 but smaller than Figure 1.96. Measurements are lengths of the cheek tooth row in millimetres.

ornaments. In male musk deer, the canines are elongated into long, slender, curved tusks which extend outside the mouth (Figure 1.91). These tusks grow persistently, but become rooted in later life. They have a teardrop-shaped section, with enamel coating only the buccal side and giving rise to a sharp ridge on the distal surface. A sharp point is also maintained by wear on the lingual side of the tip. A deciduous canine is present in young musk deer, as a tiny peg.

Family Antilocapridae
Pronghorns: *Antilocapra*[AM], *Capromeryx*[AM], *Tetrameryx*[AM], *Stockoceros*[AM]

The living pronghorn *Antilocapra* of North America is similar in size to a sheep, and has teeth which differ little from those of the bovids. There are some minor distinguishing features, particularly in the slightly swollen distal form of the upper third molar (Figure 1.96). The extinct Pleistocene pronghorns differed mainly in their size, and the form of their horn cores. *Tetrameryx* was larger than living *Antilocapra*, whereas *Stockoceros* was somewhat smaller, and the diminutive *Capromeryx* smallest of all (Kurtén & Anderson, 1980).

Family Bovidae
Subfamily Bovinae
Cattle: *Bos*[EU,AF,AS,AM]; bison/buffalo: *Bison*[EU,AS,AM]; water buffalo: *Bubalus*[EU,AF,AS]

Large, robust animals with large robust teeth (Figure 1.95). *Bison* is very similar to *Bos*, whereas *Bubalus* has a distinctive form of occlusal outline in both upper and lower third molars.

Subfamily Hippotraginae
Hartebeests: *Alcelaphus*[AF]; addax: *Addax*[ASW]; oryx: *Oryx*[AF,ASW]

Exotically horned antelopes, intermediate in size between the larger bovines and smaller caprines (Figure 1.95). *Alcelaphus* has a markedly intricate infundibulum, whereas *Addax* and *Oryx* have a prominent mesial fold on their lower third molars.

Subfamilies Antilopinae and Caprinae
Gazelles: *Gazella*[AF,AS]; Chinese gazelles: *Procapra*[ASE]; chiru: *Pantholops*[ASC]; saiga: *Saiga*[ASC,EU]; goral: *Nemorhaedus*[ASC]; serows: *Capricornis*[ASC]; mountain goat: *Oreamnos*[AM]; cave goat: *Myotragus*[EU]; chamois: *Rupicapra*[EU]; musk oxen: *Ovibos*[EU,AS,AM], *Symbos*[AM]; shrub ox: *Eucatherium*[AM]; goats: *Capra*[AS,EU,AF,AM]; sheep: *Ovis*[AS,EU,AF,AM]; auodad: *Ammotragus*[ASC]; tahrs: *Hemitragus*[ASC]; bharal: *Pseudois*[ASC]

These two subfamilies include all the small bovids, all rather similar in size, and all with similar teeth which are notoriously difficult to distinguish (Figures 1.95–1.96). *Ovibos* is larger than the other genera, approaching small cattle in size, but distinguished by the slot-like infundibulum seen in the occlusal surfaces of worn upper premolars and lower molars. In the remainder of the genera,

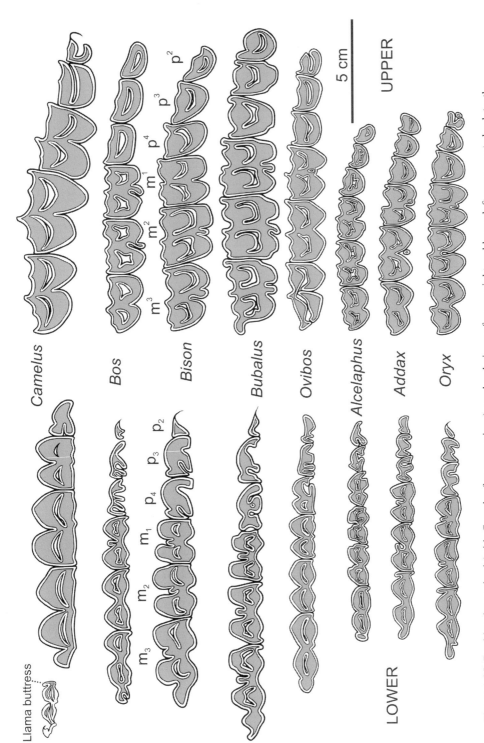

Figure 1.95 Bovidae, large sized (with *Camelus* for comparison), occlusal views of upper right and lower left permanent cheek teeth. N.B. Scale is smaller than Figures 1.94 and 1.96.

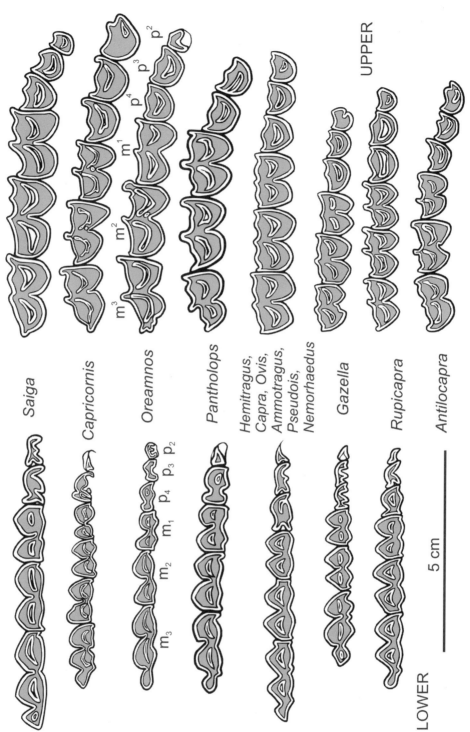

Figure 1.96 Bovidae, small sized, occlusal views of upper right and lower left permanent cheek teeth. N.B. Scale is larger than Figures 1.94–1.95.

Saiga and *Pantholops* are the most divergent forms, with reductions in their premolar row. *Saiga* has no lower second premolar, and *Pantholops* is lacking both upper and lower second premolars. Other differences are rather minor. *Capricornis* has relatively large premolar teeth, with a particularly well-developed distal element to the lower fourth premolar. *Rupicapra* has prominent buccal ridges or buttresses on its upper cheek teeth, and *Gazella* has a narrow, slit-like infundibulum in worn teeth. *Oreamnos* has a somewhat reduced upper second premolar, and some elaboration on the distal end of the upper third molar. Worn permanent dentitions of *Ovis* and *Capra* are very difficult to distinguish, but Payne (1985) and Halstead & Collins (2002) have published a number of features that consistently do this. The extinct cave goat *Myotragus* of Mallorca (Ramis & Bover, 2001) had a reduced cheek tooth row, with greatly enlarged, ever-growing lower incisors. In mature animals, there was a single incisor on each side, but this replaced another much smaller tooth and there is discussion about whether the later tooth was a permanent or a deciduous incisor (Bover & Alcover, 1999).

Suborder Tragulina

The chevrotains live today in tropical West Africa and South Asia. Their teeth follow the usual pattern of Artiodactyla, with the addition of upper canine tusks.

Suborder Tylopoda

The Tylopoda includes one family, containing the camels, llama, alpaca, guanaco and vicuña. Only the camels are found in the Old World and, although extinct forms of camels and llama were present in North America during the Pleistocene, none survives today and all the other forms are confined to South America. Very little is known about the origins and development of the important living domesticates, the two-humped Bactrian camel of Asia, or the single-humped dromedary of Asia and Africa. Camel fossils in Asia and Africa are rare; there is no living wild one-humped form, and it is not at all clear whether the very small population of wild two-humped camels in Mongolia is truly wild or, in fact, feral (Clutton-Brock, 1987). The Tylopoda ruminate, but are not ruminants and their stomach has a somewhat different structure. Their teeth are superficially similar, but the Tylopoda differ in retaining an upper incisor, and canine tusks. The vicuña has rodent-like, ever-growing incisors, unlike any other living artiodactyl.

Family Camelidae
Camels: *Camelus*[AS,AF]; yesterday's camel: *Camelops*[AM]; llamas: *Hemiauchenia*[AM], *Palaeollama*[AM]

$$\text{di}\frac{1}{3}, \text{dc}\frac{1}{1}, \text{dp}\frac{2-3}{1-2} \rightarrow \text{i}\frac{1}{3}, \text{c}\frac{1}{1}, \text{p}\frac{1-2}{1-2}, \text{m}\frac{3}{3}$$

The single upper incisor is reduced to a simple point, but the three lower incisors on each side are spatulate and arranged in a fan which bites against a horny pad over the

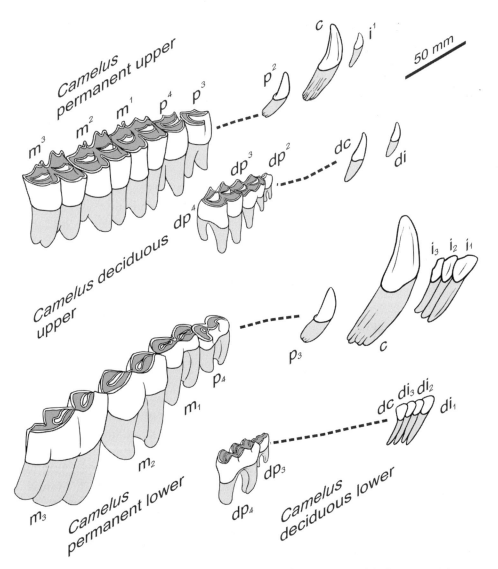

Figure 1.97 *Camelus*, upper right and lower left permanent and deciduous dentitions.

upper jaw (Figure 1.97). They are more asymmetrical than those of the equivalent-sized ruminants, and frequently wear a notch in the lower third incisor, against the single upper incisor. Camelid permanent canines are enlarged into tusks and, in males, these can be quite large, with stout roots and sharp ridges running down the crown to mesial and distal. The permanent upper second premolar is separated from the other teeth by wide diastemata, and is reduced to a small point. When worn, the permanent molars have a double infundibulum (Figure 1.95), with a similar B-shaped occlusal outline to the ruminants, but with a simpler, squarer form; they have less pronounced 'buttresses' on the buccal sides of the upper molars, *without* the additional small cusp (Figure 1.93) between the twin bulges of the lingual side (in upper teeth) or buccal side (in lower teeth). Llamas (Wheeler, 1982), living and extinct, have pronounced buccal 'llama buttresses' at the mesial end of their lower molars (Figure 1.95), but this varies between genera (Webb, 1974). The deciduous upper third and fourth premolars of camelids are like the permanent molars in form, although the third has a somewhat triangular occlusal outline. The deciduous lower fourth premolar has a triple infundibulum, as in the ruminants, but lacks the additional buccal root. *Camelops* was a large extinct Pleistocene camelid in America (Kurtén & Anderson, 1980), larger than the modern camel but with teeth otherwise similar in form. *Palaeolama* and *Hemiauchenia* are likewise large fossil llama from the Pleistocene of America, particularly Florida (Webb, 1974).

Conclusions

Teeth are amongst the most readily identifiable elements of mammal remains in archaeology. The teeth may be preserved as isolated specimens, or as partial dentitions, held in the jaws. Most mammal families show enough variation between genera, and often between species, to make a clear identification possible when the critical parts of the dentition are present. However, some of the most important archaeological families – particularly the Bovidae – show only a little variation and this makes identification difficult. Recent work has addressed some of the most difficult problems (for example the sheep/goat distinction) and the geographical and stratigraphical context of the remains can help as well. This chapter aims to act as a starting point: an introduction to the range of tooth forms. It is necessarily a rather brief treatment and each family could easily provide material for a whole chapter, or even book, of its own. In particular, it has only been practical to present most dentitions in an arbitrarily chosen 'average' fully developed state. In reality, the processes of development and tooth wear make a difference to the appearance of a dentition. The ideal situation is to have a set of reference material which covers the full eruption and wear sequence for each genus, as there was for many of the families when the drawings for this chapter were made.

2

DENTAL TISSUES

This chapter deals with dental *histology,* the study of the microscopic structure of dental tissues: enamel, dentine and cement. There are three important boundaries between them: the enamel–dentine junction (abbreviated to EDJ below), cement–dentine junction (CDJ) and cement–enamel junction (CEJ). Dental histology has considerable potential for archaeology. It can be used to determine age at death. Isolated tooth fragments from one individual can be matched up. A detailed growth sequence calibrated by a daily clock can be investigated. In some mammals, the enamel structure is so complex that it is possible to use it to unravel the course of evolution.

The main tools of dental histology are polarised light microscopy of thin sections and scanning electron microscopy (SEM). Most of the figures in this chapter come from one or the other. Increasingly, confocal light microscopy is being used as well. These are all described at the end of this chapter, together with methods for preparing specimens. The most useful unit of measurement for light microscopy is the micrometre – one micrometre (1 μm) is one thousandth of a millimetre. A human hair is about 100 μm thick, and so is the slice of tooth in most thin sections. In electron microscopy the nanometre is often used as well (1 nm is one tenth of a micrometre).

The inorganic components of dental tissues
Calcium phosphate minerals and their preservation
Living calcified tissues contain between 69% and 99% (by weight) of inorganic material. In all the dental tissues, this inorganic component consists almost entirely of calcium phosphate minerals, and most is in the form of apatite. The *apatites* are a family of minerals, based on calcium (Ca^{2+}) and phosphate (PO_4^{2-}) ions. The general chemical formula is $Ca_{10}(PO_4)_6X_2$. 'X' can be a variety of ions, the most common being hydroxyl (OH^-) and fluoride (F^-). Calcium and phosphate may also be replaced by other ions, and there is continuous variation from one form to another, with partial or complete substitution of any of the ions in the general formula (Elliott, 1997). In geology, apatites occur most often (although never abundantly) in a form approximating to fluorapatite $Ca_{10}(PO_4)_6F_2$. 'Approximating' because there are always some substitutions of other ions. Most of the apatite in bones and teeth is hydroxyapatite $Ca_{10}(PO_4)_6(OH)_2$. This is found only in animal tissues, but again is never in a pure state. Continuous variation between hydroxyapatite and fluor-apatite occurs by the substitution of some fluoride ions for hydroxyl. Intermediate

forms are known as fluorhydroxyapatites. Fluoride ions are frequently present in drinking water and substitutions take place during growth, in the mouth and finally in the ground, when water percolates through buried teeth. Other substitutions include:

- sodium (Na^+) or strontium (Sr^+) in the place of calcium
- orthophosphate (HPO_4^{2-}), carbonate (CO_3^{2-}) or hydrogen carbonate (HCO_3^-) instead of phosphate
- chloride (Cl^-) or carbonate (CO_3^{2-}) in the place of hydroxyl.

Furthermore, additional ions and molecules may be adsorbed onto the surface of the crystals. Apatite crystals in dental tissues are long, narrow and hexagonal in section, and are very small – 'crystallites'.

Although most of the inorganic component is apatite, a proportion exists in other forms. Several other crystalline calcium phosphates are found, including whitlockite, β-$Ca_3(PO_4)_2$, brushite $CaHPO_4.2H_2O$ and octacalcium phosphate $Ca_8(HPO_4)_2(PO_4)_4.5H_2O$. All of these are rare in dental tissues (although common in dental calculus; below), and most non-apatitic mineral in dental tissues is instead amorphous (non-crystalline) calcium phosphates (Williams & Elliott, 1989). Their chemical composition is variable and they are unstable, reverting easily to apatites.

Tertiary and Quaternary fossils show electron microscope evidence of crystallites (Wyckoff, 1972), similar to apatites of living tissues in organisation, orientation and size. X-ray diffraction patterns of mineral bundles in the Middle Pleistocene hominid femora from Trinil, Java, were characteristic of apatite (Day & Molleson, 1973). In a study of human dentine ranging in date from c. 2000 BC to the seventeenth century AD, Beeley & Lunt (1980) showed that calcium and phosphorus remained in similar proportions throughout, not departing greatly from that of fresh dentine. Apatites may thus remain stable, but fossils also show evidence of apatite loss and alteration, often with replacement by crystals of different form and orientation (Wyckoff, 1972) and a larger range of crystallite sizes (Sakae *et al.*, 1997). The mechanism for apatite removal is presumably solution, followed by leaching in groundwaters percolating through the burial matrix. Buried apatite is also affected by the adsorption and substitution of fluorine from groundwater. This occurs progressively, at a relatively constant rate. The resultant fluorhydroxyapatites have a lowered solubility and may be better preserved. Another factor likely to affect rate of loss is the porosity, size and structure of the buried bone or tooth itself. In a wide range of archaeological and geological contexts, however, apatites seem to survive remarkably well.

Dental tissues on archaeological sites have frequently been involved in fires, or deliberate cremations. It is possible to recognise the effects of strong heating in SEM studies and also, to some extent, in the colour of the tissue. Colours may, however, be dependent on secondary mineral deposition or adsorption of trace elements during burial – so they are not an absolutely reliable guide. Burned dental tissues (Shipman *et al.*, 1984) are usually cracked and warped, due to loss of water and organic components. Colour changes and diversifies in the region of 360–525 °C, when most of the organic component is probably lost. The crystallites increase in size at 675 °C, and sinter at 800 °C to fuse as round, smooth masses. Underlying this is a change from apatite to whitlockite between 200 and 600 °C (Sakae et al., 1997).

Non-apatitic calcium phosphates have been demonstrated in medieval samples of human dental calculus from Jämtland, Sweden (Swärdstedt, 1966). Whitlockite was a major constituent in some samples, with brushite in smaller quantities. The proportions were little different from fresh calculus (Chapter 5).

Other minerals in living and archaeological tissues
The enamel of rodents and insectivores (p. 26) may be pigmented by iron oxides. These are granular in form (Schmidt & Keil, 1971) and are deposited by enamel-forming cells (Halse & Selvig, 1974). This type of stain is preserved variably (Stuart, 1982) in Quaternary fossils. Calcium carbonate minerals may be present in dental calculus, in addition to the calcium phosphates. They are more common in the calculus of mammals other than humans and include calcite $CaCO_3$ and weddellite $Ca_2CO_4.2H_2O$ (Navia, 1977). Pore spaces are left in preserved dental tissues after decomposition of cells and proteins. These may be extended by apatite loss. Groundwaters percolating through the burial matrix may contain dissolved ions which crystallise as minerals within the spaces. Typical *secondary minerals* of this kind are calcite or the iron sulphide mineral, pyrite FeS_2. The vivid blue phosphate mineral vivianite $Fe_3P_2O_8.8H_2O$ may also be deposited under these circumstances (Limbrey, 1975).

The organic component of dental tissues
Collagen
Collagen is a fibrous protein, found in dentine, cement and bone. It has a very characteristic 'spectrum' of some 20 different amino acids, with abundant glycine, proline and hydroxyproline (which is unique to collagen, see Williams & Elliott, 1989), joined together to form three spiral chains, which are twisted together into a further spiral macromolecule. These are packed together into fibrils 10–200 nm in diameter. Under the transmission electron microscope, when appropriately stained, these show a characteristic 67 nm banding along their length. Collagen fibrils are mineralised in a very intimate way, with apatite crystals seeded into gaps within them. In anorganic preparations of bone and dentine where the collagen has been removed, details of the fibrils can still be seen, faithfully reproduced in apatite. This has been noted also in fossil bone where little collagen is preserved (Day & Molleson, 1973).

Collagen is very stable. It is insoluble in water and not very susceptible to attack by bacteria or fungi, although heating breaks it down rapidly to gelatine (Hedges & Wallace, 1978). This is how old-fashioned glues are made from hides and bones; incidentally, the word collagen comes from Greek *kolla* (glue). After burial of teeth and bones, collagen is progressively altered and broken down. The chemical process is not well understood, but decomposition is inhibited by low soil pH, exclusion of oxygen by waterlogging, by dryness and low temperature. What remains might better be put into quotation marks as 'collagen' because, although its amino acid composition may not differ much from collagen in fresh tissue, relatively little of the original macromolecule structure may survive unaltered (van Klinken, 1999). About 20% by weight of fresh bone and dentine is collagen but, in archaeological specimens, over 5% protein would be considered well preserved (Schwarcz & Schoeninger, 1991) and the figure may be less than 1%. Even so, it is only under 0.5% that significant departures from the expected amino acid composition are shown (Beeley & Lunt, 1980). The protein content of archaeological dentine shows a strong relationship with how brittle or soft it has become. Specimens with more than 10% protein are more solid and those with less are more vulnerable (Beeley & Lunt, 1980). In some cases, however, electron microscopy of well-preserved Pleistocene teeth and bones has shown collagen fibrils with the characteristic 67 nm banding (Wyckoff, 1972).

Ground substance
Dentine, cement and bone have another organic component in addition to collagen (Jones & Boyde, 1984; Frank & Nalbandian, 1989). Under the light microscope, this is made visible by selective stains (Hancox, 1972) and shows as an amorphous material, known as *ground substance*. This presumably represents the small percentage of organic material other than collagen shown in analyses of fresh dentine and bone. Discounting proteins which may be related to cell contents or serum, and organic acids which may be related to mineralisation, this includes lipids, peptides, glycoproteins and glycosaminoglycans (Elliott, 1997). The lipids are a class of compounds which are soluble in 'fat solvents' such as ether or benzene. Various proteins and peptides are present ('non-collagenous proteins' or NCPs), with long and short amino acid chains. Glycoproteins (also called mucins) are short-chain polymers of proteins and carbohydrates, more commonly known for contributing to the viscosity of saliva. Glycosaminoglycans (acid mucopolysaccharides) are also protein/carbohydrate polymers, but by contrast have long-chain molecules, with regular sequences. In archaeological bones and teeth where collagen is not well preserved, the NCPs seem to survive rather better (Masters, 1987). This may be because they are more closely bound to the mineral component.

Enamel protein
When initially formed, enamel matrix is about 30% protein (below). Later in its development, mature enamel is formed by removal of protein and water, and growth of mineral content. The fully mature tissue has less than 1% protein. About 90% of the proteins in the original matrix are amelogenins, derived from one gene with two copies – one on the X-chromosome and one on the Y-chromosome (Fincham & Simmer, 1997). The original gene product is modified to produce a number of

different smaller molecules. Amelogenins are rich in the amino acids proline, histidine, leucine and glutamine. The remaining 10% of proteins in the original matrix are a mixture. They include tuftelins (rich in glutamic and aspartic acid), ameloblastins (also known as amelins or prism sheath proteins, and rich in proline, glycine and leucine), serine protease enzymes known to modify amelogenins, and other proteases. Serum albumin is also present, but there is discussion about its role and it may be a contaminant. During maturation, the proteins of the matrix are broken down and removed (Robinson *et al.*, 1997). The tiny fraction that remains consists of small proteins and peptides, just a few amino acid residues long. They are rich in glycine and glutamic acid, but low in proline. Protein is concentrated in mature enamel at the tufts, which are 'fault planes' radiating a short way out from the EDJ, and prism sheaths (p. 156). There are similarities between the non-amelogenins of matrix and the mature enamel proteins. For example, the tuftelins are identical to tuft protein (Deutsch *et al.*, 1997), and ameloblastins are found in sheath protein. The tiny fraction of protein in mature enamel survives apparently intact for thousands, and indeed millions, of years, with size of molecules and amino acid composition like those of fresh tissue (Doberenz *et al.*, 1969; Glimcher *et al.*, 1990).

Chemistry and physics of dental tissues in archaeology
Trace element studies
Trace elements are those that are found in low concentrations in body tissues. Archaeological research has centred particularly on the chemistry of fluorine, uranium, strontium and lead, but a more recent trend is to examine groups of elements together (Aufderheide, 1989; Sandford, 1992; Sandford & Weaver, 2000). The fluoride content of groundwater, and therefore of the food animals and plants, is variable. Once ingested, fluoride is rapidly incorporated into the developing dental tissues. In a few places, fluoride concentration in the groundwater is so high that it results in enamel defects (*fluorosis*, p. 169). The fluoride content of freshly extracted teeth shows strong gradients – highest in the dentine next to the pulp, falling through the dentine towards the EDJ and CDJ, and then rising again to the surface of the enamel and the cement (Weatherell *et al.*, 1977; Robinson *et al.*, 1986). The higher the groundwater fluoride, the higher the fluoride content of dental tissues and the more marked the gradient. After death and burial, percolation of groundwater through the teeth adds further fluoride and superimposes its own pattern. Overall fluoride content of archaeological bones and teeth can thus be much higher than that of fresh tissue, and the gradient through the tissues tends to even out (Coote & Sparks, 1981; Coote & Nelson, 1987; Coote, 1988). This gradual accumulation of fluoride with time has been used as a relative dating method (Oakley, 1969; Aitken, 1990), best known in its application to the Piltdown fraud. More recently, it has been applied in association with uranium. Uranium occurs at very low concentrations in fresh dental tissues, but sometimes

much higher in fossils because, like fluorine, it accumulates from the groundwater. It tends to accumulate particularly in a thin surface layer (Henderson *et al.*, 1983), where its concentration can be measured non-destructively by radiometric methods (Demetsopoullos *et al.*, 1983). Uranium dating has been used in conjunction with electron spin resonance (below) dating for hominid sites in Israel (Grün & Stringer, 2000).

Lead enters the body through water or inhaled air. The modern world has many sources of lead, which finds its way into both people and animals. Lead is a 'bone seeking' element that accumulates in the dental and skeletal tissues. It shows a gradient, quite like fluoride (Brudevold & Söremark, 1967; Rowles, 1967). Lead analysis is of interest in archaeology because of the possibilities for monitoring the growth of pollution (Aufderheide, 1989). The difficulty is that lead levels vary between teeth from a single individual, and some specimens show such anomalously high values that they can only be the result of post-mortem change (Waldron, 1983).

Strontium has been of particular interest because of its relationship with diet (Sillen & Kavanagh, 1982; Aufderheide, 1989; Sandford & Weaver, 2000). The mammalian gut absorbs calcium more readily than strontium and the effect is increased at higher tropic levels in an ecosystem. Plants have the highest strontium levels, herbivore tissues lower, carnivores that eat herbivores lower still and carnivores that eat carnivores the lowest. Strontium is readily incorporated into the dental tissues and, unlike fluorine and uranium, does not show strong gradients. Much discussion has centred on diagenetic change to strontium in buried teeth and bones (Ambrose, 1993). It is also known that there is variation in relation to other aspects of the diet, rather than only trophic level (Burton & Wright, 1995; Burton & Price, 1999).

Preservation of DNA
Since the 1990s dentine has emerged as a tissue of choice for recovery of DNA from the individual to which the tooth belonged, although hair in mummified remains is also becoming an important source. DNA is now routinely extracted from ancient teeth and bones, and particular sequences of bases amplified using the polymerase chain reaction (PCR). This is not the place to discuss ancient DNA, which is reviewed elsewhere (Kaestle & Horsburgh, 2002). Ancient DNA is very degraded, and one of the main problems is contamination, from the organisms breaking down the bone or tooth, from the soil, from archaeologists and from laboratories. This so far outweighs the tiny surviving proportion of ancient DNA that it is very difficult to exclude the possibility that the sequence amplified actually represents an individual's DNA. A number of checks can be made (Cooper & Poinar, 2000), but one of the most convincing is when the DNA of extinct creatures is shown to contain a sequence which cannot be matched by any living one (Krings *et al.*, 1997; Ovchinnikov *et al.*, 2000).

Table 2.1 *Commonly studied isotopes in dental tissues*

Isotope	Abundance (%)	Tissue component in which analysed
Hydrogen		Collagen
^{1}H	99.99	
^{2}H (D – Deuterium)	0.02	
Carbon		Collagen and carbonate in apatites
^{12}C	98.90	
^{13}C	0.10	
^{14}C	<0.01	
Nitrogen		Collagen
^{14}N	99.64	
^{15}N	0.36	
Oxygen		Carbonate and phosphate in apatites
^{16}O	99.80	
^{17}O	0.04	
^{18}O	0.2	
Strontium		Phosphate in apatites
^{84}Sr	0.56	
^{86}Sr	9.87	
^{87}Sr	7.04	
^{88}Sr	82.53	

Adapted from Schoeninger (1995).

Isotopes and dental tissues

Most of the elements in both the inorganic and organic components of dental tissues have more than one form, or *isotope*. They differ in the number of neutrons in their nucleus, and this is denoted by a number in front of the symbol for the element (Table 2.1). Elements are incorporated into the body through a series of chemical reactions, and the isotopes of one element react at different rates, so that different quantities are incorporated (Schoeninger, 1995). This is known as fractionation. The effect is less marked for heavier elements, so strontium shows almost no fractionation and hydrogen shows a great deal. Isotope ratios are measured using mass spectrometry and are expressed as a δ-value in parts per mil (‰). So, for Table 2.1, the ratios would be δD, $δ^{13}$C, $δ^{15}$N, $δ^{18}$O and $δ^{87}$Sr (or sometimes ^{87}Sr/^{86}Sr). Teeth make a good material for extracting the organic and inorganic components from which these isotopes can be measured. The difficulty is that there are diagenetic changes to dental tissues which may change the isotope ratios that were present in life. Carbonate is readily altered in this way, so it is better to use phosphates for strontium and oxygen, and collagen for carbon. Enamel is usually less altered than dentine. Bone or dentine collagen is the source for carbon, nitrogen and hydrogen, but it is rarely intact in ancient tissues (above) and care is needed in interpretation. Because teeth were formed in childhood and are not subject to tissue turnover, they offer the possibility of comparing isotopes for different stages of an individual's life (Balasse *et al.*, 2001; Price *et al.*, 2001; Richards *et al.*, 2002).

Carbon isotope ratios give information about the plants on which the food chain is based. Four groups can be considered. C_4 plants are mostly tropical grasses, including maize, sorghum and millet, and they have high $\delta^{13}C$ values. CAM plants are mainly the succulents of arid areas and have values like C_4 plants. The majority of other terrestrial plants are C_3, with low $\delta^{13}C$ values, and marine plankton are in between these two extremes. These values are passed on up the food chain to the dental tissues of herbivores and carnivores and, although there is fractionation at each stage, the tissue values are correlated with the plant values. The $\delta^{13}C$ of collagen reflects mainly the protein in the diet (and to some extent the carbohydrate), whereas the $\delta^{13}C$ of enamel carbonate shows the diet as a whole. Nitrogen isotope ratios show differences between marine and terrestrial food chains, between nitrogen-fixing plants such as legumes and non-fixers, and at different trophic levels so that carnivores have higher $\delta^{15}N$ ratios than herbivores.

Carbon isotope ratios have been used to document the spread of maize agriculture through the Americas (maize domestication originated in Mexico around 4000 BC) (summary in Schoeninger, 1995). Nitrogen and carbon isotope ratios have indicated a change in Europe from a widespread Mesolithic use of marine resources, to a strong Neolithic dependence on terrestrial resources (Lubell *et al.*, 1994; Richards & Hedges, 1999). They have also indicated that European Neanderthals were strongly terrestrial carnivores and that, by contrast, some Upper Palaeolithic groups used substantial aquatic resources (Richards *et al.*, 2000, 2001).

Oxygen isotope ratios in bones and teeth from animals that obtain water mainly from drinking show temperature variation. Where animals obtain much of their water intake from leaves, the ratios vary with humidity. Values of δD also seem to be related to temperature and humidity, although there are also differences between C_3, C_4 and CAM plants.

Strontium isotopes, which show effectively no fractionation, represent the rocks from which groundwater came. The isotope ratios vary with the geology, from region to region, so it is possible to see if the values for the enamel (formed during childhood and then little altered) vary within one site, and if they match the values for the surrounding geology (Ezzo *et al.*, 1997). Similarly, it is possible to compare the ratios for enamel with those of bone (which turns over and therefore represents in adults the water intake over the few years before they died) and gain an idea of population movements (Sealy *et al.*, 1995; Price *et al.*, 1998, 2001).

The other important information given from carbon isotopes is radiocarbon dating (Aitken, 1990). It relies on the ratio between the radioactive isotope ^{14}C and stable isotope ^{12}C, in samples prepared from collagen (or other organic materials such as wood). Conventional dating uses β-particle counting which needs large samples (10 g minimum), and most teeth are too small to provide this much dentine. The newer technique of mass spectrometry (AMS), however, requires much smaller samples (0.5 g), and dentine, protected by a cap of dense enamel, is a good material for dating. The maximum range of radiocarbon dating, for practical purposes, is around 45 000 BP.

Electron spin resonance dating of enamel apatites
ESR dating is based on the trapping of free electrons by defects in the apatite crystal lattice. The electrons are generated by radioactive elements in bones and teeth, and the burial matrix so that, over time, the number of trapped electrons increases. An ESR spectrometer measures the signal generated by the trapped electrons in such a way that the accumulated radioactive dose can be evaluated and, from this, the date is estimated. Thick enamel is the preferred material, although the sample size is only 10–100 mg. The earliest dates proposed are around 2 million years BP, so ESR is potentially important in the Quaternary for the period beyond radiocarbon. It has been used for key hominid sites in Europe (Falguères, 2003) and Israel (Grün & Stringer, 2000).

Policy on taking samples for analysis or sectioning
Museums around the world now receive many requests each year for teeth which will be removed and sampled or sectioned in a way which results either in complete destruction of the specimen, or partial damage. Exciting new research ideas need appropriate samples but, once the tooth has been destroyed or damaged, it is not available for other researchers to follow up *their* exciting new ideas. Some might have difficulty in regarding teeth as patrimony, but they are an important source of information on our past, which should be preserved for future generations, and this resource is being eroded. Sampling is becoming a difficult issue and researchers might consider a number of questions:

1. Why is this tooth, in particular, needed and not another specimen? Is there a clear hypothesis that requires it? Could the research question be answered in another, non-destructive way?
2. Is there a good chance of success? It makes sense to carry out a preliminary pilot study on a few less important specimens before larger numbers are taken.
3. Is the tooth appropriate for the analysis or sectioning? For example, if enamel has been worn away to expose the dentine to diagenetic change, does that invalidate the results?
4. Is the most minimally destructive sampling method being used? What will happen to any parts of the tooth remaining?
5. What is the procedure for validating results? For example, how are the potential effects of contamination in DNA analysis going to be investigated?
6. If possible, the tooth should have an antimere that remains in the collection and is never sampled. If there is a choice of teeth, is there one which could be more readily sacrificed by other types of study? Does one tooth have more potential than another, say, for a study of morphology, estimation of age or pathology?
7. A good record should be made beforehand, including measurements, morphology and pathology of the whole dentition, not just the tooth that will be taken. A dental impression should be taken (p. 199), and a detailed photographic record made. It seems reasonable to suggest that this should be built into the research project sampling the material, and budgeted accordingly.

Many people, including curators, researchers and others, present and future, have a stake in ancient remains. It is important not to stand in the way of current research,

Dental tissues

but consideration does need to be given to other fields of research that may not yet have examined the material.

Dental enamel
Composition of enamel
Mature dental enamel is unique amongst mammal tissues. It is almost entirely inorganic and is acellular – effectively dead even in living organisms. By dry weight fresh enamel contains 96% inorganic material, less than 1% organic material, and the rest water (Williams & Elliott, 1989). The chemical composition of the inorganic component approximates to hydroxyapatite. Enamel crystallites are considerably longer than those of bone and dentine. Estimates vary, but they are at least 1600 nm long and maybe even more. In mature enamel, the crystallites are packed together to make a dense, very finely crystalline mass.

Amelogenesis, the formation of enamel
Enamel is formed by the internal enamel epithelium; a closely linked sheet of cells called *ameloblasts*. They are narrow and cylindrical, and packed together so that only the narrow face of one end is in contact with the developing enamel. At various stages of amelogenesis the cell end may be flat, bear a protuberance, or be convoluted into 'ruffles'. When a protuberance is present, it is known as *Tomes' process*, after the nineteenth-century anatomist C. S. Tomes. Tomes' process varies in size and shape between different stages in amelogenesis, different regions of the crown and different species. In addition, the ameloblasts are able to move in relation to one another, weaving complex patterns of enamel structure.

Amelogenesis takes place in two stages, *matrix production*, and *maturation*. Matrix production involves the formation of an organic matrix and the seeding of thin, ribbon-like crystallites into it. This matrix has the structural features of fully mineralised enamel, but is only about one third mineral, with one third protein and one third water (above). Maturation involves removing the protein and water, and increasing the size of the crystallites, to produce the heavily mineralised mature enamel. Each ameloblast is involved in all stages of amelogenesis: committed to matrix production for up to one year, then switching to maturation. Their form and physiology change accordingly. For the very first part of its active matrix-forming life, an ameloblast has no Tomes' process. After a few days, the process develops and is then present throughout the matrix production phase, until just before the end. When present, each Tomes' process fits into a tiny pit in the developing enamel matrix surface. As each ameloblast switches into maturation, it develops a ruffled cell wall and shows internal changes related to the removal of protein. The bulk of maturation is carried out by the ameloblasts but the process continues slowly after the cells have died, during tooth eruption (Chapter 3) and even when the tooth is in the mouth.

Each ameloblast is responsible for a given territory of enamel. It produces matrix for the whole thickness at this point and brings about its maturation. Different areas

156   Teeth

Figure 2.1 Enamel prisms in a human first molar from medieval York. SEM-ET image (field width 90 μm) of fractured, etched surface. The alternate swelling and constriction of the prisms is related to the cross striations.

of enamel start to form at different times. Ameloblasts at the coronal tip of the crown start first. Further down the crown sides, ameloblasts start the sequence progressively later and later, so that the cells at the cervix start last.

The prismatic structure of mature enamel
The enamel crystallites are organised into bundles about 4–12 μm wide (Figure 2.1), known as *prisms*, although they are far from being simple prismatic elements and the width varies with the plane of section. Within and between them the crystallites vary in orientation. The boundaries of prisms are formed by discontinuities, where crystallites of one orientation meet crystallites of another, very different orientation. Such discontinuities exist within the secretory territory of each ameloblast and are introduced by the Tomes' process. Much of the organic content of mature enamel resides at prism boundaries, forming a *prism sheath* (Osborn & Ten Cate, 1983). As the ameloblasts near the end of their matrix production phase, they lose their Tomes' processes and the final layer is without prism boundaries. This is known

Dental tissues

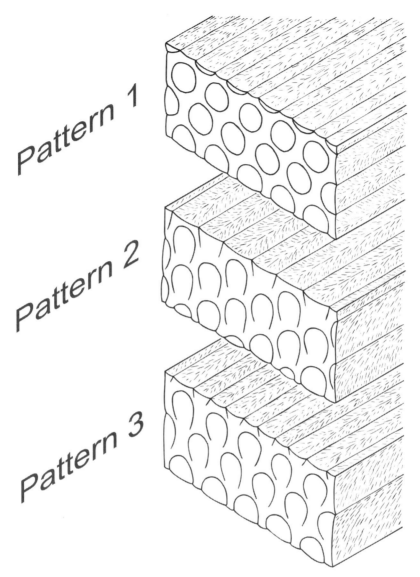

Figure 2.2 Enamel patterns 1, 2 and 3. Isometric views with the prisms sectioned transversely and longitudinally.

as *prism-free true surface zone enamel*. Soon afterwards, the ameloblasts change over to their maturation mode.

Three main patterns of enamel structure (Boyde, 1989) can be distinguished (Figure 2.2). *Pattern 1* enamel is the only form to have discrete prisms, surrounded by complete, cylindrical boundaries. The prisms fit into tunnels within an enveloping *interprismatic* enamel. *Pattern 2* enamel has incomplete cylindrical discontinuities. When viewed from the developing enamel surface, they are horseshoe shaped in outline. The prisms are arranged in vertical rows, separated by *interrow sheets* of interprismatic enamel. Each prism is connected by narrow bridges to the interprismatic enamel on either side. *Pattern 3* again has incomplete cylindrical discontinuities, horseshoe shaped in outline (at the developing surface). The prisms are also arranged in rows but, this time, the rows are horizontal and the interprismatic enamel is attached as a 'tail' below each bulbous prism 'head'. The tails of each prism row fit between the heads of the prisms in the row below, and all are connected by narrow bridges of enamel.

All three patterns can be found in most teeth, with gradations between them, but different groups of mammals generally show one more commonly than the others (Boyde, 1971, 1989; Boyde *et al.*, 1988). Pattern 3 predominates in the Primates, Carnivora, Pinnipedia and Proboscidea. Rhinoceroses and *Hippopotamus* also have Pattern 3, but most Perissodactyla and Artiodactyla have Pattern 2 as the predominant form, as do marsupials, lagomorphs and rodents (although their complex decussation hides it; p. 180). The interrow sheets are particularly prominent in Perissodactyla and Artiodactyla. The odontocete whales have a mixture of Patterns 2 and 3 in the deeper layers of their enamel, but thick layers of Pattern 1 lie above them. Pattern 1 is often found just under the surface of the tooth crown in many mammals so, although it has been described as particularly abundant in the Sirenia, Tapiridae, Chiroptera and Insectivora, further research may show other patterns.

Many mammals possess minute *tubules* in their enamel. All marsupials with the exception of the wombat (Boyde & Lester, 1967) and the honey possum (Gilkeston, 1997) have them. Each tubule is associated with a prism, and follows along its length from the EDJ right through the full enamel thickness. The tubules are continuous with the dentinal tubules (below) and range from 0.2 to 2.5 μm in diameter. Related structures are also found in at least some rodents, insectivores and bats, and in prosimian Primates (Carter, 1922; Boyde & Lester, 1967; Boyde, 1989). Short enamel tubules, just extending a few micrometres from the EDJ, are more widely found, for example in humans (Boyde, 1989).

Enamel preservation
Enamel is normally the best preserved of hard tissues. It is almost wholly mineral, so that decomposition of organic matter has little effect on it. The felted crystalline nature of enamel renders it hard and tough, and resistant to mechanical erosion. In addition, it may already have considerable quantities of adsorbed fluorine before burial, due to contact with fluorides in food and drink, and after burial, from groundwater. Its solubility may therefore be lowered in relation to the other tissues. The cap of enamel protects underlying dentine but, if this has become brittle, the poor support may cause the enamel to be fractured and eroded.

Archaeological enamel nearly always yields good microscope sections, often indistinguishable from fresh enamel. Conditions on some archaeological sites, by a process which is not understood, actually enhance the features of the enamel in

Dental tissues

sections (Antoine *et al.*, 1999). Fossil enamel sections, in particular, often show this. By contrast, there may be post-mortem pitting and cratering of the enamel surface, similar to the effects of the disease dental caries (Chapter 5).

Teeth are frequently caught up in fires on archaeological sites. Strong heating can change the appearance of enamel under the SEM (Shipman *et al.*, 1984). At 185–285 °C, the appearance of a fractured surface under the SEM is smoother than normal. At 285–440 °C such surfaces appear granular and at 440–800 °C larger glassy granules are visible, separated by pores and fissures. Between 800 and 840 °C the granules coalesce into larger, rounded, smooth globules about 0.5–1 μm across. In cremations of adult humans, enamel is usually lost completely, fracturing away to expose the surface of the dentine but, where developing teeth are protected inside the bone of children's jaws, the enamel often survives (McKinley, 1994).

Incremental structures in enamel

There is a rhythm to amelogenesis, preserved in mature enamel as regular changes in prism structure. In sections under the optical microscope these show as alternate bright and dark bands across prisms, spaced about 4 μm apart (Figure 2.3), called *prism cross striations*. Treatment of enamel sections with dilute acid accentuates them (Schmidt & Keil, 1971; Boyde, 1979), and they show as compositional contrasts in backscattered electron mode SEM (SEM-BSE) images (p. 205), suggesting that they are caused by variation in mineralisation. This may be due to changes in the ratio of phosphates to carbonates in the apatite crystallites (Boyde, 1979). Examination of fractured preparations in Everhart–Thornley or secondary electron mode (SEM-ET) imaging (Figure 2.1) shows that the cross striations correspond to a similarly spaced regular bulging and constriction of the prism boundaries, perhaps related to rhythmic variation in the rate of enamel matrix production (Boyde, 1989). The dark bands of the cross striations seen in ordinary transmitted light microscopy probably represent the constrictions of the prism boundaries seen in SEM-ET, and the regular lower density regions seen in SEM-BSE (Jones & Boyde, 1987). There has been some discussion of possible confusion between cross striations and rows of transversely sectioned prisms (which have similar dimensions), because overlain structures are difficult to resolve in the relatively thick sections of conventional light microscopy (Weber & Glick, 1975). Their existence has, however, been confirmed by confocal microscopy (Figure 2.4), which focuses on a very thin plane within the section, and by the SEM evidence. This leaves little doubt that cross striations exist, even though care is needed in identifying them with conventional microscopy (Boyde, 1989; Shellis, 1998). Some sections show a phenomenon known as 'doubling' in some areas, where small dark bands are seen at half the spacing of the cross striations. Some are probably 'artefacts' relating to the section thickness (Shellis, 1998), although some may be real features, but they can in any case be identified (and discounted) by measurements.

One combined bright/dark unit of the prism cross striations represents approximately 24 hours of enamel matrix formation (a *circadian* growth rhythm), at least

160 Teeth

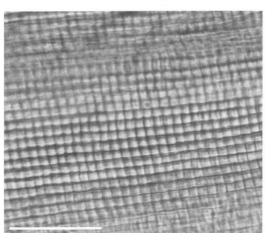

Figure 2.3 Brown striae of Retzius and prism cross striations in a human upper second incisor from the crypt of Christ Church, Spitalfields; 100 μm thick section seen in ordinary transmitted light. Scale bars are 40 μm. In the left image, the prisms run almost horizontally and are broken at regular intervals by the dark marks of the cross striations, which are about 4 μm apart. Running diagonally across the image are the shadowy lines of the brown striae of Retzius, which are not well defined at higher magnifications like this. The right-hand image has the surface of the crown on the right, with the more well-defined surface brown striae angling up towards it. The brown striae are less well defined deep to this, but the prism outlines and cross striations can clearly be made out.

in primates. Counts of cross striations from the earliest formed enamel to the end of crown formation correspond well with independent estimates of crown formation times (Asper, 1916; Gysi, 1931; Boyde, 1990). Sections of teeth from laboratory animals which had received periodic injections of chemical markers have shown good matches of counts between marker lines and intervals of injections (Mimura, 1939; Bromage, 1991; Dean, 1998a). Sections of teeth from children with historically documented ages at death have also shown a close match between these ages and cross striation counts, from the enamel formed at birth, to the enamel still being formed at the point of death (Antoine *et al.*, 1999). Cross striations therefore seem to provide regular markers that may be used to establish a reliable chronology (Dean *et al.*, 2001) of enamel formation.

Most mammals also show a coarser layering of enamel. Lines, marking out successive positions of the matrix-forming front, run through the tissue (Figure 2.3).

Dental tissues 161

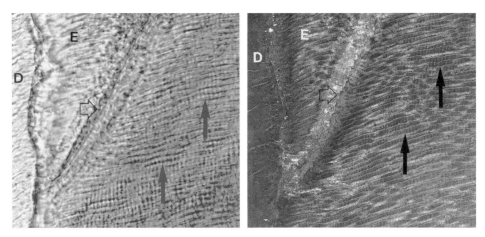

Figure 2.4 Prism cross striations in a human incisor from the crypt of Christ Church, Spitalfields; 100 μm thick section, with the same field of view seen in ordinary transmitted light (upper image) and confocal reflected light microscopy (lower image). Field width 170 μm. The confocal microscope was a Leica TCS SP (reflection mode; 488 nm laser) using an oil immersion objective with the oil applied on top of the cover-slip. Confocal microscopes can focus into the thickness of a section (in this case about 50 μm below the surface), defining a plane of focus less than 1 μm thick within the section. The images were taken by Daniel Antoine, Institute of Archaeology, UCL. The arrows in the images mark features which can be matched exactly between the two types of microscopy. On the left of both images is the dentine (marked D), with the EDJ running vertically. The rest of the images is all enamel (E). The prisms run almost horizontally, with a slight angle upwards, across the images. Regular cross striations mark the prisms at 4 μm intervals in both images. The short, open arrow shows the line of a strongly defined brown stria of Retzius and, in the confocal image, this can be seen as a sharp line of disruption to the prisms. The boundaries of the prisms appear unbroken in the ordinary light image, but it is clear in the confocal image that some (pointed to by the lower long dark arrow) are longitudinally sectioned and represent parazones (p. 178), and some are more transversely sectioned (just below the upper dark arrow) and represent diazones. There is, however, a clear difference between the cross striations pointed out by both dark arrows, and the more transversely sectioned prisms.

These were first described in detail by the Swedish anatomist Anders Retzius (1837), and so are called the *brown striae of Retzius*. In a radial section (p. 201) of a tooth, the arrangement of the striae records the history of crown formation (Figure 2.5). The earliest enamel-forming fronts are 'dome' shaped. Successive dome-like increments of enamel are formed one on top of another. When the full height of the cusp tip is formed, the ameloblasts at the apex of the dome change over to maturation. The active matrix production front then becomes 'sleeve' shaped. Successive sleeve-like increments of enamel are deposited around the 'last dome', overlapping one another towards the cervical part of the crown. The sleeves become narrower and thinner as the crown nears completion and the angle relative to the EDJ becomes nearer to the perpendicular. So the cusps and cutting edges of crowns are built up of dome-like increments and the sides are built up of sleeves. The dome-layered region

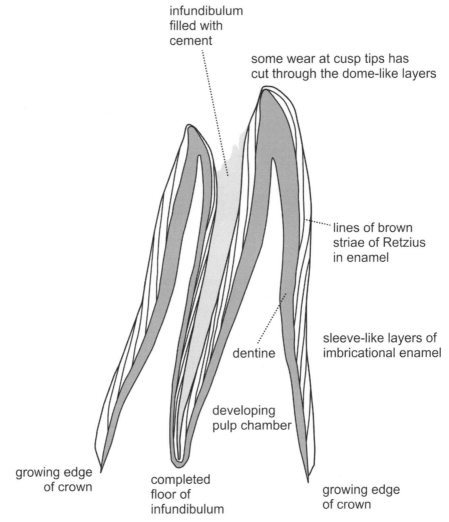

Figure 2.5 Pattern of enamel layering in a cattle molar. The lines along which brown striae of Retzius run have been traced from a section through two cusps of a crown that, whilst almost complete, was still developing. The grey-shaded dentine always develops slightly in advance of the enamel and projects as a sharp forming edge. Hypsodont teeth like this one have relatively thin enamel over the cusps, with few dome-shaped appositional or cuspal layers. As the tooth would already have been partially erupted, there is some wear at the cusp tips. The bulk of the crown height is made up of the sleeve-like imbricational layers. The pulp chamber is still developing, but already shows the tall thin projections (pulp 'horns') up each cusp, characteristic of hypsodont teeth.

is often called *appositional*, or *cuspal enamel*, whereas the sleeve-layered region is distinguished as *imbricational* or *lateral enamel*. This arrangement is characteristic of carnivores as well as Primates, but the tall crowns of ungulates do not really have cusps; the period of dome-like formation is very short and almost the whole crown is made from sleeve-like layers of lateral enamel. Brown striae in taller-crowned teeth also usually have a steeper inclination relative to the EDJ than they do in the bunodont teeth of primates. In cattle or horses, they are only a few degrees away from being parallel to the EDJ.

Brown striae may be difficult to observe. They are best seen with relatively thick sections (150 μm) under low magnifications (Figure 2.3). Often they are less distinct occlusally and deeper into the enamel, and more sharply defined cervically and near the surface. In many mammals, cross striations line up together to make prominent zigzag lines, finely spaced, which make it difficult to distinguish a coarser layering inside the enamel even though, near the surface, there clearly are regular brown striae. The enamel of deciduous teeth in humans shows them less clearly than permanent enamel.

In the region of imbricational enamel, brown striae outcrop at the crown surface. At each outcrop, the layer of prism-free true surface zone enamel (above) is breached and a shallow microscopic trough is formed around the crown circumference (Risnes, 1985a, 1985b). Viewed from the crown surface, these troughs follow one another at regular intervals down the tooth, somewhat like waves (Figure 2.6). When they appear in this way, these structures are called *perikymata*. Within the troughs, where the brown striae appear, there is no prism-free enamel, and exposed Tomes' process pits mark the point at which ameloblasts ceased matrix secretion (Boyde, 1989). Whatever causes the brown striae must also cause the ameloblasts nearing the end of their matrix production phase (and just about to produce prism-free enamel) to jump straight into maturation. The wave troughs are formally known as perikyma grooves, and the wave crests between them as perikyma ridges (Risnes, 1984). Towards the cervix of the crown in humans, the perikymata become closer spaced and change to a more sharply defined form that looks like overlapping sheets (Hillson & Bond, 1997). In some mammals, they take this form all the way down the crown side. These are known as *imbrication lines* but they represent the same layering, and all these features are generally referred to as perikymata.

If examined carefully, the enamel of the crown surface shows perikymata in most mammals. For some, the surface layer of cement has to be stripped away but, even here, regular perikymata are clearly visible in the enamel surface underneath (Figure 2.6). Spacing varies down the crown and between species. In humans, they go from 150 μm near the cusps down to 30 μm near the cervix. In pigs and cattle they are about 100 μm through most of the crown height, in horse they are 170 μm and in mammoth 270 μm (author's own observations). The largest (>1 mm) perikymata, so prominent they can be seen clearly with the naked eye and felt with a fingernail, are seen in mastodon molars.

164 Teeth

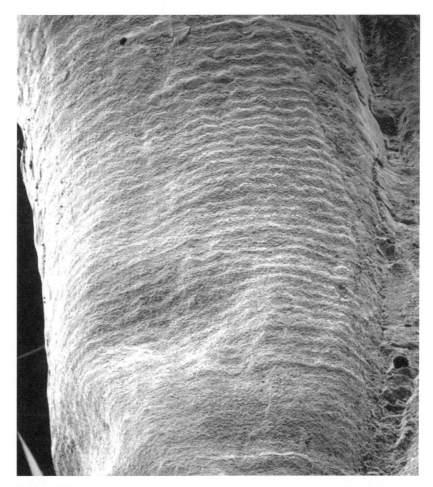

Figure 2.6 Perikymata on the enamel crown surface of a horse molar. The cement has been stripped away with a dental scaler. SEM-ET of cleaned, gold coated surface, field width 6 mm.

Variation in the brown striae of Retzius
Brown striae of Retzius are only seen clearly in thick sections (above) and under low magnifications, but cross striations are best seen at higher magnifications in thinner sections, so it is often difficult to determine the relationship between the two structures. Within each section, however, there are usually patches where it is possible to count cross striations between sharply defined brown striae. If this is done repeatedly throughout one section, and in different teeth from one individual, the count remains the same for all enamel in all teeth and this implies a constant rhythm throughout dental development. Many studies of human teeth (Dean, 1987; Fitzgerald, 1998; Fitzgerald & Rose, 2000) have shown that the count varies between individuals. The counts range from 6 to 11, but the most common

counts are 7, 8, 9 or 10. Studies on living great apes show a similar range and the modal count for great apes and humans together is 9 (Dean et al., 2001). Few counts are available for other mammals, but Fukuhara (1959) counted macaques (7–8), elephants (7–9), horse (5–6), rhinoceros and tapir (10), pig (11), cattle (10), carnivores (4–5), insectivores and bats (2).

The brown appearance of the striae is due to light scattering, perhaps related to changes in crystallite size or orientation, or enamel porosity. They are enhanced by etching sections (Schmidt & Keil, 1971), and usually show as poorly mineralised lines in SEM-BSE images (Boyde & Jones, 1983). Rarely (Gustafson & Gustafson, 1967) they may be hypermineralised. The striae are also marked in SEM images of fractured enamel by a parallel diversion of prism boundaries (see also Figure 2.4), varying from a very slight notching to a clear translocation (Osborn, 1973; Weber & Ashrafi, 1979) or an actual break in the prisms (Frank, 1978).

Striae of Retzius vary in prominence throughout a section of enamel, most noticeably in the darkness of the line, and its apparent width. The sequence of prominent striae from one tooth can be matched with other teeth (Fujita, 1939; Gustafson, 1955; Gustafson & Gustafson, 1967) forming at the same time in one individual, and the counts of cross striations between striae in the sequence remain the same in the different teeth. This shows that some systemic change during development causes the variation, but it is not clear what this would be. Prominent striae have been associated in humans with developmental defects (Figure 2.8 below) seen at the crown surface (Goodman & Rose, 1990; Hillson et al., 1999). These defects are caused by childhood fevers, nutritional deficiencies and other upsets, so it seems reasonable to suggest similar causes for the more prominent striae. Prominent striae in enamel sections have been matched with documented periods of illness in a captive colony of macaques (Bowman quoted in Dirks, 1998). One particular case of prominent striae is the *neonatal line*. This is found in those teeth forming at the time of birth (deciduous teeth and some first molars) and separates prenatal enamel in which brown striae are less prominent, from postnatal enamel in which they are more prominent (Schour, 1936; Weber & Eisenmann, 1971; Whittaker & Richards, 1978). It is particularly useful as a starting point for chronologies of enamel development (below). Gustafson & Gustafson (1967) distinguished between normal rhythmic incremental lines and pathological lines. By the latter they appear to have meant prominent striae which showed hypo- or hypermineralisation. Wilson & Schroff (1970) also distinguished pathological bands, this time as striae which were associated with 'atypical' prism forms. Rose et al. (1978) used their definition and methods to record the bands as markers of physiological stress in childhood (below). In order to recognise that the bands might not be caused by disease, Rose introduced the neutral term 'Wilson bands'.

Development of brown striae varies between mammal orders (Fukuhara, 1959; Schmidt & Keil, 1971). In some, notably Primates and carnivores, they are clearly marked. In others, such as rodents and marsupials, they are not. Further comparative histological studies are needed.

Enamel incremental structures and archaeology

In archaeology, brown striae were first noted in human remains by Soggnaes (1956), including Upper Pleistocene material from Skhūl and Tabun in Israel, predynastic material from Upper Egypt and Medieval from Norway. Falin (1961) noted both cross striations and brown striae in prehistoric teeth from Dniepropetrovsk. Hillson (1979) found both structures in predynastic to Roman period material from Egypt and Nubia, and recorded perikymata in dental impressions. At a similar time, Rose (1977; Rose *et al.*, 1978) examined Wilson bands (above) in teeth from Woodland and Mississippian sites in Illinois. He divided sections of lower canines into 'enamel units' – zones based on the tooth development sequences of Massler *et al.* (1941) – and counted the number of Wilson bands within each. Individuals in which they were frequent had a low average age at death, suggesting that Wilson bands reflected physiological stress. Cook (1981) used the alternative technique of measuring along the EDJ and counting the number of Wilson bands per millimetre as a measure of their frequency. Condon (1981) used similar EDJ measurements, from the first formed dentine, combined with published dental development times (Moorrees *et al.*, 1963) to derive a chart to determine the ages at which the Wilson bands had been initiated. It was, however, found that when Wilson bands were matched in different teeth from one individual, the estimated timings did not correspond well between teeth (Goodman & Rose, 1990). Rudney and colleagues (Rudney & Greene, 1982; Rudney, 1983a, 1983b) used similar methods for Meroitic and X Group material from Nubia, and developed a 'weight' system for Wilson bands, according to the degree of prism abnormality. Weights were summed for each crown, to give a 'growth disturbance score'.

At the same time, however, counts of prism cross striations, brown striae of Retzius and perikymata have been used directly to build chronologies of tooth crown formation and, if teeth were still forming at the time of death, to estimate age. Boyde (1963, 1990) first applied this method to a 4 to 5 year old Anglo Saxon child (Miles, 1963b). A neonatal line could be identified under the mesiobuccal cusp of the permanent first molar, and prominent brown striae sequences could be matched between this tooth and the first incisor. A total of 1692 cross striations were counted from the neonatal line to the last enamel formed on the incisor crown. The child died only just after this crown had been completed, so 1692 days (4.6 years) must have been close to the age at death. The advantage of this method is that it is independent of the assumptions inherent in using dental development standards (Chapter 3). Similar methods were used to determine the formation timing of tooth crowns in Medieval dentitions from Picardie, France (Reid *et al.*, 1998), and in a child from the eighteenth to nineteenth century crypt at Christ Church, Spitalfields, in London (Dean & Beynon, 1991). The latter site yielded the remains of a number of children whose age at death was known independently from coffin plates and parish records. This made it possible to compare cross striation counts with the age in days, and these matched well in several children (Antoine *et al.*, 1999). Something about the Spitalfields crypt has made cross striations particularly prominent, but many sections show them less clearly. They are particularly difficult to count in the 'gnarled enamel' under the cusps, where decussation (p. 177) causes a large variation in prism direction. In such cases, an alternative is to measure the length of prisms in the cuspal enamel, from the first secreted matrix at the EDJ (or the neonatal line in the mesiobuccal cusp of the first molar) up to the first perikyma groove at the crown surface, on the tip of the cusp (Reid *et al.*, 1998; Dean, 1998a). The prism length is multiplied by 1.15 to allow for decussation (Risnes, 1986). Cross striation spacing in this region is measured carefully in several places and this is applied to the prism length to give a time for the formation of cuspal enamel. Imbricational enamel formation time is determined by counting brown striae or perikymata, determining the cross striation count between brown striae, and multiplying the two counts.

Where it is necessary to match isolated teeth, this can be done with a good degree of confidence by matching brown stria sequences. The pattern of prominent striae, and the count of cross striations between them, is dependent upon individual growth history. If patterns match between teeth, there is little doubt that the specimens come from the same individual. Similar matches can be made using perikymata and hypoplastic defects on the crown surface (below).

Dental development sequences can be built for other mammals. This has been done for baboons (Dirks *et al.*, 2002), gibbons (Dirks, 1998), gorilla and orangutan (Beynon *et al.*, 1991). Unworn tall crowns of ungulate teeth may provide even better material, because they have a very thin cuspal enamel zone and pass almost directly into imbricational enamel. This would allow perikymata counts to represent practically the whole of crown growth.

Similar approaches have been used to investigate the development of fossil hominids. Discussion of hominid dental development originated with the Taung child, found in 1924 in South Africa, and the type fossil for the genus *Australopithecus*. Dart's (1925) description emphasised the human-like teeth and used modern human standards to estimate the age at death as 6 years. This assumption, however, has wider implications for human evolution. Human development runs on a different schedule from ape development – humans live longer and take longer to develop, not only physically, but also in their more complex level of cognition. Amongst Primates, the length of dental development is highly correlated with brain size (Smith, 1991a). Ape and human dental development shows characteristic differences. A chimpanzee, for example, has a short dental development sequence with much overlap between teeth in formation timings, and molars with thin enamel that form fast, and form early relative to incisors (Anemone, 1995). Humans have a long development time, less overlap between teeth in formation time, and molars with thick enamel that form slowly. One non-histological study (Mann, 1975) concluded that robust australopithecine *Paranthropus* jaws from Swartkrans followed a human-like development schedule, as had been suggested for the gracile australopithecine from Taung. Another non-histological study (Smith, 1986) found instead that *Australopithecus* and early *Homo* had an ape-like dental development, and further, that *Paranthropus* had its own development pattern unlike either apes or humans. At the same time (Bromage & Dean, 1985) perikymata counts of incisors had been used to infer ages for immature *Australopithecus*, *Paranthropus* and early *Homo* jaws that fitted best with an ape-like development. There was intense discussion (Mann *et al.*, 1987, 1990, 1991; Smith, 1987, 1991a; Mann, 1988; Conroy & Vannier, 1987; Wolpoff *et al.*, 1988). Finally, a canine from a *Paranthropus* jaw was sectioned (Dean *et al.*, 1993) and cross striation counts confirmed that the development state of the dentition fitted the ape-like pattern best. Similar histological studies (Dean *et al.*, 2001) have shown not only that all australopithecines and early *Homo* had a short, ape-like dental development, but that *Homo erectus* also did. The earliest fossil in the study with human-like development was the Tabun Neanderthal. From this work, it appears that the human pattern of development, with all that implies, is a recent phenomenon in hominid evolution.

Hypoplasia and hypocalcification
The surface of the crown frequently shows defects. Some take the form of grooves or pits, and are known as hypoplasia. They result from disturbances to enamel matrix secretion. Others consist of white opaque patches in the translucent enamel. These are called hypocalcifications and result from disturbance to maturation. Hypoplasia and hypocalcification most frequently occur separately, but can be seen together in the same defect.

Hereditary defects
Rarely, enamel defects in human teeth are inherited. Some are inherited as part of syndromes involving non-dental parts of the body, including epidermolysis bullosa, pseudohypoparathyroidism, trichodento-osseous syndrome, Ehlers-Danlos' syndrome and trisomy 21 (Pindborg, 1982). Others are defects of the teeth alone, and these are usually distinguished as *amelogenesis imperfecta* (or AI), a term first used by Weinmann *et al.* (1945). Such defects are rare; between 0.06 and 1.4 per 1000 people (Bäckman, 1997). In a small proportion of the families affected, AI can be linked to mutations of the amelogenin gene (p. 149) on the X-chromosome but, in the majority, inheritance is autosomal (not involving X- or Y-chromosomes) and the genes responsible have not been identified at the time of writing.

Characteristically, all teeth in the dentition are affected, although defects are more marked in permanent than deciduous teeth. In terms of their clinical appearance, three main subdivisions are usually recognised (Winter & Brook, 1975; Bäckman, 1989): hypoplastic, hypomaturation and hypocalcification. Hypoplastic AI is further divided into three types. Pitted hypoplastic AI has pinprick-size pits scattered randomly over the surface of all crowns, whereas local hypoplastic AI has pits and grooves concentrated in the middle of the buccal sides. Rough hypoplastic AI is characterised by very thin, rough enamel. Hypomaturation and hypocalcification AI both exhibit enamel of normal thickness but deficient mineralisation, and this is more marked in the latter than the former. Hypomaturation AI has two forms. 'Snow-capped teeth' are distinguished by opaque white cusps and incisal edges. Other hypomaturation AI teeth have mottled enamel, with a similar density to dentine, which often breaks away from the EDJ. X-linked AI defects may be hypoplastic or hypomaturation and, whereas the teeth of men are affected all over, the women show a mixture of normal and defective enamel, usually in vertical stripes down the crown.

Fluorosis

Fluoride ions are present in most drinking water and become incorporated into the enamel (above). The very low concentrations in most drinking water do not cause defects, and indeed confer some protection against the disease dental caries (Chapter 5). In some areas, however, fluoride concentrations are unusually high. At concentrations above 0.5 ppm, some people show enamel defects. If fluoride concentrations are not too elevated, hypocalcifications are produced, first as white lines running around the crown circumference, and at higher levels as larger diffuse opaque white patches (Møller, 1982). At higher fluoride levels still, pit-form and plane-form hypoplastic defects are formed (below). The defects affect molars more than incisors and may be stained yellow or brown. In archaeological material, the only published case is from the Neolithic/Chalcolithic site of Mehrgarh in Pakistan (Lukacs et al., 1985). It is, however, difficult to be sure about hypocalcifications in archaeology, where diagenetic effects may mimic them.

Developmental defects of enamel

By far the most common defects are caused by episodic disruptions during crown formation. These are best known in humans, amongst whom it is rare to see anyone without at least a microscopic defect, but any mammal can show them (Miles & Grigson, 1990). They can be hypoplasia or hypocalcification (above) and the Fédération Dentaire Internationale (FDI) has developed a general descriptive scheme for recording them (the Developmental Defects of Enamel (DDE) index, see Commission on Oral Health, 1982). In living people, the bright white marks of hypocalcifications are more prominent and are regarded as more common than hypoplasia. They are subsurface zones of poorly mineralised enamel (Suga, 1989) which scatter light back to the observer and therefore appear brighter. A distinction is made between *demarcated* or *diffuse opacities* (Suckling et al., 1989). Diffuse opacities affect just the enamel near the crown surface, whereas demarcated opacities can involve any depth. The nature of the disruption to maturation is not clear. In archaeology, hypocalcifications are rarely seen and there must, in any case, be some doubt about possible diagenetic changes.

Hypoplasic developmental defects (Figures 2.7–2.8), by contrast, are very common in archaeological dentitions. The term *enamel hypoplasia* was coined by Otto Zsigmondy (1893). It is mostly used in anthropology to describe disturbances to matrix production. The term *linear enamel hypoplasia* (LEH) may be used to emphasise that an episodic disturbance has initiated a line of defects through

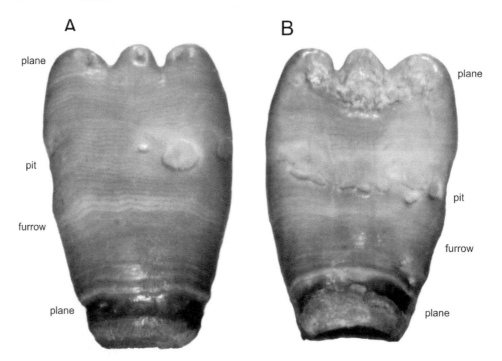

Figure 2.7 Defects of dental enamel (enamel hypoplasia) in a lower first incisor from the crypt of Christ Church, Spitalfields. Macrophotograph by Daniel Antoine, Institute of Archaeology, UCL, who also provided Figure 2.8. Image A shows the buccal/labial surface and B shows the lingual surface. The crown is still developing and the growing edge can be seen at the bottom. As the tooth had not erupted at the time of death, the mamelons (p. 17) are sharply defined. Four main zones of defect are visible: a plane-form defect involving the mamelons (best seen in image B), a pit-form defect at half crown height, a furrow-form defect just below this, and another plane-form defect just above the growing crown edge.

the dentition, unlike AI defects which affect most of the crown in all teeth. In hypoplastic defects, the enamel is hard and well mineralised, but has a local deficiency of thickness that extends around the circumference of the crown, following the line of the perikymata and is therefore clearly related to development. The nature of the deficiency varies. It may be a single fine line in the crown surface, or a broader furrow. In other cases, a large area of the crown surface is ridged across with a 'washboard' effect. There may be a band of isolated pits, large or small, or a row of pits associated with a furrow. Sometimes, a whole area of enamel is missing, to expose a plane of underlying enamel, or even the EDJ. Berten (1895) divided defects into *furrow-form*, *pit-form* and *plane-form*, and this terminology will be followed here (Hillson & Bond, 1997). It might be assumed that they represent a scale of growth disruption – least in furrow-form and most in plane-form, but this has not been established. The furrows are by far the most common and are caused by a variation in the spacing of perikymata. Different mammals vary in the pattern of perikymata spacing on their crowns but, in any one region of the crown,

Dental tissues

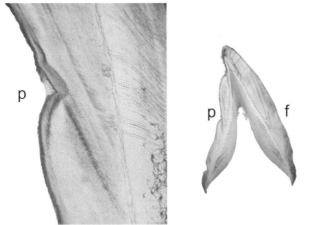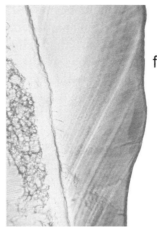

Figure 2.8 Radial section, running buccal-lingual and centred on the central mamelon, of the tooth shown in Figure 2.7. The section is 100 μm thick and viewed in ordinary transmitted light microscopy. The central image is at low magnification and shows the whole developing tooth crown which is some 6 mm across. The upper and lower plane-form defects are seen particularly clearly on the concave lingual side of the crown, as is the pit-form defect marked 'p'. This is shown at a larger magnification in the left-hand image, where the defect is clearly marked by a prominent brown stria of Retzius. The plane of section on the buccal side did not pass through the upper plane-form defect, or one of the pits, but it does show the furrow-form defect, marked 'f', well. The right-hand image shows this at a higher magnification and, once again, it appears to be associated with prominent brown striae. There are large areas of interglobular dentine (p. 192) which correspond, in particular, to the lower plane-form defect.

the perikymata are normally evenly spaced (Hillson et al., 1999). If one pair of perikyma grooves is further apart than the neighbours, the crown surface is locally depressed into a fine furrow-form defect. Where, say, 10 perikyma grooves are widely spaced, the furrow is wider and deeper. In the largest, 20 or more perikyma grooves may be involved (Hillson & Bond, 1997). The size of the most common type of furrow-form defect (but see plane-form defects below) is controlled mostly by the number of perikymata, and therefore by the duration of the growth disruption. The defect, and number of perikymata involved, can be matched on other tooth crowns being formed at the same time, showing that a systemic disturbance is responsible (if not, it implies an injury or other disturbance local to that tooth). The size of the furrow, however, also depends on its location. In an incisor, for example, the furrow may be near the cervix whereas, on the neighbouring canine, the enamel matrix formed at the same time might be in the middle of the crown. Perikymata at the cervix are normally more closely spaced than those in the middle (above), so the furrow will appear narrower even though, under the microscope, the same number of perikyma grooves can be counted (Hillson & Bond, 1997). Molars, in particular, show an abrupt change from widely spaced perikymata occlusally to narrowly spaced cervically, and this makes hypoplasia difficult to record. Most studies therefore concentrate on incisors and canines, covering a slightly reduced range of dental development from about one to six years of age (Figure 2.9 below; Reid & Dean, 2000). Furrow-form defects are most conveniently examined using high resolution replicas cast from dental impressions (below) which can be taken as a rapid record in museums or on site. It is best if the teeth are not too worn so, in humans, children or young adults provide the best specimens and modern teeth with toothbrush abrasion are of little use. Almost all studies have been carried out on humans, but many mammals show hypoplastic defects. They are very common in monkeys and apes (Miles & Grigson, 1990; Guatelli-Steinberg & Lukacs, 1998, 1999; Guatelli-Steinberg, 2000, 2001, 2003; Guatelli-Steinberg & Skinner, 2000), pigs (Dobney & Ervynck, 2000), dogs (Mellanby, 1929; Mellanby, 1930) and even the tall crowns of cattle or horses show them if the overlying cement is carefully stripped away.

Under high magnification, a whole range of defect sizes can be seen, from a single pair of perikyma grooves which result in a furrow 100–200 μm wide, up to furrows 2 mm or more wide which can easily be seen with the naked eye, or felt with a thumbnail. It is therefore difficult to use simple visual scoring to record the defects; a better approach is to use a measuring microscope to measure the spacing of perikyma grooves along a transect down the crown side (Hillson & Jones, 1989; King et al., 2002), or measure their spacing at the edge of a section (Hillson et al., 1999). Furrow-form defects can be defined as an increase in spacing, above a given limit. The age at which a disruption occurred is most precisely determined (Hillson et al., 1999) if a section can be made and an enamel development chronology built from cross-striation counts (above). This is often not possible and, in such cases, estimates can be made from perikymata counts (King et al., 2002). It is necessary to use a standard figure for the start of imbricational enamel formation (p. 163) – the age at which the first perikyma groove appears on the surface – and to assume a cross-striation count for the intervals between brown striae of Retzius/perikyma grooves. So, for example, it may be assumed that the first perikyma groove on the unworn incisal edge of human lower second incisors appears at 358 days (almost 1 year) of age and the last formed enamel along the cervix 1164 days later at 4.2 years of age (Reid & Dean, 2000). If 127 perikymata are counted, it implies a cross-striation repeat of $1164/127 = 9$ between perikymata. Perikymata counts can then be used to estimate the age at which a hypoplastic defect started and ended. If a tooth is too worn to count perikymata from the incisal edge, it may be possible to establish the cross-striation repeat from another tooth, because it should remain the same in all teeth for one individual. Alternatively, a cross-striation repeat may be assumed (usually 8 or 9) and then, after matching the defect with another tooth, perikymata counts can be used to estimate the difference in age at crown completion for the two teeth. If the difference is much larger or smaller than expected, another cross-striation repeat can be tried.

Many studies have recorded hypoplasia using the naked eye, or modest magnification. This is difficult because of the white, translucent, glossy surface and

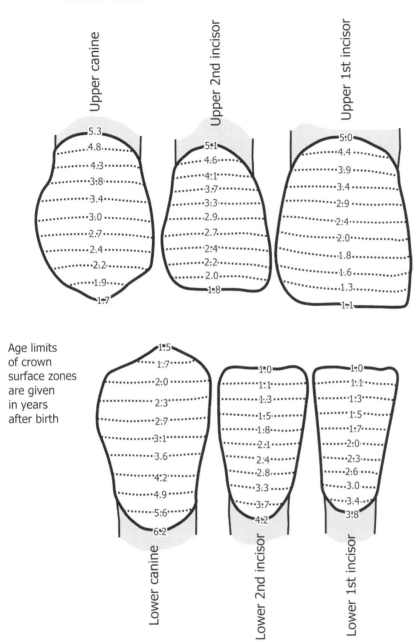

Figure 2.9 Formation times for the crown surface in human permanent upper and lower incisors and canines. Redrawn from Reid & Dean (2000). It is apparent that the incisal part of the crowns grows more slowly than the cervical part. The timing of enamel defects is estimated by matching them across the different teeth. Altogether, these tooth crown surfaces provide a sequence from one year after birth, to about six years.

because the smallest defects cannot be seen. There is no definition of a lower limit for the smallest furrow-form defect recorded and it is up to the individual observer, so there can be little comparability between studies (Hillson & Bond, 1997) in hypoplasia frequencies. There is also the problem that the same disruption may produce different-sized defects in different teeth, and it is essential to match defects across the dentition. Not all studies have done this in the past. The most difficult problem, however, is estimating the age at which defects were initiated. Swärdstedt (1966) pioneered a method in which measurements were taken from the CEJ to the defect, using needle point calipers, and converting to ages using a table derived from mean crown heights and ages scaled from a diagram in Massler *et al.* (1941). This has been widely used (Goodman *et al.*, 1984a; Goodman & Rose, 1990) and is recommended as a standard method (Buikstra & Ubelaker, 1994). Unfortunately, several mistakes were made in the original table; in particular the ages for the first enamel formed at the crown surface are incorrect (Hillson, 1992b). This makes a difference of up to a year in the occlusal part of the crown. Other problems are that calipers make an imprecise instrument to measure the tiny defects and the chart assumes a constant unworn crown height, when this is known to vary by several millimetres. As an alternative rapid method, attempts have been made to subdivide the crown surface into zones of an equal proportion of the height (Figure 2.9), and work out their ages from cross-striation counts (Hillson, 1979; Reid & Dean, 2000). Here, the age is estimated as a best fit, by matching the defects between zones in different teeth, rather than measuring their position (Hillson, 1996).

Pit-form and plane-form defects (Figures 2.7–2.8) are even more difficult to interpret (Hillson & Bond, 1997). Each pit represents a group of ameloblasts that stopped matrix secretion early, leaving the exposed plane of a brown stria in the floor of the pit. Examination in the SEM confirms this because the characteristic Tomes' process marks are present. The pits may follow a furrow-form defect, in which case they represent large irregularities in a perikyma groove and the growth disruption must have been the same as that causing the furrow. Where the pits form an isolated band, with normal enamel around them, the brown stria associated with the disruption is deeper inside the crown, and the position of the pits on the surface is not related to the age at which the disruption occurred. Also, unlike furrow-form defects, the width of the band may bear no relationship to the duration of the defect, because all the pits may originate in just one brown stria. Even wide bands probably represent very short disruptions. Plane-form defects represent a whole band of ameloblasts which ceased matrix secretion. The exposed brown stria plane may occupy half the crown height, but still relate to a momentary disruption to amelogenesis. It is better not to record pit-form and plane-form defects in the same way as furrow-form and, fortunately, they are much less common. Histological studies of enamel sections are the only way to determine their chronology.

Growth in general is affected by many different factors including diet, disease, altitude, climate and socioeconomic factors (Bogin, 1999). It is to be expected that any of these might cause a disruption to enamel matrix secretion and many factors have been documented (Pindborg, 1982). Experimental evidence has also accumulated. In a series of classic experiments with beagle dogs, Mellanby (1929, 1930, 1934) showed that deficiency in vitamins A or D could cause enamel hypoplasia. Similarly, Kreshover & Clough (1953) showed that artificially induced fever in pregnant rats caused a wide range of disturbances in the enamel formation of their offspring, as did inoculation with pathogenic viruses (Kreshover et al., 1954) and bacteria (Kreshover, 1944). Kreshover (1960) concluded that defects of enamel formation should be seen as 'non-specific' in the relationship with possible growth-disturbing factors. They have been used in this way as an important indicator of Selyean stress in bioarchaeology (Goodman et al., 1988; Goodman & Rose, 1990; Larsen, 1997).

In fact, very few clinical studies have dealt with permanent teeth, which are by far the most studied in archaeological material. The earliest studies noted the relationship of deciduous incisor defects to the neonatal line (Schour & Kronfeld, 1938; Kronfeld & Schour, 1939) in children with birth injury. Sweeney et al. (1969, 1971) showed that deciduous incisor defects in a similar position were more common in Guatemalan children who had been hospitalised for malnutrition than in those who had not. This effect is seasonal and reflects social factors (Infante & Gillespie, 1974). For permanent teeth, Goodman et al. (1991) showed that a control group of malnourished Mexican children had more defects than a group who had received a powdered milk supplement. Lindemann (1958) used Massler et al.'s (1941) original figure to estimate the age of disruption for hypoplastic defects in permanent teeth. It was possible to match the approximate age of disruption with periods of disease in the majority of cases. Sarnat and Schour (1941, 1942) used similar means to determine the age of disruption in extracted teeth from 60 children. In over half of the children, they found it difficult to match defects with medical histories including chickenpox, diarrhoea, diphtheria, measles, pneumonia, scarlet fever and whooping cough. There is a need for research using histological methods to determine the age of growth disruption with more precision.

One special case is congenital syphilis in humans, where a foetus is infected by *T. pallidum* from the mother. Half the babies die at birth and half the survivors show signs including saddle nose, frontal bosses, sabre shins, Hutchinson's incisors, Moon's molars or mulberry molars (Jacobi et al., 1992; Condon et al., 1994; Hillson et al., 1998). The last three are all unusual hypoplastic defects in parts of the teeth which were being formed just after birth. Hutchinson's permanent upper first incisors have a prominent notch in the incisal edge, and drawn-in sides which make the crown look a little like a melon seed in outline. Moon's permanent first molars have no visible sign of a defect on the surface, but the cusps in the occlusal surface are much closer than normal. To continue the botanical analogy, the crowns are sometimes called 'bud-form'. Mulberry permanent first molars (plants again) have a sharply developed plane-form defect around each of their cusps. This leaves little nodules of enamel where the cusps should be, and these often break away to leave crater-like hollows. With wear, both mulberry molars and Hutchinson's incisors can be difficult to recognise.

Archaeological studies of enamel hypoplasia
Some studies have focused on fossil hominids. Robinson (1956) reported that hypoplasia was more common in *Australopithecus* from Sterkfontein than in *Paranthropus* from Swartkrans, whereas White (1978) found the opposite. Guatelli-Steinberg (2003) showed that there were indeed more LEH defects on *Australopithecus* canines than on *Paranthropus*, but also suggested that the closer spacing of perikymata on *Australopithecus* teeth might have an effect. Brothwell (1963c) pointed out the presence of defects in photographs of Chinese *Homo erectus* teeth in Weidenreich (1937), and noted defects in Neanderthal, Upper Palaeolithic and Mesolithic jaws from Europe. Ogilvie *et al.* (1989) found that hypoplastic defects in the Krapina Neanderthals were common and suggested that this represented a high level of physiological stress. Hutchinson *et al.* (1997) by contrast found that, although common, the figures lay within the overall range of later assemblages from North America, including hunter-gatherers and agriculturalists. Studies of the hunter-gatherer/agriculturalist transition in the Levant (Smith *et al.*, 1984) and North America (Sciulli, 1977, 1978; Cook, 1984; Larsen, 1995; Goodman *et al.*, 1980, 1984a, 1984b) have suggested that hypoplasia became more common with the adoption of agriculture. This is explained by the increased risk involved in reliance of fewer food resources, and the increased risk of infectious disease resulting from sedentism and the growth of settlement size.

The course of the prisms through the enamel
Prisms almost never run straight through the enamel, radiating out perpendicularly to the plane of the EDJ. Most are inclined to occlusal; they also deviate laterally to one side or another and often 'wriggle' sinusoidally or even helically along their length. This is achieved by a coordinated movement of ameloblasts within the internal enamel epithelium. It makes the enamel stronger and gives the worn surface particular characteristics that enable it to function in grinding or cutting. There is a wide range of variation in the weave of enamel prisms and this is seen as an adaptation to the varying role of the dentition in different mammals (von Koenigswald, 1982; Boyde & Fortelius, 1986; Rensberger, 1997; von Koenigswald & Sander, 1997b).

At its simplest, the deviation may be a simple occlusal inclination, relative to the EDJ plane. All the prisms at a particular level in the enamel layer are affected at once. Von Koenigswald (1997b) has termed this radial enamel because, in transverse microscope sections (p. 201), prisms radiate out perpendicularly to the EDJ. In radial microscope sections, the prisms are inclined relative to the perpendicular at varying angles, from just above 0° to almost 90° (i.e. nearly parallel to the EDJ). In evolutionary terms, radial enamel is seen as the most primitive mammalian enamel form. One departure from this arrangement is a simultaneous deviation of prisms to one side, as seen in the transverse microscope sections. Von Koenigswald has called this tangential enamel. Such deviation tends, in any case, to occur at corners of the crown because the enamel layer is curved sharply round and there is a mismatch between the curve of the EDJ and the curve of the crown surface (Boyde, 1989). Large areas of tangential enamel are, however, a particular feature of marsupials (von Koenigswald & Sander, 1997a).

Much more common in placental mammals is a division of prisms into alternating zones or bands, running around the circumference of the crown in an orientation parallel to the sequence of development, and therefore also approximately parallel to the perikymata (above). Each zone contains a ring of prisms, all 'doing roughly the same thing' (Boyde, 1989). The neighbouring zones above and below, by contrast, contain prisms doing exactly the opposite thing. The simplest form that this takes (although under the microscope it looks extremely complex) is in the enamel of many rodents. Here, the alternating zones are just one prism deep and each zone deviates laterally (in effect like tangential enamel – but just for one row of prisms). The zones above and below deviate in the same way, but in the opposite direction. Prisms of different zones cross one another at about 90°, described by Sir Richard Owen (1845) as *decussation*, from the Roman numeral *decus* (ten, written 'X'). There is still interprismatic enamel, because rodents are based on Pattern 2 (above), but it is confined to the interstices of the prism crossings, and the crystallite direction is almost at right angles to both sets of prisms. This is what makes it look so complex in the SEM.

The rodent arrangement is known as *uniserial enamel*, because each zone contains one row of prisms. Next in complexity is when each zone contains 3, 4 or 5 rows of prisms, still with an abrupt transition between zones. This is called *pauciserial enamel*. It is found particularly in hystricomorph rodents (below) and is once more based upon Pattern 2 but, because they are less disrupted, interrow sheets of interprismatic enamel are visible within each zone (Boyde, 1969). In many carnivores, the zones contain 6 to 10 rows of prisms and, although the transition is still abrupt, there is a transitional row of prisms running more or less radially. Carnivore enamel thus occupies a transitional position. The most complex arrangement is *multiserial enamel*, which characterises most placental mammals. Here, the zones contain 10 to 20 rows of prisms, and the transition between zones is much more gradual. No one row of prisms crosses another at a large angle. In human enamel, for example, it takes 10–13 rows of prisms to go from a 20° deviation one way to a 20° deviation the other way (Boyde, 1989). This makes a total decussation angle of 40°, but no single row of prisms crosses another at that angle. Another feature of multiserial enamel is that, superimposed over this gradual decussation, the prisms may follow a sinusoidal path which takes them out of one zone and into another. This adds another level of complexity which is very hard to visualise as a three-dimensional model.

The zones may be seen in radial, transverse or tangential microscope sections (below). In addition, they are often visible at the surface either by etching and viewing in the SEM, or under reflected light microscopy (the different angles at which the prism approaches the surface are respectively more or less light scattering at different viewing angles). The view provided at the surface is effectively a tangential one in which prisms are all relatively transversely sectioned and the zones are circumferentially arranged, one above the other down the crown. Often, zones can be seen to divide into two, in a 'Y'-junction, or to terminate. In radial (sagittal in rodent incisors; below) microscope sections, they are seen as alternating zones of more transversely and more longitudinally sectioned prisms (Figure 2.4), radiating out from the EDJ. These were respectively called *diazones* and *parazones* by Preiswerk (1895). In polarising microscopy of thin sections, the diazones appear darker and the parazones brighter (p. 205). This appearance is named after Hunter (1771) and Schreger (1800) as Hunter–Schreger Bands (HSB). This abbreviation is often used to describe the decussating part of the enamel.

Organisation of enamel structure in different mammals
The different components of radial, tangential, uniserial, pauciserial and multiserial enamel are arranged into layers within the thickness of the enamel (*schmelzmuster*, see von Koenigswald, 1982). This arrangement varies widely between different groups of mammals.

Insectivora: Insectivores have been described as having Pattern 1 prism packing and no decussation, with the prisms running straight through the enamel layer (Boyde, 1969, 1971).

Chiroptera: Bat enamel was also described by Boyde (1971) with Pattern 1 prisms, and no decussation. Since that publication it has transpired that, although Pattern 1 does characterise the outer one third of the enamel layer, the inner two thirds is dominated by a Pattern 2 in which the interrow sheets are not strongly developed, but with some Pattern 3 (Lester & Hand, 1987). Near the EDJ, irregular Pattern 3 predominates, with thin prism-free enamel layers actually at the EDJ and crown surface.

Carnivora and Pinnipedia: Most carnivores have Pattern 3 prism packing, and strongly developed pauciserial HSB, with 6 to 10 rows of prisms in each zone and an abrupt change from one orientation to another (Boyde, 1969). There is relatively little undulation of individual prisms, so they lie more or less parallel to the axis of the parazones. The zones are almost perpendicular to the EDJ and run through most of the enamel layer, passing into radial enamel and then irregular areas of prism-free enamel near the crown surface (von Koenigswald, 1997a). In tangential section, or as seen from the crown surface, the zones run horizontally with a slight undulation, 'Y'-shaped splits into two zones and occasional termination of zones. In some carnivores, the undulations become more prononouced, into 'acute-angled' HSB and even more strongly, to 'zigzag' HSB (Stefen, 1997). Undulating

HSB predominate in the Procyonidae, Viverridae, most Mustelidae, and Phocidae. Canidae, Ursidae, Felidae, Hyaenidae and Otariidae show a transition from undulating HSB near the cervix to acute-angled and/or zigzag HSB more occlusally. In some Hyaenidae (*Crocuta*), zigzag HSB predominates throughout, and this form is seen as an adaptation to the forces generated by bone crushing (von Koenigswald, 1997a).

Odontoceti: Boyde (1969, 1971) characterised odontocete enamel as Pattern 1 and lacking decussation – radial enamel grading into prism-free enamel near the crown surface (von Koenigswald, 1997a). In some living odontocetes there may be no prisms at all. More recently (Sahni & von Koenigswald, 1997), decussation has been found in some fossil whales, and in the river dolphin *Platanista*.

Primates: Humans and apes have predominantly Pattern 3 enamel through the underlying enamel layers and towards the cervix of the crown, with occasional Pattern 2 (Boyde & Martin, 1982). Pattern 1 is found at the surface, particularly occlusally (Boyde & Martin, 1987). There is also a thin layer of prism-free enamel next to the EDJ and at the crown surface. Multiserial HSB are moderately prominent in the inner half of the enamel thickness, with gradually merging zones in which it takes 10–13 prism rows to pass from one extreme angle to the other. The prism boundaries cross the parazones at a slight angle, suggesting they undulate along their length. Macaque and marmoset monkeys have been shown to have predominantly Pattern 2 enamel, with some Pattern 1, and ring-tailed lemur to have predominantly Pattern 1 (Boyde & Martin, 1982).

Artiodactyla: The majority of Artiodactyla are predominantly Pattern 2, although *Hippopotamus* is largely Pattern 3 (Boyde, 1989). The high crowns of Bovidae have a thin layer of 'modified radial enamel', with thick interrow sheets, next to the EDJ (von Koenigswald, 1997a). The multiserial HSB are well developed, contain 14 or more prism rows and merge into one another without abrupt transitions. Each prism undulates along its length, because the prism boundaries are angled slightly relative to the main axis of the parazones (Kawai, 1955). Grine *et al.* (1986, 1987) have suggested that there may be small differences in the packing of prisms between *Ovis* and *Capra* which could help with the identification of teeth from these difficult genera (Chapter 1).

Perissodactyla: Most Perissodactyla have Pattern 2 enamel. Horses have well-developed horizontally arranged multiserial HSB with zones containing many prisms, like those of Artiodactyla, gradually merging into one another. The prisms must undulate along their length, because they cross the parazones obliquely (Kawai, 1955). Next to the EDJ is a layer of radial enamel with thick interrow sheets, also like hypsodont Artiodactyla. Rhinocerotidae are unique amongst living mammals (but not fossil) in having vertically zoned HSB in the inner half of their enamel layer (Boyde & Fortelius, 1986). The zones contain many prisms, characterised by Pattern 3 enamel, and show abrupt transitions with a high decussation angle. Decussation continues into the outer zone of enamel, but with a much smaller decussation angle and a pronounced inclination of all prisms to occlusal. The vertical decussation results in grater-like ridges in the worn enamel surface.

Proboscidea: Recent elephants do not have a real HSB structure, and their prisms are irregular in orientation (Boyde, 1969; von Koenigswald, 1997a). Most enamel is Pattern 3, with some patches of Pattern 1.

Sirenia: Pattern 1 is the predominant enamel form (Boyde, 1971). The prisms are inclined occlusally through most of the enamel, but near the surface are more perpendicular to it (von Koenigswald, 1997a).

Lagomorpha: The characteristic of lagomorph enamel is an irregular multiserial HSB structure. The broad, irregular broad parazones and diazones, containing many prisms, grade into one another. Individual prisms follow a strongly undulating course. The overall effect is 'like a raging flame' (Kawai, 1955) under the polarising microscope. In the incisors, this structure occupies the whole enamel thickness. In molars and premolars it is found only in an outer belt of the enamel, with radial enamel in the inner layer. The predominant prism-packing type is Pattern 2.

Incisor enamel in rodents
Rodents are traditionally grouped into three suborders, on the basis of skull, jaw and tooth morphology (Chapter 1). These are the Sciuromorpha, Myomorpha and Hystricomorpha. Each is defined on a complex of features, including the microstructure of enamel, which is so complex that it is given a separate section here. The ever-growing incisors have a different enamel organisation from molars and premolars. Their enamel is divided into pronounced inner and outer layers (*Portio Interna* and *Portio Externa*, see Korvenkontio, 1934; PI and PE, see Martin, 1997). The PI contains the decussating enamel, and the PE is composed of radial enamel. The width of the PE (Figure 2.10) can be expressed as a percentage of the total enamel thickness (*externalindex*), and the inclination of the HSB zones in the PI, or prisms in the PE, can be measured relative to a line perpendicular to the EDJ (Korvenkontio, 1934).

Hystricomorph incisor enamel (Figure 2.11) is pauciserial in the PI, with HSB zones normally 3–5 prism rows deep (Boyde, 1978) and showing abrupt changes from one to another. The zones are inclined to occlusal 10–50° away from a line perpendicular to the EDJ (Korvenkontio, 1934). The externalindex varies between about 20% and 40%. In the PE, the prisms of the radial enamel run straight and parallel, inclined occlusally at 50–75° to the perpendicular. There is copious

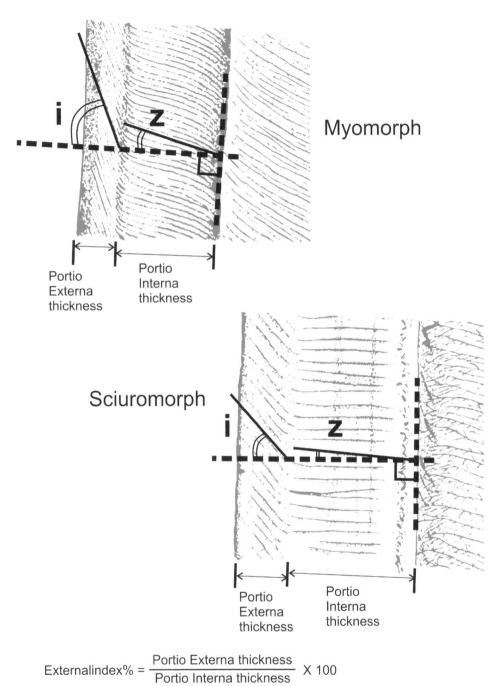

Figure 2.10 Rodent incisor enamel measurements. Images of myomorph and sciuromorph longitudinal incisor sections modified from Korvenkontio (1934).

182 Teeth

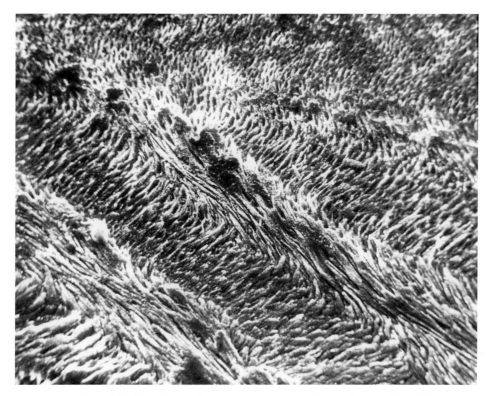

Figure 2.11 Hystricomorph enamel in a porcupine incisor from the Neolithic escargotière site of Doukhanet al Khoutifa, Tunisia. SEM-ET (field width *c*. 75 μm) of polished, etched surface. The parazones and diazones of the inner layer run from lower right to upper left. The outer layer of non-decussating prisms starts towards the top of the image.

interprismatic enamel, arranged in sheets. Another hystricomorph characteristic is spaces without crystallites, due to irregularities in the packing of prisms (Boyde, 1978).

Sciuromorph incisor enamel (Figure 2.10) is uniserial in the PI, which occupies about one half (Boyde, 1978) or less (Korvenkontio, 1934) of the total enamel thickness. The externalindex is thus usually 40% or more. In sagittal sections (p. 201), the HSB zones of the PI are relatively straight and arranged roughly perpendicular to the EDJ, or less than 10° away from the perpendicular (Boyde, 1978). The PI has very little interprismatic enamel. In the PE radial enamel, the straight prisms all run parallel, inclined occlusally to an extent that varies between taxa. About one half of the PE is interprismatic.

Myomorph incisor enamel (Figure 2.10) is uniserial in the PI, which occupies about three quarters of the total enamel thickness (Boyde, 1978). The externalindex is 30% or less (Korvenkontio, 1934). In sagittal sections of the PI, the HSB zones have a slight sigmoid curvature and are markedly inclined towards the occlusal edge of the tooth. In upper incisors the zones are inclined 20–30° away from perpendicular (to the EDJ). In lower incisors, the angle is 40–50°. There is separate interprismatic enamel in the PI, grouped into thin sheets and scattered bundles. The PE contains prismatic and interprismatic enamel, continuous with that of the PI. Prisms in the PE radial enamel are more strongly inclined occlusally than in the Sciuromorpha, 'typically within 15 degrees of being parallel' to the EDJ (Boyde, 1978).

Hystricomorphs are distinguished by their pauciserial enamel, but share few other similarities (Korvenkontio, 1934). Most Sciuridae clearly have sciuromorph enamel, but some Sciuromorpha (in terms of skull and mandible morphology), such as *Castor*, do not. All the Muridae that Korvenkontio studied fit well into myomorph enamel. Other myomorph families (in skull morphology) are less clear. The Cricetidae are rather variable. Most microtines fit reasonably well, although some genera have a thin PI. The gerbillines also fit, but in cricetines the HSB zones are not strongly inclined. Zapodidae fit reasonably well, as do the Dipodidae, Heteromyidae and Geomyidae. *Spalax* is intermediate between myomorph and sciuromorph enamel organisation, but is myomorph in skull form. Some genera of the Gliridae (also myomorph skull form) have sciuromorph enamel, and others myomorph.

On this basis, there seems to be some possibility of identifying incisors by a combination of histology and size, at least to family level (Boyde, 1971), which is potentially useful because isolated rodent incisors are common finds and are otherwise difficult to identify. There are, however, difficulties. Korvenkontio examined few specimens and the measurements show some variation when applied to large samples of single species. Wahlert (1968) measured sections of 96 upper and 93 lower incisors from brown rat *Rattus norvegicus*. Total enamel thickness varied over a range of 46 μm in upper and 55 μm in lower incisors. Inclination of HSB zones in the PI varied over 9° in upper incisors and 19° in lower. Measurements of the relative thickness of PI and PE were difficult, because Wahlert could establish no clear line between them. The estimates ranged between 20% and 39% for Korvenkontio's 'externalindex' (Figure 2.5) in upper incisors and had a range of 14–27% in lower incisors. Variation in rat incisor enamel was also noted by Risnes (1979a), although the overall structure was highly regular. The irregularities noted by Wahlert and Risnes do not invalidate rodent incisor enamel as a taxonomic tool, but highlight the need for further work on variation.

Rodent molar enamel
Hystricomorph molar enamel has pauciserial or multiserial structure. In *Hystrix* this is distinct only in the middle of the enamel layer. In *Ctenodactylus*, virtually the whole thickness of the layer is multiserial, but in the related genus *Massoutheria* this is true only of thinner enamel patches. There is little detailed information

on molars of the Sciuromorpha. Von Koenigswald (1980) described multiserial structure in elements of the molar enamel of Sciuridae, Geomyidae, Heteromyidae and Castoridae.

The general pattern for myomorph molars which are not persistently growing is for uniserial enamel to be confined to a ring at the cervix of the tooth. In the rat (Risnes, 1979b), uniserial enamel forms an inner layer or PI, against the EDJ, occupying about 65% of the total enamel thickness in the cervical region. This decreases to 45% halfway up the crown and becomes absent towards the occlusal surface. Prisms in neighbouring HSB zones of the uniserial enamel cross at 90°, and the zones are inclined occlusally at 35–90° (average 55°) to the EDJ. In the PE, the enamel is radial, with prisms running straight or slightly wavy courses and separated by sheets of interprismatic enamel. The whole molar structure is more irregular and less sharply defined than that of the incisors.

A similar organisation is found in the molars of Gliridae and most of the Cricetidae. The persistently growing molars of the Microtinae, however, are very different, because uniserial enamel is found the full height of the crown and is exposed in the occlusal surface. Von Koenigswald and colleagues (1980, 1982; von Koenigswald *et al.*, 1994) used SEM to study etched occlusal surfaces in both living and fossil forms. There were differences between the mesial and distal sides of the prismatic elements that make up microtine molars (Chapter 1). Mesial 'leading edges' in many living genera have uniserial enamel in the inner layer, and radial in the outer layer. Distal 'trailing edges' often have radial enamel in the inner layer and tangential enamel in the outer layer, although uniserial enamel spreads round the angle of the prismatic element in some genera. It is possible to follow evolution from the Pliocene genus *Mimomys*, which initially had only radial enamel in the occlusal part of a relatively low, non-continuously growing, crown. As more hypsodont forms evolved, tangential enamel was added to the trailing edge all the way up the crown, and later, uniserial enamel was added to the leading edge. In the Pleistocene, *Mimomys* evolved into *Microtus*, with its ever-growing molars. The trailing edge became thinner, and lost its tangential enamel (von Koenigswald, 1997b).

Dentine
Composition and structural components of dentine
Dentine is a mineral/organic composite: 72% inorganic by dry weight, 18% collagen and 2% other organic material (Williams & Elliott, 1989). The majority of mineral is apatite, with crystallites much shorter than those in enamel at 20–100 nm in length. Some amorphous calcium phosphates may also be present. One of the dominant features of dentine structure is the collagen, which is secreted in mats of fine fibres. Within this organic matrix the crystallites are seeded and their orientation is a further main determinant of dentine structure. Unlike enamel, dentine is a living tissue and the cells of dentine (*odontoblasts*) line the sides of the pulp chamber. Odontoblasts are long and narrow cells. One end of each cell is in contact with the developing predentine and out from it grows a long odontoblast process which

passes through the full thickness of dentine, and remains throughout the life of the tissue. Odontoblast processes occupy closely spaced tunnels called *dentinal tubules* (one main tubule per cell). Once dentine has been formed, it does not 'turn over', unlike bone, in which cells constantly remove, replace and remodel tissue. Secondary dentine formation does continue after the initial formation, however, on the walls and roof of the pulp chamber.

Formation of dentine
Primary dentine formation takes place in two phases: secretion of an organic matrix (*predentine*) and seeding crystallites into it. The organic matrix remains in the fully developed tissue and, unlike enamel matrix, it is not removed. Sections of developing dentine show a predentine formation 'front' next to the layer of odontoblasts and a mineralisation front 10–40 μm deeper into the tissue. The first predentine matrix formed, under the growing tooth crown, is *mantle dentine*. At this early stage, the odontoblast processes are finely branching and sit in finely branching dentinal tubules. Mantle dentine in the crown also has coarse collagen fibre bundles called the *fibres of von Korff*, running parallel with the dentinal tubules (Jones & Boyde, 1984). After some micrometres of mantle dentine matrix have been deposited, the main *circumpulpal dentine matrix* is laid down. It has two components: intertubular and peritubular. *Intertubular matrix* consists of mats of collagen fibres, deposited more or less perpendicular to the odontoblast processes at a rate of approximately 4 μm per day. There must also be some ground substance, although this is difficult to demonstrate (Jones & Boyde, 1984). In many mammals, the dentinal tubules are lined with *peritubular matrix* which consists entirely of ground substance.

Dentine mineralisation involves the seeding of crystallites in microscopic cavities, called *matrix vesicles*. From a single seeding point within each vesicle, the crystallites grow out radially, rupturing the vesicle walls. Each small body of mineralised dentine is spherical, with crystallites fanning out in all directions from the centre. These bodies are called *calcospherites* (Schmidt & Keil, 1971), and they grow through the matrix until they mutually limit one another's size, to result in a complex of radially crystalline, intersecting hyperboloid bodies. The calcospheritic pattern is unaffected by the dentinal tubules, the peritubular matrix, or the collagen fibres, and it is assumed that the crystallites are formed within the ground substance of the intertubular matrix (Jones & Boyde, 1984). Mineralisation of the peritubular matrix proceeds independently, resulting in a much heavier degree of mineralisation.

In each tooth, dentine formation always starts a short while before enamel. The first predentine is laid down under the centre of the cusps, followed a short while later by the first enamel layer. The dentine builds up as a series of conical layers, stacking one inside another. With each new layer, more odontoblasts on the fringes come into production, so that layers increase in size to fill the contours of the crown. After some time, the cells at the apex of the active cone stop production. Thereafter growth continues as a series of sloping sleeve-like layers, leaving a space in the

186   Teeth

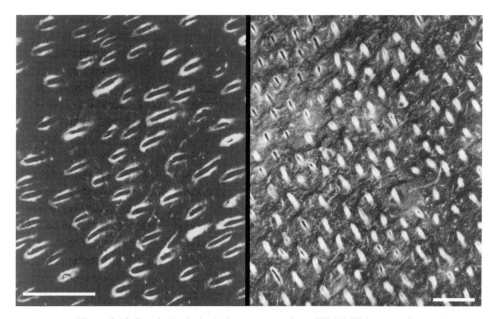

Figure 2.12 Dentinal tubules in human premolars. SEM-BSE images of cut and polished specimens taken by Sandra Bond, UCL. Scale bars 20 μm. The left image shows dentinal tubules sectioned obliquely, with their heavily mineralised peritubular matrix appearing very bright in contrast with the much less heavily mineralised intertubular matrix. The bright specks in the intertubular matrix represent the centres of the calcospherites. The right image shows an area of root dentine which is becoming sclerosed. About half the tubules in the image have been infilled with high density material. The dividing line between open and filled tubules is irregular.

middle. This space becomes the pulp chamber and, as the root grows, the diameter of the space narrows to become the root canal.

The structure of dentine
Under the microscope, the dominant feature is the dentinal tubules (Figure 2.12). These extend from the pulp cavity to the EDJ, branching occasionally and becoming finer as they go. Near the pulp chamber they are 2–3 μm in diameter and spaced 4–5 μm apart. Out near the EDJ, they are less than 1 μm in diameter, but 7–8 μm apart. Along their length, they have tiny branching lateral tubules which contain offshoots extending out of the main odontoblast processes. Most dentinal tubules stop at the EDJ, although some pass through into the enamel to join enamel tubules (p. 158). In light microscopy, dentinal tubules are visible because of refractive index contrasts between the tubule lumen (the space originally occupied by the process but now containing air or mounting medium), peritubular dentine and intertubular dentine (above). In backscattered electron mode (SEM-BSE), they are clearly seen in topographical images of fractured teeth and there is a strong compositional contrast between the intertubular and peritubular dentine, which makes a bright

edge to the tubules (Figure 2.12). Underlying the cusps, the tubules run a more or less straight course. Elsewhere, they follow a gentle 'S' curve. Superimposed on this 'primary curve' are many small wriggles in the tubules' course. Some of these are particularly pronounced and may be related to incremental structures in the dentine. The general direction in which the tubules run is angled up to occlusal. This angle is exaggerated in the high-crowned ungulates.

Peritubular dentine shows some variation between different mammals. It is scarce in the insectivores, rodents, lagomorphs or bats, but is present in Primates, carnivores, hyrax, ungulates, whales, Sirenia and elephants (Bradford, 1967). It is particularly thick in cow, horse and elephant (Jones & Boyde, 1984). In the ungulates it commonly forms only on the side of the tubules facing the EDJ (Boyde, 1971). In the roots of human teeth, the lumen of dentinal tubules is progressively filled in by peritubular dentine (Figure 2.12) from middle adulthood onwards. Presumably, the odontoblast process must remain vital for this to occur, as the tubule narrows around it (Jones & Boyde, 1984). Eventually, however, the tubules may be completely infilled (Vasiliadis *et al.*, 1983b; Frank & Nalbandian, 1989). This so-called sclerosis of the dentine starts around the root apex and gradually extends up the root towards the cervix. It is irregular, with neighbouring groups of filled and unfilled tubules, giving a complex three-dimensional structure. The sclerotic zone can clearly be seen in thick sections in the light microscope, because the unfilled tubules scatter light and appear dark, and the filled tubules are translucent. Examination of serial sections (Vasiliadis *et al.*, 1983a, 1983b) makes it possible to measure the volume, of sclerotic root dentine, relative to the overall root volume, and this is highly correlated with age. It provides one of the most promising age estimation methods for adult humans (Chapter 3).

Superimposed over the outlines of the tubules, the calcospheritic mineralisation is best seen under the polarising microscope (Figure 2.13). Parts of the hyperboloid outline of the calcospherites are shown as shadowy, arc-shaped structures. The collagen fibre mats in intertubular dentine matrix, together with their associated mineralisation, are intermittently visible as shadowy lines which mark out the progress of dentine matrix formation. In most mammals, coronal mantle dentine is distinguished by the direction and coarseness of von Korff's fibres, but it is difficult to see in a routine dentine section. In the Odontoceti, coarse fibres run right through the dentine, along the walls of some dentinal tubules (Boyde, 1980). At the sides of the root, in many mammals, the most prominent feature is a thin layer in which the calcospherites have failed to meet, leaving unmineralised predentine between them. This is the *granular layer of Tomes*. Next to it is either the CDJ, or a thin intervening layer called the *hyaline layer of Hopewell Smith* which, when present, is heavily mineralised and homogeneous, giving it a glassy appearance under the light microscope and a strong compositional contrast in SEM-BSE. Some regard this as equivalent to radicular mantle dentine, whereas others consider it part of the cement. In any case, there is some intermingling of tissues at the CDJ (Jones & Boyde, 1984).

Figure 2.13 Andresen's lines and calcospheritic structure in dentine from a human molar from Kerma, Nubia (*c.* 1720–1550 BC). Polarising microscopy of intact section (field width 100 μm). The dentinal tubules run almost vertically and the calcospherites can be seen as dark, cloud-like masses. The Andresen's lines curve across this structure.

In some mammals, root formation is completed soon after the initial eruption of the tooth crown. High-crowned teeth, however, are erupted gradually and are often still forming their crowns when they first come into wear. Their roots start to form later, and may even then grow persistently. In addition, dentine formation continues slowly inside the completed parts of the pulp chamber and root canal. Odontoblasts remain active and layers of *secondary dentine* are continuously deposited, lining the pulp chamber walls. This type of secondary dentine usually has dentinal tubules dispersed regularly through it (*regular secondary dentine*), and may be difficult to distinguish from the primary dentine. Secondary dentine is also deposited as patches over the pulpar ends of tubules in which odontoblasts have been killed. This is usually due to exposure of the outer ends of the tubules by attrition or dental caries (Chapter 5). The secondary dentine patch covering the pulpar end of these tracts may be distinguished by the irregular spacing of its tubules (Scott & Symons, 1974; Frank & Nalbandian, 1989), and is described as *irregular secondary dentine*. The sealed-off tubules of the primary dentine are recognised as a 'dead tract' in mounted thin sections – black when viewed in transmitted light and white under reflected light. This is because the irregular dentine patch prevents the mounting medium from entering the tubules, and the trapped air provides a strong refractive index contrast (p. 204). Tubules at the edge of such dead tracts may be filled in by large crystals of apatite or octacalcium phosphate (Osborn, 1981; Jones & Boyde, 1984). These infilled tubules are referred to as 'sclerotic dentine', but they differ greatly in origin and development from the root dentine sclerosis described above. The pulp chamber and root canals gradually fill with secondary dentine, particularly in heavily worn teeth. Attrition exposes the dentine in the occlusal wear facet (Chapter 3) and the secondary dentine is usually recognisable as a differently coloured patch (often darker) at the centre of the primary dentine.

There is some variation between mammals in microscopic dentine structure. The most notable example is the striped appearance of dentine in the tusks of hippopotamus, pig, warthog and elephant (Schmidt & Keil, 1971). This is due to variation in the course of collagen fibrils and shows as bands of alternately bright and very dark dentine under the polarising microscope. Even in hand specimens of elephant and hippopotamus ivory, it is possible to make out the pattern of intersecting arcs – like 'engine turning' (Penniman, 1952; Smith, 1972a). In hippopotamus ivory, the arcs are more closely packed, and there is often a surface layer of enamel. Walrus ivory is clearly distinguished by a granular core of secondary dentine which includes denticles (below). This has, for example, enabled the identification of walrus ivory in the twelfth-century AD chessmen from Lewis, in the Outer Hebrides islands of Britain (King, 1983). There is also variation between mammals in the packing and form of von Korff's fibres (Schmidt & Keil, 1971), although it would be difficult to use this for classification.

Occasionally, small isolated bodies of dentine form within the pulp chamber. These *denticles* (Pindborg, 1970) may be loose within dried specimens and rattle about on shaking. In the odontocete whales (and walrus tusks), the pulp chamber is often filled by denticles (Boyde, 1980) embedded in dentine – sometimes with blood capillaries included (vasodentine).

Dentine preservation
Fresh dentine is a very tough material, but archaeological dentine is frequently softened or brittle. This may result in tooth crowns fracturing, or enamel layers

breaking away near the EDJ. Dentine is weakened by collagen loss (Beeley & Lunt, 1980) but, even at low protein contents, the remaining mineral component may differ little chemically from fresh dentine. Sometimes, archaeological dentine shows all the microscopic structures of the fresh tissue (Poole & Tratman, 1978). Falin (1961) found evidence of odontoblast processes still in place in the dentinal tubules of Bronze Age human teeth, and Werelds (1961) even found 'mummified' pulp tissue adhering to the dentine in corners of the pulp chambers of teeth from seventh- to ninth-century AD Belgium. It is not clear how widespread this type of preservation may be, but most specimens show considerable alteration. Roots are affected most, especially at the apex. Inside, an altered zone extends around the pulp chamber and root canal, but usually stops short of the EDJ, leaving a zone of intact dentine under the crown (Werelds, 1961). Macroscopically, the alteration is seen as a discoloration and change in the texture of the root tissues. Using compositional SEM-BSE imaging, it is possible to see foci of diagenetic change (Bell, 1990; Bell *et al.*, 1991, 1996). Most are around 20–150 μm across, with a clear boundary which is usually more heavily mineralised than the surrounding unaltered dentine, and therefore showing bright in SEM-BSE (Figure 2.14). Inside this, there is a zone of lower density, or mixed low and high density, with voids. A network of these foci connects together, and may invade most of the tissue. Most early work described them as irregular canals, branching and anstomosing (Soggnaes, 1950, 1956; Werelds, 1961, 1962; Clement, 1963), but there seems to be a variety of patterns. It is assumed that the foci are made by invading micro-organisms, probably quite soon after death. Light microscope sections which are badly affected can have little original dentine structure remaining, and may be opaque. Thus, dentine condition is highly variable. On one hand, even Oligocene fossils may yield dentine that looks little different from fresh tissue (Doberenz & Wyckoff, 1967) but, on the other hand, Medieval teeth may be heavily affected (Werelds, 1962). The effect varies between teeth from one site, and even within one jaw. Dentine histology is limited by this destructive process, although careful selection of specimens for light microscopy may help. SEM-BSE allows patches of intact tissue to be examined between diagenetic foci, which may slightly extend the range of usable material.

Where teeth have been caught up in fires, strong heating changes the SEM appearance of dentine (Shipman *et al.*, 1984). At 185–285 °C the peritubular dentine shrinks, splitting away from the intertubular dentine. At 285–440 °C, fractured surfaces under the SEM are smoother, with the peritubular/intertubular dentine boundary obliterated. Also in this temperature range, tubules become oval in section. At 440–800 °C, fractured surfaces have a granular appearance and the section of the tubules is even more elliptical. Between 800 and 900 °C granules coalesce into larger, rounded, smoother globules, some 0.5–1 μm across. In human cremations, the covering enamel usually fractures away to leave the conical outlines of the EDJ. Roots also fracture into drum-like segments. Generally, these are well preserved and can to some extent be identified, with care.

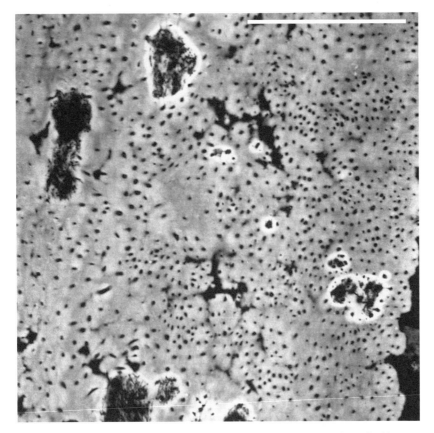

Figure 2.14 Diagenetic foci in the dentine of a sheep molar. The specimen had been buried for 32 years in the Overton Down experimental earthwork (Hillson & Bond, 1996). SEM-BSE image of cut and polished specimen taken by Sandra Bond, UCL. Scale bar 75 μm. The dentine was still in an early stage of mineralisation at death, with growing calcospherites. The front of mineralisation is the dark area to the lower right of the image. Remaining interglobular spaces are present throughout. Transversely sectioned dentinal tubules are scattered through the image – their peritubular matrix is not yet heavily mineralised. The diagenetic foci are shown as rounded areas, around 30 μm in largest dimension. Some have a bright (heavily mineralised) edge, but others do not. Inside, they show a variable degree of mineralisation.

Dentine incremental structures
Laminar structure is at least faintly visible in most dentine sections in the light microscope. It is, however, often patchy and difficult to resolve into clear layers. In humans and other Primates, it is only an occasional section that allows layers to be counted or measured. This may partly be due to the plane of section, which needs to cut directly through the main axis of the tooth, but there are other unknown factors as well.

Two types of finer layering are visible in most mammals studied (Dean, 1995): more prominent 'long period' lines spaced in apes and humans about 20–30 μm apart, between which are fainter and less well defined 'short period' lines at around 2–4 μm (up to 16 μm in the fastest growing regions of some mammal teeth). Both were described by Andresen (1898) and von Ebner (1902), and their names have been attached to them, but it is not clear which name should be applied to which line spacing. Usually, the long period lines are known as *Andresen's lines*, and the short period lines as the *lines of von Ebner* (Dean, 1995, 1998b). They are seen in polarising and ordinary light microscopy, both in intact sections and in sections where the mineral has been removed chemically (Schmidt & Keil, 1971). This demonstrates that they are due to variation in the organic component and it seems most likely that they represent cyclic changes in the proportions of collagen and ground substance (Jones & Boyde, 1984; Dean, 1995). If sections are stained with silver, however, it is possible to see a concentric layering, which appears to represent the growth of mineralisation in the calcospherites (Kawasaki *et al.*, 1980; Dean, 1995, 1998b). This may be superimposed over the linear pattern incremental structures in some areas of dentine, but the relationship is not well understood. The long period lines appear to correspond to the brown striae of Retzius in the enamel of the same tooth: the count of short period lines between long period lines matches the count of cross striations between brown striae, and prominent lines can be matched with prominent striae (Dean, 1995; Reid *et al.*, 1998). As dentine formation is always in advance of enamel, the dentine lines are positioned slightly further down the EDJ than their enamel equivalents. Labelling experiments confirm that the short period dentine lines represent a circadian growth rhythm, like the cross striations of enamel (reviewed in Dean, 1995). A neonatal line (p. 165) can be recognised in dentine (Schour, 1936), but it is difficult to use dentine layering to determine the timing of events in dentine development, because most sections show clearly only occasional patches of layering. Instead, it is better to use the cross striations of enamel to calibrate dentine layering (Kawasaki *et al.*, 1980; Dean & Scandrett, 1996). If an estimation of age, based on developing root length, is needed, then it is best to calibrate it by the relative amount of crown growth achieved in other forming teeth from the same individual (Dean *et al.*, 1992a).

Other developmental features may also appear in dentine. The *contour lines of Owen* (Owen, 1845) are parallel deviations in the course of the dentinal tubules, which must involve coordinated movements of the whole sheet of odontoblasts during development. Together they form lines, some tens or hundreds of micrometres apart, which are parallel to the successive fronts of predentine formation. They are not regular in most mammals, but seem to represent some occasional disruption. In Odontoceti, there may be large deviations of this kind, making very prominent layers. In addition, the Odontoceti show local dilations in the width of the dentinal tubules, or in the number of branches, which further accentuate layering (Boyde, 1980). Another feature, occasional in most mammals, is linear zones of calcospherites which have failed to grow large enough to mineralise all the predentine matrix. The intervening patches of predentine are called *interglobular spaces*. In effect, the structure is similar to the granular layer of Tomes, but they follow the incremental lines of dentine, and the contour lines of Owen, in their orientation. Most tooth sections show a few patches of interglobular spaces. In some individuals, there are prominent layers of them (Figure 2.8). There is a strong relationship between such layers and the hypoplastic defects of enamel. Laboratory experiments with dogs show that vitamin A and D deficiency cause both (Mellanby, 1929, 1930) and rickets in humans has been linked with interglobular spaces (Seeto & Seow, 1991). Experimentally induced fever has also produced them in rabbits (Bermann *et al.*, 1939), so a similar range of growth-disrupting factors seems to be responsible. Lines of interglobular spaces are not associated with every hypoplastic defect of enamel, but they do often seem to be associated with a prominent brown stria of Retzius (Soggnaes, 1956). Cross striations in the enamel are therefore likely to be the best way to determine the age at which the disruption occurred.

Odontoceti (Boyde, 1980) and Pinnipedia (Scheffer & Myrick, 1980) often show large areas of interglobular dentine and these may be related to annual variations in food intake, and to breeding and migration (Chapter 3). A whole pattern of longer interval layering is visible in many mammals (Klevezal, 1996), relating to seasonal and annual variations. Often, they do not involve interglobular dentine, but are seen in demineralised, haematoxylin-stained preparations as an alternation between darker and lighter staining regions of dentine matrix (see, for example Figures 13 and 22 of Klevezal, 1996). In effect, they seem largely to be variations in the prominence of Andresen's lines, but there has been little research done on the relationship between finer and coarser patterns of layering. A good place to start would be such teeth as the persistently growing canines of Pinnipedia, in which strong seasonal dentine layering might be calibrated with cross-striation counts in the enamel.

Cement

Dental cement (or cementum) is a highly variable tissue. In some mammals it coats only the roots, whereas in others it envelops the whole tooth including both crown and root. It may form a layer only 20 µm thick, or may be several millimetres thick in some species. The thickness also varies with age and with the part of the tooth being examined.

Composition and structural components of cement

The overall chemical composition of cement is similar to that of bone, which has 70% by dry weight inorganic, 21% collagen and 1% other organic components (Williams & Elliott, 1989). The collagen fibres arise from two sources: large *extrinsic fibres* that are incorporated from the periodontal ligament, and small *intrinsic fibres* that are laid down by the cement-forming cells themselves. Ground substance can also be demonstrated in sections, and living cells, the *cementocytes*, are present. The composition of cement varies between different layers, different parts of the tooth and different mammals.

Formation of cement

Cement acts as a tissue of attachment for the periodontal ligament that holds the tooth into its socket. In some mammals, this attachment is to the root only but, in hypsodont teeth, it is both to the root and the crown, so a cement layer covers the whole tooth before wear. Cement is formed on the surface of enamel and dentine by *cementoblasts*, positioned at the edge of the periodontal ligament. These manufacture first an organic matrix, *precement* (or cementoid), which is later mineralised. In actively growing cement, a layer of precement covers the surface, itself overlain by a sheet of cementoblasts. These occasionally become entrapped in the developing tissue to become cementocytes. Intrinsic fibres are formed by the cementoblasts themselves. They are typically 1–2 μm in diameter and are laid down as a series of mats, organised into overlapping patches (domains) of similar fibre orientation. Extrinsic fibres are formed by the fibroblasts of the periodontal ligament. They are progressively engulfed by the growing precement and enter it roughly perpendicular to the growing surface. These fibres are variable in size, but in humans are 6–7 μm or more in diameter (Jones, 1981). They may be incorporated as dense groups, or singly. In cement as a whole, the degree of mineralisation varies with rate of formation. The intrinsic fibres mineralise more rapidly than the extrinsic fibres, which may end up unmineralised at their cores (Jones, 1987). When cement is growing rapidly, the extrinsic fibres are less well mineralised and, when growing more slowly, better mineralised. This can be seen in SEM pictures of anorganic preparations (below) of the developing cement surface where, if it is rapidly growing, the ends of the extrinsic fibres are represented as pits because of their unmineralised cores. A more slowly growing, or resting, cement surface shows the better mineralised extrinsic fibres standing up as projections.

Cementocytes occupy spaces (*lacunae*) of variable size – 7–20 μm in diameter – and irregular shape. Processes extend out from the cell walls, mostly towards the cement surface, inside tiny tunnels (*canaliculae*). The processes provide the cementocytes with nutrition and, as the cells become more deeply buried, they become less viable. The density of cementocytes, and their lacunae, varies greatly through the cement and contributes to the layered structure (below). They are incorporated in greater numbers when the cement is growing rapidly (Jones, 1981) and this can be seen in anorganic preparations of the developing surface (Boyde, 1980), which show the developing lacunae of cementocytes that have just been incorporated.

Root development is often delayed in the high-crowned herbivores. Eruption is continuous as the tooth wears rapidly and the crown does the job of the roots for much of the time (Weinreb & Sharav, 1964). The whole surface of the crown is covered with cement in the cow and horse, including the deep infundibulum (Chapter 1) present in their cheek tooth crowns. Cement deposition follows reduction or disintegration of the enamel organ (Chapter 3). Enamel matrix formation is finished beforehand, although maturation may not be complete (Listgarten, 1968). The cement filling the infundibulum contains clusters of cells that may be derived from the disintegrated enamel organ. In the horse (Jones, 1974), enamel is resorbed before cement deposition takes place, leaving some patches of enamel intact but removing others entirely. Cement is deposited directly on this resorbed surface. The cement is cellular and also contains loops of vascular tissue (blood vessels), housed in canals. These may be regularly arranged at 100–200 μm intervals. Cementocyte lacunae are arranged concentrically around the canals – a structure reminiscent of osteons in bone. Many other mammals have extensive coronal cement coatings, including most ungulates, elephants and many rodents. They are also common amongst the odontocete whales and frequently include vascular tissue (Boyde, 1980).

Cement deposition continues throughout life. Teeth constantly erupt and change position to adjust for the effects of wear on the crown. This is partly brought about by remodelling of the bone in the alveolar processes around the tooth, but also by remodelling of the collagen fibre network in the periodontal ligament. Extrinsic fibres in the cement are thus cut off and new fibres are entrapped by the growing precement. It is possible to reconstruct the sequence of changes in tooth position by tracing the changes in extrinsic fibre orientation through the cement layer (Gustafson & Persson, 1957; Sims, 1980; Lieberman, 1994). Deposits of cement build up particularly in the areas of maximum vertical displacement including the apex of the root, and the fork of the roots in multirooted teeth. High-crowned herbivores, whose teeth wear particularly rapidly, build up considerable thicknesses of cement in this way. A thick 'molar pad' of cement is formed between the roots of their cheek teeth (Beasley *et al.*, 1992). Abnormally thick cement (*hypercementosis*) may occur around the apices of roots, for example in humans where it is seen particularly in older people, in association with heavy tooth wear or with chronic periapical (Chapter 5) inflammation (Pindborg, 1970; Soames & Southam, 1993). In some high-crowned herbivores (Jubb & Kennedy, 1963) the opposite condition of too little may occur, with a deficient cement coating to the crown. Small spherical masses of cement (*cementicles*) may occur within the normal cement layer, within the periodontal ligament or alveolar bone.

Types of cement
Cement may be classified by the presence or absence of cementocytes, or the presence and origin of collagen fibres (Jones, 1981):

1. *Afibrillar cement.* Heavily mineralised ground substance only. No cementocytes or collagen fibres.
2. *Extrinsic fibre cement.* Mineralised ground substance, with fully mineralised extrinsic fibres, closely packed and parallel. No cementocytes.
3. *Mixed fibre cement.* Mineralised ground substance, with both extrinsic and intrinsic collagen fibres. Cementocytes may or may not be present.
4. *Intrinsic fibre cement.* Mineralised ground substance with intrinsic fibres only. Cementocytes present.

Most cement sections contain all four types. Afibrillar cement is produced early in the development of the cement coating of the tooth, and is often found as a very

196   Teeth

thin and intermittent layer which cannot be seen by light microscopy. In any case, it may be difficult to distinguish at the CDJ from the similarly heavily mineralised hyaline layer of Hopewell Smith (p. 187). Extrinsic fibre cement is also found early in cement development, before the teeth erupt into full occlusion. Mixed fibre cement usually characterises the main cement formation, is variable in structure and composition and is often strongly layered (below). Small patches of intrinsic fibre cement may be found near the root apex of older individuals.

Incremental structures in cement

Cement frequently has a layered structure (Figure 2.15), with the layers following the circumference of the roots. They are best seen in demineralised preparations, sectioned with a microtome and stained with haematoxylin. Layering can also be seen as compositional contrasts in SEM-BSE images (Jones, 1987), but no research has yet been done to match these images with stained sections. It is also possible to make out layers in intact sections under ordinary or polarised light microscopy, although fine cement layers are usually harder to see, partly because of the thickness of the section. The layers result from variation in:

- the density of cementocyte lacunae
- density of collagen fibres versus ground substance
- density of extrinsic fibres versus intrinsic fibres
- degree of mineralisation of fibres
- orientation of extrinsic fibres.

It is likely that they represent a variation in formation rate of mixed fibre cement. Faster forming cement incorporates more cementocytes, more extrinsic fibres with poorer mineralisation and more intrinsic fibres relative to ground substance. Such a cement layer stains less heavily with haematoxylin (because ground substance is the basophilic component) and shows as a darker zone in compositional SEM-BSE images (because of its poorer mineralisation). The larger number of light scattering interfaces it contains (from the collagen fibres and lacunae) make it darker in transmitted light microscopy of intact thin sections and lighter in reflected light microscopy. Slower forming cement incorporates fewer cementocytes, fewer extrinsic fibres, might include a zone richer in ground substance, and would be more heavily mineralised. It stains more heavily, shows brighter in SEM-BSE images, brighter in transmitted light and darker in reflected light. Variation in the orientation of extrinsic fibres could make contrasts in polarised light microscopy, and could also make minute differences in the resistance to polishing of cement which would superimpose topographic contrasts in SEM-BSE. Layering must also be affected by the presence of the different types of cement, which are mineralised to a different extent. It is not necessary for cementocyte lacunae to be present for layers to be visible, because they are seen in extrinsic fibre cement, which has no cementocytes (Jones, 1981).

Dental tissues

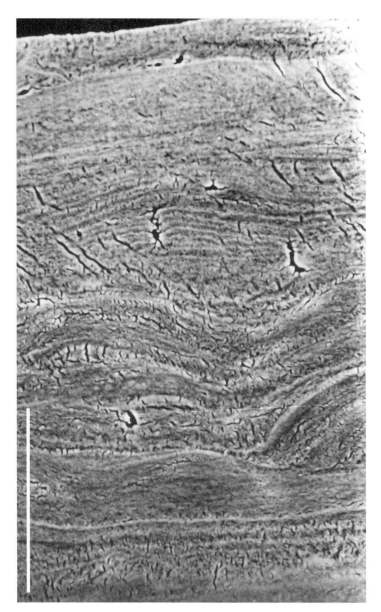

Figure 2.15 Cement layering. SEM-BSE image of cut, polished surface, taken by Daniel Antoine, UCL. Scale bar 200 μm. The specimen was a third molar, which had been extracted from a 48-year-old human. At the bottom edge of the image is the CDJ and, at the top, the surface of the cement. The layers are shown as bright/dark alternations, running approximately horizontal. It is apparent that the layers were not laid down evenly over the root. The black, rounded patches are the cementocyte lacunae, filled with resin. The diagonal black marking is typical of cement in this form of preparation and are due to cracking along the line of the extrinsic fibres.

The layers vary in thickness, from just a few micrometres to 100 μm or more, and in definition. In some mammals, the lines are very strongly and sharply defined, especially in demineralised, haematoxylin-stained preparations (see figures in Klevezal, 1996). They are seen, with incremental lines, in dentine (above and Chapter 3) as 'recording structures' which allow the age and season at death to be determined. This technique is an important tool in zoological studies of living mammals.

Structure and preservation of cement

Like dentine, cement is variably preserved on archaeological sites. This is presumably due in particular to progressive destruction of the organic component. Cement is invaded by diagenetic foci in much the same way as dentine (p. 190). Specimen preparation by demineralisation and microtome sectioning therefore yields variable results. Spiess (1976) successfully applied the technique to protohistoric and historic seals from Labrador and to 20 000–35 000 BP teeth from Abri Pataud, France. He did, however, observe that preservation needed to be good and reported 'complete dissolution of the samples' when trying to demineralise material from 18 000 BP Lukenya Hill, near Nairobi. Similarly Coy *et al.* (1982) used the technique to section a large number of Saxon cattle molars from Britain. Only about one half yielded satisfactory results. Ground, polished sections of intact tissue can be prepared, but they are often rendered opaque by diagenetic foci and, in any case, they are really too thick to see the finest layers. Compositional SEM-BSE offers a possible solution, but there has not yet been sufficient research relating these images to their conventional haematoxylin-stained counterparts.

Resorption of root and crown

A certain amount of remodelling of the attachment to the periodontal ligament may take place, by resorption of cement, dentine and enamel. All the dental tissues are avascular, so the resorption can only start from the pulp chamber, root surface or (in high-crowned mammals) the crown surface. It is carried out by cells identical to the osteoclasts of bone, known as *odontoclasts*. Each cell removes tissue across a variable front, to create resorption hollows called *Howship's lacunae* (Jones, 1987; Jones & Boyde, 1988). They vary in size from a few micrometres to hundreds of micrometres and have a characteristically scalloped *reversal line* at their edges. In most mammals, resorption of the root is most marked in the exfoliation of deciduous teeth (Chapter 3). Even with the naked eye, the resorbed surfaces have a characteristic rough 'acid etched' texture which results from many thousands of intersecting Howship's lacunae (Jones & Boyde, 1972, 1988). Resorption may also occur later in the life of the root, around the apex. In horses, strong resorption of the crown surface precedes the deposition of the thick cement layer (above).

Preparation techniques
Tooth surfaces
The surfaces of the crown and root are practically always free of the neighbouring soft tissues in archaeological specimens. In effect they are already anorganic preparations (normally prepared with 5–7% NaOCl, Boyde, 1984) from which superficial predentine or precement have long since been lost. They may, however, require cleaning. Soil can be removed by picking with a hardwood stick and, in many specimens, by simple washing with water. If they are soaked and then too rapidly dried, teeth can break up, so it is best to test a less important specimen first. Fine cleaning is best achieved with acetone, using cotton buds on hardwood sticks (roll the bud along to pick up dirt – do not scrub). This rarely causes scratching and can also be used to remove most consolidants or adhesives which might have been applied.

Enamel may be etched or polished to bring out prism structure in the SEM. Etching may be accomplished by immersion in 0.5–2% H_3PO_4 for 30 seconds to 2 minutes (Boyde, 1984). Polishing requires a soft lap to bring up the prism boundaries.

Impressions and replicas
Ideally, the specimen itself should be the subject of examination, but often in archaeology this is not possible because the material cannot be taken out of the museum, or because the tooth is in a skull which cannot be fitted inside the SEM specimen chamber. In such circumstances, a high resolution replica can be made of the exposed surfaces (Hillson, 1992a). Dental impression material (at the time of writing Colténe President gives the best resolution) comes in two forms: a putty and a fine body material. Dentists would use a two-stage technique in which an initial impression is made with the putty, to provide a tough outer casing, and the fine body material then takes the detail. For archaeological material, it is gentler just to apply the fine body material, which is more flexible and can be pulled away more easily. Caution is needed, because teeth can be broken. Spaces around teeth larger than about 1 mm need to be plugged with small pieces of crumpled aluminium foil, so that the impression material does not get stuck when it solidifies. When set, the edge of the mould should be released gently, using a wooden stick. Afterwards, epoxy resin is poured into the mould (a readily flowing resin such as Epotek 301) and allowed to harden. Typically, structures much less than 1 μm across are preserved in these replicas, although there must be some dimensional changes due to the flexing of the mould.

Sectioning
Before sectioning a tooth, a full description, measurements, photographs and a dental impression should be taken. For archaeology, intact sections (i.e. not demineralised) are normally used. These are usually cut and polished down plane parallel

to a thickness of 100–150 μm – they are therefore quite thick, even though they are often called 'thin sections'. It is possible to make them thinner, down to 60 μm or even 30 μm, but archaeological specimens can easily collapse and structures such as brown striae of Retzius require a thicker section anyway. For the same reason, it may be safest to impregnate archaeological specimens with a hard resin (Hillson, 1996) before cutting, although experienced histologists may do without this step (Reid *et al.*, 1998). A hard resin means that the specimen will polish flatter and the hardest available is polymethymethacrylate (Perspex or PMMA). Methylmethacrylate monomer is also so fluid that it passes into the finest spaces in the tissue and thoroughly impregnates it. The plane of section is crucial, so it is essential to mark it on the tooth before impregnating. After the resin has hardened, the specimen is cut and polished. This is very difficult and a lot of practice (one month or more) is required before good sections can be made. Most of the difficulties experienced in dental histology result from sections which are not good enough to see what needs to be seen. A good cutting machine is essential. The very best, and most expensive, have a very thin rotating annular blade, shaped like a doughnut, which cuts against its diamond-coated inner edge. This allows very thin cuts to be made, losing less material. Many laboratories, however, have a simpler rotating disc which cuts against its outer edge. This is perfectly adequate, but makes thicker cuts. The other essential equipment is a good lapping machine, with a heavy polishing jig to hold the specimen. These also are expensive. The procedure is to cut the specimen twice, isolating a slice 250–500 μm thick. This is temporarily fixed with sticky wax to a glass slide, which is held in the jig, and one side of the slice is polished. It is then cleaned, turned over and fixed to a slide with cyanoacrylate adhesive, before being polished down to 100 μm. Very fine abrasives are used throughout, to avoid scratching. Once the tooth has been sectioned, it is possible to remove the methylmethacrylate by dissolving it in acetone and if needed the tooth can be reassembled with a sheet of plastic to replace the lost slice.

For SEM-BSE imaging, a very flat section surface is needed. The best way to produce this is by micromilling (Boyde, 1984), but the equipment is expensive and not widely available. In practice, most specimens are prepared by cutting and polishing. The specimen needs to be impregnated with methylmethacrylate (including 5% styrene to give it stability under the electron beam) and then a single cut made, using a machine as above. The cut surface is then polished, using a series of abrasive papers, on a hard surface, with water as lubricant. It is usually best to polish as little as possible, just enough to remove the scratches of the saw. Once again, it requires much practice. The softer components of the specimen always wear more rapidly than the harder parts, producing some surface relief, but this can be minimised. The technique can be less destructive than preparing a 'thin' section, because only one narrow cut needs to be made and the methylmethacrylate can be removed with acetone afterwards.

Dental tissues

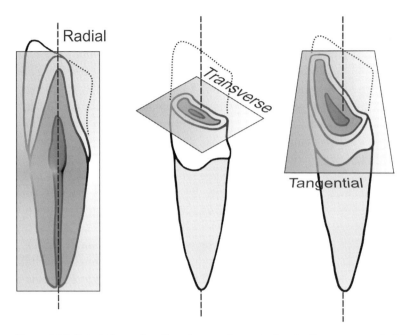

Figure 2.16 Orientations of section planes, using an incisor tooth as an example. The axis of rotation of the tooth is marked as a dashed line.

Orientation of sections and surfaces
Sections of individual teeth are described with reference to an axis of rotation passing from coronal to apical (Figure 2.16). In a multicusped tooth, each cusp has its own axis of rotation. *Radial* sections are cut with this axis at their centre, radiating out in any direction. This is the most commonly used orientation. *Transverse* sections cut the tooth in any plane perpendicular to the rotation axis and *tangential* sections are those cut in any other plane. Sections may also be related to the anatomical planes of the body as a whole. *Sagittal* sections are parallel to the median sagittal plane of the body. This is particularly useful in describing sections of rodent incisors, which do not have a clear axis of rotation. Similarly, the term *mesiodistal* may be used to describe a section cut in a plane tangential to the dental arcade, and *buccolingual* for a section perpendicular to it. The plane of section makes a big difference to what can be seen, and what cannot. For incremental structures in enamel and dentine, in particular, it is essential to centre the section on the highest point of dentine in the EDJ under each cusp. Failure to do this is one of the most common errors in section making.

Cameras and light microscopes
Cameras and scanners
Macrophotography is a useful tool for studies of teeth, and a proper camera system with macro lens, tripod and cable release is needed. Small scales can be drawn

in a computer graphics program and printed on to matt photographic paper. Many modern digital cameras have a good close-up facility, making it possible to fill the frame with a tooth 1 cm across. They have an advantage over film cameras in convenience and because the results can be checked instantly, which is useful in the field. Several researchers have found an ordinary computer scanner works well. The resolution can be set high over small areas and there are fewer problems of parallax, which can distort conventional camera close-up pictures.

Lenses and low power stereomicroscopes
An ordinary hand lens can resolve objects 10 μm in size, but has to be held close to the specimen. It is also difficult to light the specimen brightly enough, and to hold it steady. A stereomicroscope is much more practical and can be regarded as an essential tool for work with teeth. Careful adjustment of the two eyepieces is needed but, when this has been correctly done, such microscopes give a good impression of depth which is useful for the three-dimensional shapes of small teeth. Magnification ranges from ×10 to ×50, although the depth of focus is best at lower magnifications. They need a good cold light source and a fibre optic lamp is best. Forceps are needed for handling small jaws and it is helpful to have a small tray which can be moved around under the microscope. A good way to hold the jaw steady is to fill the tray with clean dry sand. One useful attachment is a drawing tube, which allows the observer to trace the image on a paper pad by the side of the microscope. Most of the original drawings for the figures in this book were done that way. It is possible to attach cameras to most stereomicroscopes, although they rarely give enough depth of focus. For objects up to around 1 cm across, macrophotography is better and, for smaller objects, an SEM (below).

Conventional transmitted light microscopes
This is the usual type of laboratory microscope, with a built-in light source in its base, a condenser lens under the section to focus light on it, an objective lens to form the primary image and eyepieces to magnify the image into the eye. For dental histology, quite low magnifications are used: ×100 up to ×400. Sometimes, structures may be made more clearly visible by very oblique illumination. This can be arranged by deliberately off-centring the condenser lens, but it is better to use dark field illumination. A dark field microscope has an opaque disc fitted underneath the condenser, so that only a ring of light enters, and shows sharp refractive index (below) changes brightly in an otherwise dark image.

Most histology of dental tissues has, however, used a polarising microscope. This is distinguished by having polarising filters above and below the stage, arranged so that their planes of vibration are at right angles to one another. A purpose-built instrument will have these filters built in, but they can be improvised by cutting discs of polaroid sheet and fitting one into the eyepiece and one into the filter carrier below the condenser. The eyepiece is rotated until the image goes completely dark, which is the point at which the vibration planes of the filters are perpendicular. A purpose-built instrument has a rotating stage which allows extinction angles to be determined (below), and this is more difficult to improvise. Polarising microscopes are a powerful tool in dental histology. It is possible to determine the direction of crystallites and fibrils in the tissues, and also the extent and type of mineralisation. Details of structure can be seen that would not be possible in any other type of microscopy. The standard text on dental polarising microscopy is Schmidt & Keil (1971).

Confocal light microscopes
The principle of a confocal microscope is to focus a point of light on to a narrow plane within the section (or indeed an unsectioned specimen), and then to make it possible to observe it through an aperture which excludes light transmitted or reflected by any other plane in the section. The point of light is rapidly scanned across the plane of focus to build up the image. A tandem scanning reflected light microscope (TSRLM or TSM) both scans points of light and observes them through apertures in a spinning disc. A confocal laser scanning microscope (CLSM) uses vibrating mirrors to scan a laser beam across the plane of focus. The laser scanning unit is often attached to a conventional microscope, so it is possible to find the field of view quickly before switching to confocal mode. Both the TSM and CLSM are often used in reflected light microscopy, although many other configurations are possible. The plane of focus is only 1 μm thick, and can be positioned several hundred micrometres inside a translucent specimen like enamel. Many confocal microscopes have image analysis systems, including frame stores which, coupled with automatic control of focus, make it possible to record multiple planes through the specimen. The information from these can be combined to create a three-dimensional model of the structures in the specimen, or of its surface.

Formation of images in the light microscopy
Images are formed by six phenomena: absorption, fluorescence, reflection, refraction, diffraction and polarisation (Hartley, 1979). In dental histology, absorption and fluorescence apply mostly to the action of stains in demineralised preparations. In transmitted light, a stain absorbs some wavelengths of light and transmits others, so that those parts of the tissue which pick up a greater amount of the stain show a particular colour. Some stains, such as the tetracyclines, fluoresce in reflected ultra-violet light, and so can be used as markers. In intact sections, however, features of the dental tissues are made visible by optical contrasts, caused by differences in the transmission rate of light (refractive index or RI). Apatites have an RI of 1.63, whereas air is 1.0, water 1.33, Canada balsam mounting medium 1.54 (quite close to glass), polymethylmethacrylate (PMMA or Perspex) embedding material 1.49, Epotek 301 epoxy resin 1.54 and cyanoacrylate adhesive 1.45. A boundary between materials of two different refractive indices causes a beam of light to bend – a process called refraction – and this bending is greater with larger differences in RI. This may bend parts of the beam of light to one side as it passes through the

section, so the boundary appears darker to the observer. It may also cause a dark fringe, called a Becké line, to parallel the boundary. If the boundary is strongly enough inclined to the beam of light, the whole beam may be reflected to one side and, once again, the boundary will appear darker. Diffraction is the scattering of the edges of the beam of light as it passes a sharp margin in the specimen. It is probably responsible for much of what is seen in optical microscopy of dental tissues and causes light to be scattered when passing through tissue which contains pore spaces. This is what makes the brown striae of Retzius visible. They are brown because the blue end of the spectrum is scattered more readily than the red end. In reflected light, they are brighter than the surrounding enamel because more light is scattered back to the observer. Much of dental histology using the light microscope revolves around trying to arrange the maximum degree of optical contrast. It is best if sections are not mounted permanently, but are examined in air (which provides the maximum contrast) or mounted in water (which is still a large contrast). Thin sections usually need to be cemented to a glass slide, but if a thick, quick-setting cyanoacrylate adhesive is used, not much will enter the fine pore spaces. Canada balsam, by contrast, takes a long time to set, and as the section 'clears' by balsam entering all the pore spaces, provides only a small RI contrast with apatite. For the same reason, embedding in polymethylmethacrylate (above) is not ideal from an optical contrast point of view, but at least the contrast is larger than with Canada balsam. A relatively thick section provides a longer path through which light may be scattered and this is why such sections are specially prepared to view brown striae of Retzius. In conventional light microscopy, the image is formed by the cumulative effect of structures within a relatively thick plane of focus inside the section. This is important in understanding such features as prism cross striations because several layers of prisms contribute to the image. Normal histological practice is to prepare a section thinner than the smallest structures being imaged. In enamel and dentine, these might be 4 μm, and intact sections this thin are just not possible. One solution to this difficulty is to use confocal light microscopy (above), which focuses on a very thin plane, typically 1 μm thick.

Images are produced in a polarising microscope by the phenomenon of *birefringence* – or double refraction. Most minerals and many large organic structures refract a beam of light into two separate beams. Each emergent beam has a different refractive index, and is polarised. In the apatites and collagen of dental tissues, one emergent beam is polarised parallel to the optical axis of the crystal or molecule, and the other in a plane perpendicular to the axis. This arrangement is called uniaxial. The difference in refractive index between the beams is the birefringence. In apatites this ranges from -0.003 to -0.004. Birefringence of collagen fibrils is positive, but the actual value is difficult to determine for collagen on its own. The birefringence of a mineral or organic structure also varies with orientation. In a uniaxial refracting medium, birefringence is zero when looking directly down the optic axis, but rises to a maximum in any orientation perpendicular to this. The optical axis in apatites and collagen follows the long axis of crystals and molecules. So apatite crystallites and collagen fibrils are dark when seen end on – no birefringence means no image. Conversely, they are brightest when seen from the side.

The brightness of the image also varies with the rotation of the microscope stage, relative to the polarising filters. When the vibration planes of the refracted beams are parallel to the vibration planes of the filters, both are cancelled out and a dark image is produced. This is called *extinction* and it allows the orientation of crystals and fibres to be measured, relative to structures in the tissue. In other positions of rotation, the brightness and colour of the image depends on the birefringence of the tissue components and the thickness of the section. In 100 μm thick sections of intact enamel, the colours are greys and bright yellows of 'first order' type (Kerr, 1959). By themselves, apatites and collagen show only their own *intrinsic birefringence*. In dental tissues, however, they form an optically mixed body. Collagen fibrils have apatite crystallites within their structure, and spaces may be filled with water, air, or mounting medium. These mixed bodies add another component – *form* or *textural birefringence*. The form birefringence of collagen composites is positive and aligned along the axis of the fibrils. In enamel which has become demineralised – with enlarged pore spaces that imbibe mounting medium – a positive form birefringence is also apparent. Low negative birefringence of enamel may be modified by demineralisation to a low positive birefringence. This is responsible for the changes seen in microscopy of carious lesions (Chapter 5). Similar changes may be seen as a result of poor preservation.

Scanning electron microscopy

A scanning electron microscope (SEM) is now an essential tool in dental histology. It can be used to examine both surfaces and sections, and can be operated in ways that emphasise the composition of the tissue or the topography of the surface. Relatively low magnifications are often used for routine dental histology – they might be ×1500, ×100 or even as low as ×10. The main advantage of the SEM is its large depth of field, which allows much of a high relief surface to be seen in focus at the same time. Specimens are given a conducting coating of gold, palladium or carbon and then placed inside a specimen chamber which, in most instruments, is pumped to a high vacuum (some now can operate at near atmospheric pressure). The specimen is scanned by a fine electron beam and images on an analogue or digital display are produced by measuring the effects of the interaction of this beam with the specimen surface. Most electrons from the incoming primary beam come to rest inside the specimen, but a proportion are scattered back out again. The yield of such *backscattered electrons* is dependent upon the degree of mineralisation and chemical composition of the inorganic component, together with the topography of the surface. Many incoming electrons also initiate *secondary electrons* from collisions with atoms of the specimen and some of these escape the surface as well. The yield of secondary electrons is controlled largely by the surface topography. A backscattered electron detector (SEM-BSE) can be operated in a way that emphasises topography or, with a very flat section surface arranged perpendicular to the primary beam, in a way that emphasises composition. In general, heavily mineralised areas show as bright in the image. Magnification should be arranged so that each pixel on the image display represents no more than 1 μm on the specimen surface, because that is the minimum resolution for compositional imaging of this type. Backscattered detectors are not available on all microscopes, and many images are produced using an Everhart–Thornley detector (SEM-ET), which measures mostly secondary electrons, with a few backscattered electrons. It therefore shows largely topographical contrasts. The image has to some extent the appearance of a very obliquely illuminated scene, viewed down the line of the primary electron beam with the apparent light coming from the direction of the Everhart–Thornley detector. For more details see Boyde (1984, 1989), Boyde & Jones (1983) and Goldstein et al. (1992).

Conclusion

Excellent preservation, under a wide variety of conditions, is one of the most useful features of dental tissues. It makes them good specimens for biochemical investigations. It also makes it possible to study them as a variety of preparations, in a range of microscopes. Two further attributes, their layered structure and lack of tissue turnover, make it possible to reconstruct the processes of development which formed them. Dental development is one of the most complex developmental sequences found in mammals, and this complexity makes it possible to retrieve a high level of detail. Enamel is the jewel of dental tissues from this point of view. It is almost wholly mineral in life and so survives better than any other part of the body after death. Its tiny organic component appears to be preserved little altered for millions of years. The weave of enamel prisms is so intricate and variable that it is possible in some orders of mammals to distinguish families on this basis. The prism cross striations, which represent a 24-hour growth rhythm, make it possible to calibrate not just the development of enamel, but also the growth of the whole teeth and, by extension, the growth of the associated skeleton. The potential of enamel histology for studies of mammals as a whole, rather than just the Primates, is only just beginning to be tapped.

3

TEETH AND AGE

Growth

Many age estimation methods are based on growth of the dentition and skeleton. Growth follows a consistent sequence, broadly comparable between different individuals, and clear-cut changes occur over a short period so, even taking into account variation within species, age can be estimated with reasonable precision from the state of development. Growth unfolds in different ways for different parts of the body. General or *somatic* growth is rapid in foetal and early postnatal life, but then slows until the 'growth spurt' of puberty, during which any marker of overall body size increases rapidly. On the attainment of sexual maturity, growth rate falls back as the individual reaches adult size. *Genital* organs and most secondary sexual characters grow little before puberty, but very rapidly during the somatic growth spurt. *Neural* structures, by contrast, grow rapidly before birth and during infancy, after which they grow more slowly. Teeth and jaws have their own pattern related to the two dentitions. The deciduous dentition is associated with the shorter face or snout of young animals. Establishment of the permanent dentition is accompanied by elongation of the snout and the eruption of the last teeth marks attainment of adult face proportions.

Skeletal structures obey one or a mixture of growth patterns. Long bone length follows the somatic pattern, whereas the skull is a complex mixture of growth curves (Moore, 1974; Osborn, 1981; Goose & Appleton, 1982; Enlow & Hansen, 1996; Meikle, 2002). The cranium, orbital and upper nasal region of the skull follow the neural growth pattern and reach mature proportions early in life. In the lower part of the face, the jaws and their supporting structures, two phases of growth may occur, associated with deciduous and permanent dentitions. During sexual maturation, additional remodelling occurs in the skull to produce sexual dimorphism in features such as muscle size, robusticity, horn development, facial protruberances and other 'ornaments'.

Even when the main phase of growth is complete, skeletal structure is not 'fossilised' because, like most body tissues, bone undergoes tissue turnover. Turnover is where cells continually break the tissue down and replace it with new tissue. Bone structures are dynamic and actively maintained, so any alteration in the balance of factors that maintains a particular form may bring about change in shape. Remodelling of the skull and jaws takes place all the time, in response to functional changes. Whilst this goes on around them, the teeth stay little altered apart from the effects of wear and disease, because the dental tissues do not turn over (Chapter 2).

Rate of growth, and the timing of spurts or more gradual changes in speed, varies from individual to individual. Some control is genetic – witness differences between populations and differences between males and females within one population. A proportion of the control is, however, environmental. Nutritional plane, dietary deficiencies, incidence of disease and even psychological stress are all controlling factors (Tanner, 1973; Bogin, 1999). The genetic influence on growth is seen as a potential size or shape, which may or may not be attained. Environmental controls dictate the extent to which this potential is fulfilled.

Tooth growth

The mouth arises out of a complex of infoldings in the head region of the embryo (Ten Cate, 1985). It is lined with a layer of cells known as epithelium, overlying tissue called mesenchyme that will eventually become the muscle, cartilage and bone of the jaws. As the mouth of the embryo grows, an arch-shaped band of epithelium extends into the developing jaw tissues. This is the *dental lamina* and it follows the line that will eventually be taken by the dental arcade. On the edge of the lamina grow small swellings; proliferations of epithelial cells. These are the beginnings of *tooth germs*, within which the enamel and dentine will be laid down.

The earliest phase of tooth germ development (Figure 3.1) is called the *bud stage*. Mesenchyme cells proliferate around the bud of epithelial cells, to become the *dental papilla*, responsible for dentine and pulp formation. The epithelially derived cells are the start of the *enamel organ*, which will deposit enamel. As the bud-like enamel organ grows, it becomes indented on the side away from the dental lamina, and the tooth germ then passes into its *cap stage*. At the indentation, there is a transient group of cells, the enamel knot, which acts as a centre of control for the developing tooth germ. The edges of the indentation in the enamel organ continue to grow and the tooth germ finally enters its *bell stage*, during which hard tissue deposition actually starts. Whilst this has been going on, the tissues within the germ have become more and more differentiated. The epithelial cells of the enamel organ resolve into a number of layers. Infoldings in the organ start to follow the eventual form of the crown, with multiple indentations for cusps and ridges. Within the enclosing bell of the enamel organ is the dental papilla and the whole germ is itself enclosed in a layer of mesenchyme-derived cells called the *dental follicle*. Around the outside of the germ, bone is developing, eventually to become a *crypt*, containing the growing tooth within the jaw.

The first dental tissue to be laid down is dentine. Cells on the edge of the papilla differentiate into *odontoblasts*, with processes going towards the enamel organ. They lay down the first layer of dentine matrix, and tooth formation has begun. A short time later the epithelial cells lining the inside of the enamel organ (the *internal enamel epithelium*) differentiate into *ameloblasts* and deposit the first dome-shaped

Teeth and age

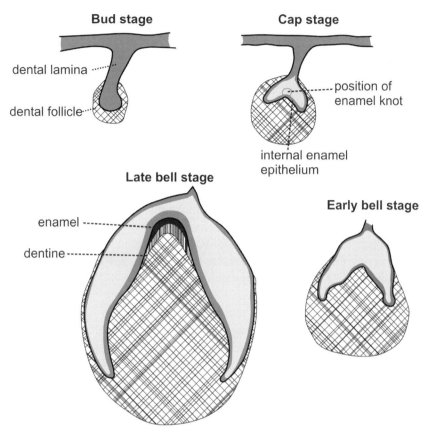

Figure 3.1 Development of tooth germs.

layer of enamel matrix. This process starts in the deepest infoldings of the enamel organ. The enamel formed here is the base for the main cusps and ridges of the crown. Cusps grow by the apposition of layer upon dome-shaped layer of enamel (Chapter 2). More ameloblasts come into action at the periphery of each layer, so the domes increase in size. Cusps merge into one another as infoldings interconnect. Where they are separated by deep folds of the enamel organ, ameloblasts continue to work until they are back to back, and then stop, leaving deep fissures. When the ameloblasts over the cusp tips come to the end of their matrix production phase, enamel is no longer deposited as dome- or cap-shaped layers, but as sleeve-like layers, overlapping down the sides of the crown towards the cervix of the tooth.

Dentine formation can also be visualised in layers – conical layers stacking one inside another. To form the pulp chamber the odontoblasts in the apex of the cone cease predentine formation. This forms the roof of the chamber and the odontoblasts remain inside to form the lining. Thereafter, the conical increments of dentine all lack their tip. The sides of the pulp chamber are formed as successive rings of odontoblasts on the inside of each layer cease predentine production. More odontoblasts become active around the apical edges of the conical layers, so that the dentine grows as a series of overlapping sleeves. In this way, the foundation for the enamel of the crown is built, together with the main part of the pulp chamber. After the crown is complete, the roots are formed by further conical layers of dentine. Where more than one root is being constructed, the band of active odontoblasts separates into several bands – one for each root – each laying down separate conical increments until the roots are complete.

Many forms of crown or root are produced by this mechanism. Bunodont molars are formed by the coalescence of several low cusps. The elongated canine of carnivores is simply a large single cusp in crown formation terms. To produce the hypsodont crowns of ungulates, the enamel organ is large and very deeply indented. The infundibulum represents a wide fissure between particularly tall cusps. Even the laminae of elephant molars can be explained in this way. The ever-growing incisors of rodents retain an active tooth germ at their base, in effect continually forming the side of an infinitely tall, one-sided cusp.

Age determination from tooth growth
Where modern comparative data are available, the state of development of crowns and roots can be used for age determination in archaeological material. For humans, many radiographic studies of living children have been carried out, and these are used as reference groups. Far fewer studies are available for non-human mammals. Only the mineralised tissue is likely to be preserved in archaeology, so it might seem logical to suggest that radiographs would be directly comparable. The difficulty is that radiographs record structures only when they are sufficiently mineralised to appear as a radio-opacity. Enamel and dentine are both initially deposited in a poorly mineralised form (Chapter 2), so it is to be expected that the age at which a particular stage of development is apparent in a radiograph is slightly delayed in comparison with a histological preparation of a dissected tooth. Depending upon the circumstances of preservation, archaeological specimens may also be missing recently formed enamel and dentine but it is difficult to argue that they are therefore closer to the radiographic image because of all the other factors that may be involved.

In studies of human children, girls reach most tooth development stages earlier than boys, with the exception of the last formed teeth, the third molars. This is in accord with somatic growth, in which girls are consistently in advance of boys until puberty, and then fall behind slightly. In addition, it is known that some human populations are in advance of others for dental development stages. This type of variation is less well established for other mammals, but occurs between different breeds of domestic animals for which there has often been deliberate selection for rate of development. Plane of nutrition, and disease in the young, are known to affect growth rate and the timing of developmental stages, but experiments with pigs have suggested that dental development is much less affected than the skeleton

(McCance *et al.*, 1961). In addition, many growth disruptions are recorded as defects in the enamel (Chapter 2) so it is possible to recognise teeth which may have been affected.

For these reasons, in archaeology there is always some uncertainty about the suitability of the reference group. It is difficult to devise ways to check this directly. Confidence would be increased if the age ranges were similar in different populations, breeds and conditions of management of living animals, and if the dental and skeletal development of each individual matched. Dental histology has been used to provide independent evidence of dental development timing in humans, apes, monkeys and in extinct hominoids (p. 166). This approach has been successful, and it is just starting to be applied in other mammals.

Initial eruption and establishment of the periodontium

Whilst tooth formation is taking place within the crypt, resorption and redeposition of bone around it allows the crypt to migrate through the growing jaw, to bring it into position in the dental arcade. Bone between the crypt and the alveolar crest is resorbed and the crown makes its way upwards. Eventually the crypt communicates with the bone surface, and the crown penetrates the mucosa of the mouth to emerge through the gingivae. Further tooth growth and bone remodelling bring the crown into the occlusal plane, while the soft tissues grow around it to form the gingival cuff and interdental papillae (Figure 1.1). This is the process of *eruption*. It is continuous. Even after the tooth arrives in the occlusal plane, constant remodelling and development are needed to keep it there. At no point can the tooth be truly said to have 'erupted'. What point in the process do you choose? Appearance of the crypt in the bone surface, gingival emergence, or arrival in the occlusal plane? All have been used as definitions, but all are a matter of degree, rather than fixed events.

In many mammals, there are two distinct eruption phases for the establishment of the deciduous and permanent dentitions. The roots of the deciduous teeth are narrow and widely spaced, in order to fit around the developing permanent tooth crypts. As the latter migrate up through the jaw, the deciduous roots (and often part of the crowns as well) are resorbed along with the bone. Root resorption is an important guide to the recognition of deciduous teeth, which often have a similar crown morphology to the following permanent teeth. Eventually, so much support is lost that the deciduous teeth become loose and are shed – a process called *exfoliation*. Details of the order of eruption and exfoliation vary with the parts of the deciduous and permanent dentitions which are present but, in general, the deciduous teeth erupt from anterior to posterior. Deciduous incisors are followed by canines and then by premolars. Size often increases with the order of eruption and the fourth deciduous premolar is usually almost as large as the permanent first molar, which erupts just to distal of it. The permanent second molar erupts behind the first, and the third behind the second – usually the last tooth to erupt. Permanent incisors often erupt around the same time as the permanent first molar, and the canines/premolars about the same time as second molars. There is, however, considerable variation in this general order.

While the tooth is growing, its attachment must also grow. This is a function of the dental follicle. Cells next to the recently formed crown surface or developing root dentine surface differentiate into cementoblasts which will lay down a cement covering (Chapter 2). Other cells differentiate into fibroblasts, responsible for producing collagen fibrils. Yet further cells of the follicle differentiate into osteoblasts and are involved in the deposition of alveolar bone. These are all parts of the *periodontium*. The collagen fibres become embedded in cement and bone as extrinsic (Chapter 2), or Sharpey's fibres. They form the periodontal ligament that binds the tooth into its socket. Little is known about changes in the periodontal ligament during eruption. Continuous remodelling must take place, with resorption and replacement of alveolar bone, ligament and cement. These processes must also continue when the tooth reaches the occlusal plane, as part of normal tissue turnover and to accommodate slight changes in tooth position. As teeth wear, continuous eruption is needed to maintain occlusion. In animals where roots have finished growth, cement apposition usually accompanies this process. Hypsodont teeth continue root (and often crown) formation for some time after crowns arrive in the occlusal plane.

Initial eruption and age estimation
Gingival emergence is the most convenient reference point for eruption in the living and is consequently the most used. It is not a precise marker, but it is still possible to demonstrate considerable variation in eruption. There is continuous variation in timing and order of eruption between sexes and in man between ethnic groups. Different breeds of domestic animals come out consistently as late or early developers. If population, sex or breed cannot be defined, then wide ranges have to be given for ages of emergence of individual teeth.

Gingival emergence is clearly an unsuitable marker for eruption to use in archaeology. It does represent the first point at which the cusp tips are exposed to wear (although this is slight until the tooth arrives in the occlusal plane), but much still has to be left to guesswork and the gingival emergence tables of veterinary texts cannot be used on their own to derive age estimates. Data need to be collected on different stages of the eruption process, using markers related to the occlusal plane and bony surfaces of the jaws. For archaeology, eruption needs to be seen as part of tooth formation and development as a whole, and it is possible to define a series of stages which can be applied to most individuals of one species, irrespective of their actual timing. The *stage* of development is usually directly comparable. It is the *age* attached to it that is more uncertain.

Dental age
Chronological age as such says less than might be expected about the biology of a growing individual. There is variation in rate of growth so that dental development,

skeletal maturation, size and shape can be slightly different in two individuals of similar chronological age. This has led, in growth studies of children, to the concept of developmental age. The development of, say, the skeleton is studied in wrist radiographs of a large sample of children and from this a standard sequence is set up (Cameron, 1998). Mean state of development can be expressed in terms of *skeletal age* – the age at which a given maturation score is on average achieved in the control sample. Skeletal age can then be determined for any child. It is only loosely related to real, chronological age, but says a lot about the process of maturation. In a similar way, standards have been set up for *dental age* in children (Demirjian *et al.*, 1973; Harris, 1998a). The dental age of any one child is an expression of the state of development and eruption of his or her dentition. Because teeth and the skeleton develop in different ways, dental age may differ from skeletal age. Similarly, neither may equal chronological age.

All this is primarily of medical value, but the concepts are also useful in archaeology and in animals other than man. Teeth and parts of the skeleton are usually the only evidence available. Ages determined from archaeological material of this type are therefore *always* dental ages or skeletal ages. What is seen is the biological development of the individual – not the chronological age. Because of the unknowns involved in archaeology, ages in years and months are often not particularly useful, so dental age is usually expressed in terms of scores for dental development. The ranges of eruption ages that are often quoted can then be seen clearly as suggestions for when a particular stage of development might occur in similar modern populations. In any case, age in months may be less important for an archaeological interpretation than the season of death, or the pattern of year groups culled during hunting so, where an animal has a strong seasonal pattern of reproduction and birth, it may be of more concern to relate the state of development to the sequence of seasonal changes.

The dentition after initial eruption
The forces of eruption and mechanical relationships between teeth interact to arrange the dentition into *occlusion*. This is the way in which teeth fit together. In normal occlusion, cusps, blades and surfaces interlock. As they work against each other, these surfaces become worn and the nature of occlusion changes. The height of the tooth is reduced, sometimes very rapidly, requiring equally rapid eruptive changes to keep the dentition in occlusion. Many mammals have persistently growing, tall-crowned teeth, and these continue to grow even after the main eruptive phase which brings the teeth into occlusion. For some, the roots become 'closed' with a completed apex that ends further growth. For other mammals, the crown or root continues to grow throughout life. In still others, the roots are completed soon after initial eruption. Once in occlusion, the teeth start to wear, and eruption continues slowly as the jaw remodels to keep the teeth in occlusion with their reduced height. This *continuous eruption* is an important factor in the dentition of mature individuals. As well as wear on external surfaces, changes occur within the tooth,

including secondary dentine deposition, mineralisation of root dentine tubules and cement apposition. These processes are age progressive and all have been used for estimation of age at death in individuals whose main phase of growth has been completed.

Tooth wear

Once a new tooth emerges from the gingivae, it starts to wear. Teeth are designed to be worn and may indeed be unable to function properly before they do so. The wear is caused by grinding of teeth against one another, and contact with food, cheeks and tongue. This may occur during chewing, but it also occurs at other times – even during sleep. The effects of wear are seen most on the *occlusal* surfaces, where teeth of upper and lower jaws meet. But, at the points where teeth in the same jaw come into contact, there is also *approximal* wear. Parts of the crown outside these areas are also worn to some extent, so that the microscopic features of the enamel surface are gradually obliterated (Scott *et al.*, 1949). As the teeth wear, adjustments must be made so that occlusion is maintained. Some rodents solve this problem with ever-growing, ever-erupting teeth. Other high-crowned animals, such as the ungulates, have roots which continue to grow for a long time. Elephants erupt their six cheek teeth one after another, over a period of up to 50 years.

It is usual to distinguish between attrition and abrasion. *Attrition* is the formation of well-defined wear facets where teeth meet, often with fine parallel scratches resulting from the abrasives in food. The direction of scratches is related to movements of the jaws during chewing (Rensberger, 1978) and the pattern may be related to diet. *Abrasion* is a more diffuse wear, with scratches randomly orientated. This occurs on tooth surfaces which do not come into contact with other teeth. Where attrition penetrates the enamel, the soft underlying dentine is worn more quickly, to produce a depression in the facet. This effect is part of the way in which hypsodont teeth function. The hard enamel stands up as sharp-edged ridges that cut against one another like shears. In elephants and rodents this is developed to a remarkable extent, with multiple enamel blades cutting against each other as the mandible moves back and forth.

Rate of wear must depend on many factors. One is the overall morphology of the crown: the disposition, height and depth of cusps and fissures; the area of the occlusal surface; the internal structure; and the thickness and microstructure of enamel. Developmental defects (Chapter 2) may lower resistance to wear. Structure of dentine and cement, and reactions such as secondary dentine inside the pulp chamber are also involved. Then there is the chewing mechanism itself, the nature of the diet, and non-feeding behaviours which bring the teeth together forcibly such as 'crib biting' in horses, or 'bruxism' in stressed humans. As attrition proceeds, the changing shape of the occlusal plane will have its own effect. Even the changing form of the occlusal facets, as different parts of the internal structure are exposed, will be involved. So, basic anatomical, physiological and behavioural differences between species ensure differences in rate of wear. A constant rate of wear cannot be assumed even within one species. The problem is exacerbated in domestic animals, where diet may be highly variable and there may be variation in dental morphology between different herds (Moran & O'Connor, 1994). In humans, there is the further factor of deliberate tooth mutilation and use of teeth as 'tools' in manufacturing processes and other activities (Milner & Larsen, 1991). All of this causes problems if wear is to be used for age determination, but it is also the basis on which diet and behaviour can be investigated.

Recording wear
Attrition is most often recorded from the pattern of dentine and cement exposed in the occlusal facet. In bunodont teeth, as the enamel at the cusp tips is lost, dots of yellower dentine are exposed. With continued attrition, these dots join up and finally the whole occlusal surface consists of a shallow, dished area of dentine, surrounded by a raised rim of enamel. In the hypsodont crowns of herbivores, attrition exposes complex loops and pleats of enamel which change in pattern as attrition exposes deeper layers of tooth structure. Schemes for recording occlusal attrition in this way are given below. Any such scheme assumes that the pattern and depth of infoldings, infundibulums, fissures and pits, and height of cusps and ridges, are relatively constant throughout a collection of teeth. Internal crown morphology varies within species (Chapter 4), so there are some hidden uncertainties in such methods.

For the common domestic species, and humans, charts have been developed to summarise the expected progress of attrition in the dentition. Each jaw is matched against the pattern of the occlusal attrition facets in these charts, and assigned a score. For some deer, a refined recording technique has been developed (Brown & Chapman, 1990, 1991a). This involves a score table (below). Points are awarded for the exposure of dentine in different places on each cusp of each molar and premolar, and the points are summed to provide a wear score for each tooth. These can then be summed for the whole cheek tooth row, to provide a total wear score, which shows a clear relationship with age. This approach could be applied to any bovid or cervid dentition, and similar score tables could be devised for other mammal groups. Another method is to measure the heights of crowns, from the cervix to the occlusal attrition facet (usually on the buccal side). Increasing occlusal attrition can be followed by the lowering of the occlusal surface (below).

Another possibility for measurement is to record the occlusal attrition facet photographically – nowadays with a digital camera (a scanner would also be excellent if the occlusal surface can be laid flat on the glass). Image analysis programs can be used to measure, in the image, the area of dentine exposed. This can then be expressed as a proportion of the total occlusal area (including the enamel as well). In theory, a threshold can be imposed upon the tones in the image, to divide the dentine from the enamel but this is difficult in practice, so it is usually easiest just to draw around the outlines with a series of mouse-clicks. This approach has already been used with some success in human dentitions, and there is no reason why it should not be equally successful with other mammals.

Determining age from wear
If rate of wear in one homogeneous population is reasonably constant, then extent of wear is a function of age. Even if the rate itself is unknown in absolute terms, the age at death of individuals within the population could be compared on the basis of the stage of wear reached at death. Crown and root development of different teeth can be used to calibrate the sequence of wear stages, until the eruption of the last tooth in the dentition (often the third molar). This results in a combined set of dental development–eruption–wear stages. The validity of this procedure depends on three main factors:

- variation in wear rates within the population
- the way in which wear rate changes through life
- variation in the age at which teeth erupt and come into function.

Wear tends to be faster in younger individuals and the greatly diminished rate in older age classes may make them difficult to differentiate (Spinage, 1973). Extent of wear also becomes much more variable in older age groups (Brown & Chapman, 1990, 1991a), so there is always greater uncertainty in ageing more elderly individuals. Standard sequences of eruption and wear stages have been determined for most of the common mammals on archaeological sites (below). A mandible, for example, with a partially preserved dentition quadrant can then be assigned to a stage in this series. These dental development stages are essentially 'dental ages' (above) and so record state of maturity rather than chronological age.

It is possible to assign true ages to such dental development stages by fitting comparative data on dental development and attrition from living populations. Some domestic mammals have the range of variation well established. This procedure does, however, add another assumption – that the living animals of today represent dental development in the archaeological material. Even when comparative groups are carefully chosen for similarity of skeletal structure to ancient forms, direct comparability is not guaranteed. In practice, archaeology has to live with such assumptions. It is mainly important to realise when they are being made. The wide limit given to most attrition-based age estimates reflects the uncertainty involved. Where actual ages are applied to archaeological material, it is important to see them as 'suggested ages' (Payne, 1973) rather than absolute ones.

If the relative ages of eruption for several teeth are known, wear rate in the earlier erupting teeth can be estimated from the stage of wear reached at the eruption of the later teeth. From this, an estimate of the *functional age* (the time it has been in occlusal function) for that tooth can be made. If it can be assumed that other teeth were worn at a similar rate, it is possible to calibrate later wear stages, but the assumptions of this method are difficult to test.

Crown height methods

Attrition has most frequently been estimated by visual assessment of the patterns of cement, enamel and dentine exposed, but an alternative approach is to monitor it by the decreasing crown height. One of the earliest applications of this idea was Spinage's (1972) study of attrition in zebra. Crown height of first molars decreased regularly with age, and an exponential curve line, fitted to the scatter of points (wear rate was faster in younger animals and slower in older), could be used to estimate age in unknown specimens. Equids have very tall cheek tooth crowns which do not change greatly in morphology through much of their height, and have rapid wear. Crown height is often the only practical alternative in archaeological material. Levine (1982, 1983) therefore extended Spinage's method to Upper Palaeolithic horses, using a small group of known-age New Forest ponies which matched well in size. Once again, an exponential line fit was used to produce an age estimation table. This has also been applied to Middle Palaeolithic horses (Fernandez & Legendre, 2003).

Klein (1978) developed techniques for constructing age distributions of fossil bovids from the early Upper Pleistocene sites in South Africa. Crown height was measured from the neck of the tooth to the occlusal surface, for the most mesial cusps on the buccal side for lower teeth, and lingual side for upper. Rate of crown height reduction was established, for some species on modern material of known age, and in the rest by inference. The starting height for unworn crowns was measured in the combined sample, and maximum values were used as a baseline. Worn heights were subtracted from this and, using known age specimens, a rate of crown height reduction was calculated for each tooth. In each species, crown height distributions were divided into ten age classes, each representing the crown height lost during 10% of the animal's life. This allowed age distributions for sites and species to be compared.

A further development was to fit a line to the scatter of points in a plot of age against crown height. Klein and colleagues (Klein, 1981; Klein *et al.*, 1983; Klein & Cruz-Uribe, 1984) converted the Spinage formula to make a standard equation allowing age to be calculated for any species with hypsodont permanent teeth:

$$AGE = AGE_{PEL} - 2(AGE_{PEL} - AGE_E)(CH/CH_0) + (AGE_{PEL} - AGE_E)(CH^2/CH_0^2)$$

where AGE_E is age of eruption for the tooth in question, AGE_{PEL} is the potential ecological longevity (i.e. theoretical maximum age attained by that species), CH_0 is the original unworn crown height, and CH is the crown height as measured at death.

This is known as the Quadratic Crown Height Method (QCHM). Although originally developed for fully hypsodont teeth, it has been used for the 'semi-hypsodont' teeth of red deer *Cervus elaphus* (Klein *et al.*, 1981). In a test on bison teeth, Gifford-Gonzalez (1992) found that it tended to underage young individuals and overage older individuals. Pike-Tay and colleagues (2000) tested the QCHM in a large group of caribou *Rangifer* jaws. The method assumes that initial, unworn crown height is equivalent for all individuals in the series, and that worn crown height does not vary greatly within any age cohort in the series. The study, however, found that height varied by as much as 10 mm, which, translated by the QCHM into age, would represent over 100 months of difference. One component might be variation in unworn crown size, so Pike-Tay *et al.* followed the suggestion of Ducos (1968) that crown diameters (mesiodistal and buccolingual, taken both at the cervix and occlusal surface; Chapter 4) might be used to standardise the crown heights by calculating an index: height/breadth of crown. Gifford-Gonzalez (1992) had already found little evidence for a relationship between crown breadth and crown height in bison, and Pike-Tay *et al.* found that calculation of an index did little to improve age estimates in their caribou sample. A tall crown was just as likely to be present on a small tooth as on a large tooth. They therefore used unaltered crown heights and, overall, found that the QCHM formula produced age at death distributions that did not match the known ages well. The regression line fitted to the real distribution of caribou age versus crown height did not match the one predicted by the QCHM. It might therefore be better to use the real regression line for age estimation, but Pike-Tay *et al.* argued that a more generalised formula would allow adjustments to be made for different populations. They replaced AGE_{PEL} with AGE_{MAX} (the average maximum life span, or the age reached by individuals reaching maturity and dying of natural causes). This produced a better fit.

Similar measurement methods have been used in a variety of mammals (below). Crown height techniques suffer the same problems as any other attrition-based age determination methods. 'The use of measurements is only pseudo-objective, it may remove observer bias, but it can not correct for varying rates of wear. If wear is accentuated for age, the measurements will give the same false information as visual appraisal' (Spinage, 1973).

Patterns of dental wear
Unless age at death is independently known, it is seldom possible to establish attrition rate confidently. Instead, it may be better to examine the pattern of wear within the dentition. That is, the relative wear of different teeth. In general, the earliest erupting teeth are the most worn within any one dentition and the latest erupting are the least worn, with others arranged in series between these two extremes. For the permanent teeth, the first molars are usually the first to erupt and there is a gradient from this tooth to the second molars and the third molars. The nature of the gradient can be seen by plotting first molar attrition scores against second molar, and so on. Similarly, attrition score in any tooth can be standardised by expressing it as a proportion of the first molar score – in effect an attrition index. This approach has been followed in studies of human dentition (Murphy, 1959a, 1959b; Molnar, 1971; Lunt, 1978; Hinton, 1981; Hinton, 1982), showing that attrition gradients

vary between populations in a way that suggests they reflect differences in diet and the use of teeth as tools.

Microwear

Abrasion and attrition are caused by repeated pressure between opposing teeth, and abrasive particles in the mouth. Such abrasives may be from the diet. Clear candidates in grazing herbivores are the silicate phytoliths, which form part of plant skeletons. Other plant-borne abrasives include cellulose and lignin. The polishing properties of either material can be seen in the blunting of scissors by paper. Meat has fewer abrasive particles, but collagen in tendons, skin and blood vessel walls may be involved. The chitin of insect exoskeletons, and any bones chewed along with meat, may have their own abrasive effect. Abrasive particles may be added to the mouth from atmospheric dust, or soil adhering to the food, particularly if animals forage near the ground. Leek (1972) found mineral particles, apparently dislodged from millstones, in ancient Egyptian bread, along with small quantities of wind-blown sand. In modern human populations there is the additional effect of tooth brushing. Many toothpastes contain abrasives which, when applied vigorously to the crown surface, remove not only plaque and calculus (Chapter 5) but also enamel.

The processes by which dental tissue is lost during the course of wear are not well known, but a great deal is known about the wear of ceramic bearing surfaces in the branch of engineering known as tribology. Loss of material from a bearing surface may occur in three ways: sliding wear, abrasion and erosion. *Sliding wear* can occur in the absence of any intervening abrasive particles. It is due to microscopic high points (asperities) on the surface, which catch against each other as one bearing surface slides against another. The asperities deform elastically – that is, they spring back again after catching an opposing asperity – and repeated deformation causes cracks to propagate, particularly along crystal boundaries. *Abrasive wear*, in a tribological sense, is caused by hard particles (i.e. abrasives) between the bearing surfaces. The load is concentrated on a tiny point, so that the force applied is magnified so greatly that it causes even a tough crystalline material like enamel to deform plastically (it does not bounce back after the load is released). This deformation affects an area only a little wider than the abrasive particle at its centre, so it is seen as microscopic scratches which track the particle as it is dragged along. *Erosive wear* also involves hard particles but, this time, they are carried in a fluid and impact on the surface, to remove material either by plastic deformation or by fractures following repeated elastic deformation. In dental enamel, these processes together produce a pattern of scratches and pits, ranging from 1 μm to 30 μm or so in width. A worn enamel surface often develops a surface relief which follows the arrangement of the long thin crystallites exposed (Chapter 2). Those that are nearer parallel to the surface are removed more rapidly than those at a larger angle, so a depression is produced (Bell *et al.*, 1991). This is one of the problems in specimen preparation for microscopy.

Attrition-type wear produces polished facets with a clear margin, but dental wear also occurs outside well-defined facets. This is distinguished in dental anthropology as abrasion (above), although this is slightly unfortunate because it might potentially involve all three wear mechanisms outlined above. Wear away from attrition facets can be monitored by the progressive obliteration of enamel surface features (Chapter 2). This occurs gradually, with age, and may indeed be used as an approximate age estimating method (Scott & Wyckoff, 1949a, 1949b). The approximal crown surfaces retain their original form longest.

The fine structure of worn surfaces, or *microwear*, can be studied using optical or electron microscopy. Most studies use SEM-ET examination of replicas prepared from dental impressions (Chapter 2). One of the difficulties of SEM for examining three-dimensional structures is that their visibility can depend strongly upon the rotation of the specimen relative to the detector. Gordon (1988) found substantial differences in counts of pits and scratches for the same specimen rotated into two positions. It would be better to use techniques that map the microscopic surface in three dimensions, such as (Chapter 2) confocal microscopy (Ungar *et al.*, 2003) or optical interferometry (Walker & Hagen, 1994).

Microwear has been studied almost entirely in enamel, particularly on the surface of attrition facets. Standard SEM images are made, with a fixed magnification in a carefully selected position which is thought to be homologous on all the teeth studied. Software is now available in which the longest and shortest axes of the features can be marked by mouse-clicks, and measurements calibrated from the settings of the SEM ('Microware 4.01', http://comp.uark.edu/~pungar/index.html). Different criteria can be used to distinguish between scratches and pits, although a commonly used ratio is 4:1 longest to shortest axis of the feature. It is also possible simply to present the length (longest axis) and width (shortest axis) measurements directly, or as a proportion of length to width. Microwear feature density is determined as the count of features in each standard view. Experiments in people and other primates have shown that the pattern of scratches and pits at any one location on a wear facet changes rapidly, over a timescale of days or weeks (Teaford & Oyen, 1989a, 1989b; Teaford & Tylenda, 1991; Teaford & Lytle, 1996). This is the 'Last Supper' phenomenon (Grine, 1986). Abrupt changes in diet cause corresponding changes in the microwear features, so the technique does seem to yield information about the past few days of diet, rather than an accumulation of wear over months or years. Teaford & Walker (1983) found that microwear on the teeth of stillborn guinea pigs (some teeth erupt *in utero*) was different from that of mature guinea pigs, implying that it was abrasive wear rather than sliding wear (above) that caused the usual microwear features. Presumably the abrasive particles derive mainly from the diet, with soil particles or other dirt that comes with it, although it is possible that some also come from dust. The residence time of abrasive particles in the mouth is not clear, and it is possible that relatively few particles could cause scratches on many surfaces. Wear is likely to take place between meals as well as during them, although herbivores tend to feed for much of their waking hours. In humans, far greater forces are exerted between teeth when people are worried, or grind their teeth in their sleep (bruxism), than during eating. In addition, people often use their teeth as tools, or a third hand, and this must add its own patterns of wear.

Studies of microwear in non-Primates have concentrated on the difference between browsing and grazing ruminants (Solounias & Hayek, 1993). Grazers are able to metabolise more fibrous fodder, such as grass. Browsers are not able to tolerate such large amounts of fibre and carefully select more fleshy leaves, usually higher off the ground. Other ruminants are intermediate in their requirements. Giraffe are prime examples of browsers, with a long neck to prove it, and enamel wear facets on their molars show far fewer scratches, relative to pits, than wildebeest, which are grazers. Grant's gazelle are intermediate feeders and fit in between, although they are difficult to distinguish from the others (Solounias *et al.*, 1988). These differences between grazers and browsers have usually been attributed to the greater abundance of phytoliths in grass (Walker *et al.*, 1978). Domestic sheep are also intermediate feeders, and they vary in microwear pattern. Mainland (2003) found clear differences in a flock of Gotland lambs kept at Lejre, in Denmark. Half had been kept in woodland, and half on grassland. The grassland lambs had significantly more microwear scratches relative to pits than the woodland lambs. These results compared well with upland grazing sheep from Orkney, but not with similar animals from the Scottish borders and Greenland, which showed lower densities of scratches. An examination of the sheep dung from the Lejre flock showed no differences in the phytolith content between the woodland and grassland animals, but significant differences in grit from the soil. Mainland (2003) suggested that the level of feeding was the main cause of the microwear features, as it is ground-feeding animals that usually ingest soil along with their food.

Another focus of microwear studies has been on living and extinct Primates, with the aim of reconstructing feeding behaviour. Most living primates eat a variety of different plant and, to some extent, animal foods. Some, however, are primarily fruit feeders (frugivores), and some leaf feeders (folivores). Fruits are often encased in hard shells, or contain phytoliths. Folivore molar attrition facets have a substantially higher ratio of scratches to pits than frugivores, and the 'hard-object specialists' have relatively more pits than the other frugivores (Teaford & Walker, 1984; Teaford, 1988; Rose & Ungar, 1998). There are differences between members of the same species, living in different environments, and within the same environment in different seasons, showing that small changes in diet may have a recognisable effect. This complicates interpretation, but the variation is relatively small compared with overall differences between folivores, frugivores and hard-object feeders. Microwear has also been investigated on the sides of molars, away from attrition facets, where features are present most often for frugivores eating tough fruits containing phytoliths, and for Primates that feed near the ground, and therefore ingest more grit from the soil (Ungar & Teaford, 1996). Incisor microwear has been investigated both on the attrition facets, and the sides of the crown just below the facet (Ungar,

1994). Density of pits seems to show little variation, but density of striations does relate to use of incisors for processing tough foods, particularly the lower incisors which move more than the upper incisors, which act as anvils.

Such studies on living and recent primates act as a reference series against which to compare extinct fossil Primates (Rose & Ungar, 1998; Teaford & Ungar, 2000). Some fossil genera have high microwear scratch densities, including *Micropithecus*, *Rangwapithecus* and *Oreopithecus*, which suggests they were primarily folivorous. High pit densities suggest that *Griphopithecus* and *Ouranopithecus* were hard-object frugivores and intermediate patterns suggest that the majority of genera including *Gigantopithecus*, *Dendropithecus*, *Proconsul*, *Dryopithecus* and *Sivapithecus* were soft-fruit feeders. The robust australopithecines (*Paranthropus*) have higher microwear pit densities and shorter, wider and more randomly orientated scratches (Grine, 1986, 1987) on their molar attrition facets than gracile australopithecines (*Australopithecus*). *Paranthropus* fits best with the hard-object feeders. *Australopithecus* molar microwear fits best with the soft-fruit feeders, but their incisor microwear suggests they took a greater variety of foods, including hard objects, than *Paranthropus* (Ungar & Grine, 1991; Ungar, 1998).

Microwear has been used to investigate behavioural and dietary transitions in more recent human remains. At Tell abu Hureyra in Syria, Mesolithic and early Neolithic teeth had larger microwear pits than later Neolithic teeth (Molleson *et al.*, 1993). By contrast, agricultural intensification between the seventh and third millennia BC in the Indus Valley was marked by an increase in coarseness of molar attrition facet microwear features (Pastor, 1994). Ungar & Spencer (1999; Spencer & Ungar, 2000) took a different approach, studying incisor wear to the labial side of the attrition facet, rather than actually on it. They compared three groups. Recent Aleutian islanders (after AD 1700), who existed almost entirely on fish, molluscs and marine mammals, also made great use of their teeth as a third hand. Late Woodland (AD 600–900) native Americans from Illinois included plant foods in their diet, with a variety of meat and shellfish. Arikara (AD 1600–1700) from the American Plains were bison hunters, although they gathered a variety of plant foods and also hunted smaller animals. There were consistent differences in microwear, with lowest densities of features and shorter, wider scratches in the Aleuts, and highest densities with longer and narrower scratches in the Woodland group. The Arikara fitted in between. Schmidt (2001) compared molar microwear between Early/Middle Woodland (1500–2000 BP) and Late Archaic (4000–4500 BP) people from Indiana. Late Archaic diet seems to have included a larger proportion of tubers and wetland plant foods, and Woodland more nuts and starch or oil rich seeds. The Woodland group molar attrition facets showed more (and larger) pits relative to scratches (which were also shorter) than the Archaic group. Preliminary work by Teaford (1991) found, in remains from the Georgia and Florida coast, that indigenous people before European contact had more microwear pits and larger scratches on molar attrition facets than postcontact people. In a larger study, however, Teaford *et al.* (2001) discovered that differences between inland and coastal groups were greater than differences between early and later periods in the precontact group, or between precontact and postcontact.

Age estimation from dental development, eruption and wear in different orders of mammals

Most common ageing methods do not separate dental development and attrition, but use a score combining the two. Eruption is difficult to use on its own, although by far the majority of data collected from living animals concern gingival emergence. As observed above, it is only the alveolar emergence that can be seen in archaeology, but gingival emergence must mark the first wear on the cusp tips and so is related to the sequence of attrition.

It is important to remember the difference between a dental age, based on the development of a reference population, and a chronological age (above). Males differ from females in their development schedules and there are differences between human populations but, for domestic animals, variation between breeds may make larger differences. Breeding is often based upon selection specifically for rapid development, so a dental development table based on living domestic animals may be particularly unsuitable for archaeological specimens. Most recent work in archaeology has therefore confined itself to dental age, based on schemes for recording and comparing eruption and wear as a system of scores, without assigning chronological ages.

For accurate assignment of scores, it is first necessary to identify teeth correctly for both the permanent and deciduous dentitions. This can be difficult, particularly in worn teeth. In some animals, notably horse, the deciduous tooth crowns may be similar in size to their permanent successors and there are considerable possibilities for confusion. X-rays are helpful in these situations and are essential for a detailed study. Development scoring systems, for use on X-rays, are available for several mammals.

Sometimes age estimates form the basis of life tables like those employed in the demography of living people (Angel, 1969; Buikstra & Mielke, 1985). There is a debate about the validity of this (Bocquet-Appel & Masset, 1982; van Gerven & Armelagos, 1983; Buikstra & Konigsberg, 1985; Konigsberg & Frankenberg, 1994, 2002). In addition to the uncertainties of ageing methods, there are problems inherent in the nature of archaeological evidence itself. The central assumption is that the accumulation of human remains recovered from an archaeological site represents some once-living population. Unfortunately, archaeological age at death distributions rarely resemble those of any recent recorded population. For one thing, not everybody was necessarily buried in the same place. In many cultures, young children were buried in other locations where they are less readily recovered. Different socio-economic groups may be buried in different parts of a cemetery. Men may be separated from women. It is also a mistake to assume that, even within these divisions, individuals represent a random selection of that cohort in the population. Many biases operate, from burial in the site, to preservation, recovery and storage.

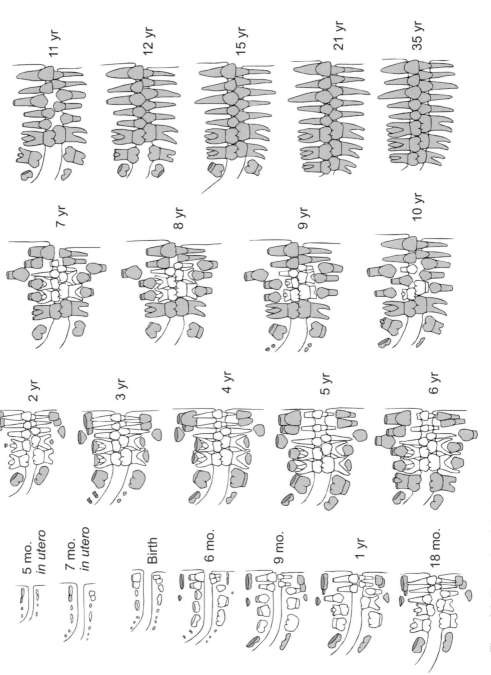

Figure 3.2 Human dental development in one year stages. Traced from Schour & Massler's 1944 chart as reproduced in Brescia (1961). Ages for stages are given in Table 3.1.

In any case, a death assemblage is a special subset of any population, because it contains those individuals that died over a given period. Every creature dies for a reason. In the case of humans, it is disease or injury that brought them into the death assemblage and this group of people is likely to include a particularly large number of those who suffered illness or deprivation in childhood. Growth rate is affected by such factors and it is likely that modern reference studies are not good standards with which to compare ancient populations. Most assemblages of other large mammals on archaeological sites represent either hunting or the culling of domestic herds. Both activities involve strong selection for particular age groups and, because the majority of large mammals reproduce in an annual cycle, the age at death distribution will be greatly affected by the season of site occupation.

Primates
Apes and monkeys
Dental eruption ages have been collected for many Primates (Smith *et al.*, 1994). The length of different dental development stages is only moderately correlated with body size, but strongly correlated with brain size of different species. Detailed dental development chronologies have been established for the apes (Beynon *et al.*, 1991; Anemone, 1995; Anemone *et al.*, 1996; Kuykendall, 1996; Winkler *et al.*, 1996) because of their relevance to debates on the development of fossil hominids (p. 167).

Humans
Dental development and eruption are both well known for living humans, although attrition rate has been so much slower over the past 200 years than in the rest of human history that few comparisons of wear are possible. Three main schemes of dental development and eruption are in common use. The oldest is the chart of Schour & Massler (1941, 1944), which divides development into annual stages, with extra stages at 6, 9 and 18 months (Figure 3.2; Table 3.1). It was originally intended just as a guide for dentists, but it has seen wide use for age estimation in archaeology. To assign a dental age, the development of crowns and roots, and state of eruption, in the specimen are matched against the diagrams in the chart. It is rare to find a perfect match, but the closest stage is chosen. Where tests have been carried out with children of independently known age, the Schour & Massler chart performs well (Miles, 1958; Gray & Lamons, 1959; Liversidge, 1994), particularly in early childhood. The origins of the age ranges given in the chart are slightly mysterious (Smith, 1991b), although they probably lie in the studies of Logan & Kronfeld (1933), who made histological sections of the jaws of a relatively small number of children. They had died of a variety of diseases and, for archaeological assemblages which also represent sick children, could therefore be considered a better standard than X-rays of healthy children. They also represent direct observations of the developing teeth, whereas X-rays only show those parts of the developing tooth that are well enough mineralised to cause a radio-opacity. Ubelaker (1978) revised the

Table 3.1 *Age ranges for stages of human tooth development*

Stage	Schour & Massler[a] range (±months)	Ubelaker[b] range (±months)
5 months *in utero*	–	2
7 months *in utero*	–	2
Birth	–	2
6 months	2	3
9 months	2	3
1 year	3	4
18 months	3	6
2 years	6	8
3 years	6	12
4 years	9	12
5 years	9	16
6 years	9	24
7 years	9	24
8 years	9	24
9 years	9	24
10 years	9	30
11 years	9	30
12 years	6	30
15 years	6	36
21 years	–	–
35 years	–	–

Notes: [a] Schour & Massler (1941, 1944),
[b] Ubelaker (1989). See Figure 3.2.

Schour & Massler chart, particularly by increasing the age ranges quoted for each stage. This new chart is also widely used, and is a recognised standard (Ferembach *et al.*, 1980; Buikstra & Ubelaker, 1994). In practice, it makes little difference which chart is used, because it is always necessary to keep in mind the distinction between developmental dental age and true chronological age. The chart really allows the archaeologist to assign a specimen to a particular stage of development, and any inference drawn about chronological age of the individual is hypothetical.

Gustafson & Koch (1974) published an alternative graph-based method, which is regarded as a standard in forensic work. Start of crown formation, crown completion, root completion and eruption are summarised in lines which represent each tooth. A ruler is laid across the graph, moving it around until a best fit is achieved for the developmental state of each tooth in the specimen. An age is then read off from the scale at the side. Once again, this works reasonably well when tested against jaws of children whose age at death is independently known (Crossner & Mansfield, 1983; Hägg & Matsson, 1985; Liversidge, 1994). It yields ages similar to the Schour & Massler chart (it was derived from similar data) but, because of the small scale, the Gustafson & Koch graph distinguishes less sharply between stages.

When recording archaeological specimens, it is convenient to note down the state of development of each tooth using a set of codes which were originally developed for scoring X-rays in growth studies (Hunt & Gleiser, 1955; Moorrees *et al.*, 1963). Their results were only presented as graphs, which are difficult to use for age estimation. Smith (1991b) scaled mean ages for each stage in each tooth, from the large growth study of Moorrees, Fanning & Hunt (Moorrees *et al.*, 1963). An overall mean age is calculated from the ages derived for each tooth. Studies of independently aged children (Liversidge, 1994) show that the method yields reasonable results, but they are not as good as for the Schour & Massler chart. Harris & Buck (2002) rescaled the means and standard deviations for enlarged versions of the Moorrees, Fanning & Hunt graphs.

Third molars are a problem in age estimation, because they vary so much in development timing. For this reason, they are not included in the Gustafson & Koch graph. Unfortunately, they are the last teeth in the dentition to develop, and cover an important stage in the late teens and early twenties. The largest study of third molar development is that of Mincer *et al.* (1993). They devised a set of grades for third molar formation, scored them on thousands of radiographs, and derived formulae for estimating age in people under 25 years (which is particularly useful for forensic purposes).

Where teeth are loose and can be measured directly, or in an undistorted X-ray, most researchers find it easier to take a measurement of the size of the developing tooth, rather than worry over which score to assign. Care is needed in taking the measurement, but different observers are usually quite consistent in their results. Regression formulae have been developed (Liversidge & Molleson, 1999) using the teeth of 76 known-age children from Christ Church, Spitalfields in London (Table 3.2). They can be used to estimate age from tooth measurements, although the confidence intervals are still quite large.

Providing the rate of attrition is the same throughout a population, the extent of attrition can be used to estimate age at death in remains of adults. Attrition rate is so slow in recent people that this method cannot be used. In Medieval and earlier material, the attrition rate is fast enough, but there are no individuals whose age is independently known, so it is difficult to calibrate the method. Attrition is usually recorded using the system of Smith (1984), which assigns a score based upon the pattern of dentine exposed. Age is estimated from the three molars. Brothwell (1963b, 1989) derived a method based on British skeletal material from Neolithic to Medieval contexts, in which the pattern of wear was divided into ten-year stages. It was calibrated largely by ages estimated from the skeleton, such as from the pubic symphysis and, where tested by similar means in non-British material, similar alternative ages match well with the attrition ages (Hillson, 1979). Brothwell's system is simple and quick, and is the most widely used.

Miles (1958, 1962, 1963a, 1978) developed a method in which the rate and pattern of attrition in molars was calibrated against eruption. Working with Anglo-Saxon skulls from Breedon-on-the-Hill, Leicestershire, England, he selected a baseline

Table 3.2 *Formulae for calculating age from measurements of developing human permanent teeth*

	3rd molar	2nd molar	1st molar	4th premolar	3rd premolar	Canine	Upper 2nd incisor	Lower 2nd incisor	1st incisor
b0	8.177 500	0.119 800	0.125 800	2.232 600	1.614 000	0.064 400	−0.448 600	1.601 600	1.062 700
b1	0.666 600	1.604 900	−0.199 200	0.560 400	0.535 500	0.253 000	0.652 000	−0.869 700	−0.565 400
b2	0.000 000	−0.114 100	0.129 700	0.000 000	0.000 000	−0.006 100	−0.008 000	0.224 900	0.151 800
b3	0.000 000	0.003 410	−0.008 320	0.000 000	0.000 000	0.009 620	0.000 000	−0.012 850	−0.007 650
b4	0.000 000	0.000 000	0.000 170	0.000 000	0.000 000	−0.000 724	0.000 000	0.000 233	0.000 120
b5	0.000 000	0.000 000	0.000 000	0.000 000	0.000 000	0.000 015	0.000 000	0.000 000	0.000 000
s	3.0	1.5	1.3	2.0	1.8	1.4	1.6	1.1	1.2
age range	11.6–22	3.4–15	0.1–9	3–14	2.5–13	0.4–13	1.1–10	0.6–9	0.5–9

Source: From Liversidge & Molleson (1999).
Note: Age is calculated as follows:
$y = b0 + b1.x + b2.x^2 + b3.x^3 + b4.x^4 + b5.x^5$
where y = age in years, x = tooth height in mm.
95% confidence intervals are determined as follows:
Lower limit = $y - s$. Upper limit = $y + s$.
where s = square root of residual mean square, multiplied by 2.03.
Teeth are measured with one jaw of the calipers flat along the developing edge of the crown or root, and the other jaw at the highest cusp tip.

group of dentitions in which at least one of the permanent molars was still erupting. This was used to estimate the number of years taken to reach different attrition stages in the first and second molars relative to the third molar. So, for example, if the third molar is assumed to erupt at 18 years, and the first molar at 6 years, there would be 12 years of wear on the first molar. Within this baseline group, Miles estimated that there was a wear gradient of 6:6.5:7 (m1:m2:m3) so that the wear state reached by 12 years of first molar wear represented 14 years in the third molar. At this stage, the individual would be $18 + 14 = 32$ years old. The attrition state of the first molar in an individual at this third molar stage would therefore represent 26 years of first molar wear, which could then in turn be applied to the third molar. In this way, the initial calibration in the baseline group could be extended. There are clearly several untestable assumptions, and Miles himself (2001) later highlighted a number of difficulties. The method, however, has the advantage that it is calibrated by internal data from the group under study, rather than external standards, and it has yielded good results in several very different contexts (Nowell, 1978; Kieser *et al.*, 1983). The basic concept of seriating eruption and attrition for the assemblage being studied is a fundamentally useful one, and is applied to several different mammals (below).

Another possibility for recording attrition in humans is to measure crown height (Mays *et al.*, 1995). This has, however, shown only a moderate relationship with independently known age in a nineteenth-century archaeological collection from the Netherlands (Mays, 2002) but, even amongst people 60 years or more of age, there was relatively modest wear. This has also been a problem in a study of crown height and wear in similar material from St Bride's Church, in London (Walker, 1996). The relationship may be stronger in more heavily worn Medieval and older material, but there are then difficulties of age calibration, in the standardisation of measurements, and in the assumption of comparable initial unworn crown height (Walker *et al.*, 1991). It might be expected that these difficulties would be less significant for herbivorous mammals with very tall crowns and extremely rapid wear.

Artiodactyla
Sheep and goats
A series of sheep jaws at different stages of development and wear is shown in Figure 3.3. Sheep deciduous premolar crowns are mostly complete before birth and are in eruption, with roots forming, at birth (Weinreb & Sharav, 1964). The permanent first molar crown is about half formed at birth, and two thirds complete when it starts to erupt at three months. The crown is complete by nine months and is fully in occlusion, its high sides acting as temporary 'roots'. Growth of the true roots then starts, continuing until about four years of age. The second permanent molar starts crown formation soon after birth and continues to one year of age – about the time the tooth erupts – when root growth starts, and continues until five or six years. Permanent third molar and all premolar crown formation does not even

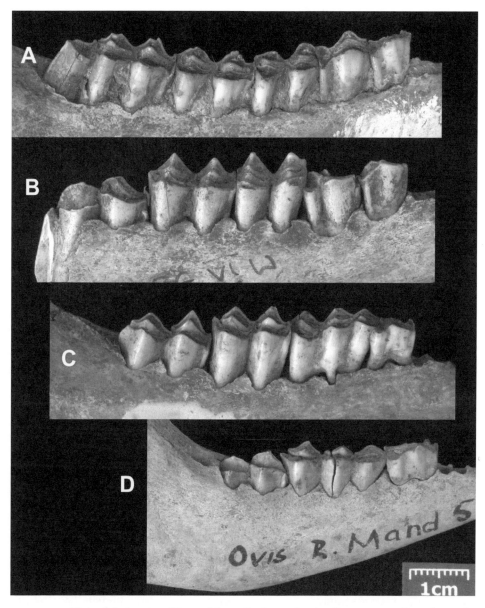

Figure 3.3 A series of jaws of domestic sheep, showing the sequence of eruption and wear. A: full set of permanent cheek teeth, with the third molar well worn (this is the oldest animal in the series). B: permanent cheek teeth almost completed, with third molar in eruption. C: mixed dentition, with deciduous third and fourth premolars still present, permanent first molar fully erupted and in wear, and second molar still erupting – but also in wear. D: deciduous dentition, with deciduous third and fourth premolars fully erupted and in wear, and permanent first molar in early stages of eruption (this is the youngest animal in the series).

begin until one year of age. It is completed by around two years of age, when these teeth erupt. Premolar roots are completed by four years, and third molars by five to six years. Incisors are of little use for archaeological purposes, because they usually fall out of the jaw, although they are often used by veterinarians because they are easier to see in live animals (Moran & O'Connor, 1994).

In the first edition of this book (Hillson, 1986), tables of eruption were given for both goat and sheep. These are no longer given, because it is apparent that there is considerable variation, and the reader needs to consult the original sources. Studies include Silver (1969), Noddle (1974), Bullock & Rackham (1982), Deniz & Payne (1982, 1979) and Moran & O'Connor (1994). The most reliably established events (Moran & O'Connor, 1994) appear to be eruption of the permanent first molar (one that is in wear is probably from an animal older than three to four months), the second molar (if in wear, the animal was probably older than 12 months) and third molar (if in wear, older than 24 months). There is some variation between sheep and goat, but this seems relatively small (Deniz & Payne, 1979). In addition, there is variation in eruption timing between living sheep breeds (notably in Soays, see Moran & O'Connor, 1994), and between male and female goats (Deniz & Payne, 1982). Castrates may have a different eruption timing too (Clutton-Brock et al., 1990).

Attrition in sheep and goats seems to be even more variable. It is particularly affected by the presence of gritty soil particles (Healy & Ludwig, 1965; Healy et al., 1967) in pasture. Moran & O'Connor (1994) suggested that no attempt should be made to assign any specimens to chronological age groups past three years of age. The alternative is to arrange mandibles (maxillary dentitions are usually too fragmentary) from archaeological assemblages into a series of eruption and attrition stages, without reference to age. Ewbank et al. (1964) were the first to do this, assigning letter codes to developmental stages and using them to seriate mandibles in a table. Payne (1973, 1987) developed a set of symbols and scores to record the wear state of premolars and molars. These can then be used to allocate mandibles from an archaeological assemblage into a series of attrition stages, to which approximate ages can be attached using the reference data quoted above (Payne, 1973). Grant (1982, 1975b, 1975c, 1978) developed a series of attrition diagrams for the deciduous and permanent fourth premolars, permanent first, second and third molars (Appendix A). Each tooth is scored, and the scores can be summed for the whole lower dentition to give a Mandibular Wear Score (MWS). This can be used as the basis of comparison, instead of assigning an age to the animals. Developments of this idea have been used in deer and pronghorn (below) but, as yet, the Grant system has continued to be used unaltered for sheep/goat. One of the difficulties of an overall MWS is that it does not summarise variation in attrition stage within the dentition (Moran & O'Connor, 1994) and one way to develop it would be to calibrate attrition in one tooth against eruption in another, like the Miles system used in human jaws (above). Deniz & Payne (1979) have also suggested that the most useful way to categorise attrition is by the sequence of 'erasure' or

Table 3.3 *Development stages of lower teeth in cattle*

	Stage								
	1	2	3	4	5	6	7	8	9
Deciduous dentition									
1st incisor		RC			L				
2nd incisor		RC				L			
3rd incisor		RC					L		
Canine		RC						L	
2nd premolar	RC					L			
3rd premolar	RC				L				
4th premolar	RC					L			
Permanent dentition									
1st incisor		CS	CC		RC				
2nd incisor			CS	CC			RC		
3rd incisor				CS	CC			RC	
Canine					CS	CC			RC
2nd premolar			CS	CC			RC		
3rd premolar		CS	CC				RC		
4th premolar		CS	CC				RC		
1st molar	C1/3–CC	RC							
2nd molar	CS	CC			RC				
3rd molar		CS		CC			RC		

Source: Brown *et al.* (1960).
Notes: CS, crown formation starts
CC, crown complete
RC, roots complete
L, tooth evulsed
Approximate ages: 1 (birth to 0.5 year), 2 (0.5–1 year), 3 (1–1.5 years), 4 (1.5–2 years), 5 (2–2.5 years), 6 (2.5–3 years), 7 (3–3.5 years), 8 (3.5–4 years), 9 (4–4.5 years).

wearing-out of the infundibula in the fourth premolar and molars. Their age ranges for this sequence are very variable, but it offers clearly defined points which could be recorded consistently. Hamilton (1978, 1982) compared the results for Ewbank, Payne and Grant systems on Iron Age sheep from the Ashville Trading Estate site, Oxfordshire, and all three produced reasonably similar distributions of age.

Cattle
At birth, cattle deciduous incisors and third deciduous premolars are usually erupted already (Table 3.3 and see Brown *et al.*, 1960). By two weeks after birth the deciduous dentition is complete except for the canine, which may come into occlusion a little later. Permanent crown formation starts with the first molars, which are already one third formed at birth. Next to start are the second molars, at one month of age, and then the first incisors, at six months or so. Formation of second incisors and second premolars starts around one year of age, followed by third incisors and canines.

Teeth and age

Table 3.4 *Dental eruption and wear stages in cattle lower jaws*

Stage	Definition	Suggested age(s)
1	1st molar not yet erupting	<6 months
2	1st molar erupting	6 (9–12) months
3	1st molar in wear, 2nd molar not yet erupted	6–18 (13–18) months
4	2nd molar erupting	18 months
5	2nd molar erupting, 3rd molar not yet in wear	18–27 (21–24) months
6	3rd molar erupting	2.25–2.5 years
7	3rd molar erupting, 4th premolar erupting	2.5–3 years
8	3rd molar slightly worn	>3 years
9	3rd molar medium worn	>3 years
10	3rd molar heavily worn	>3 years

Source: Grigson (1982a).

The permanent first molars erupt between six months and eighteen months and are followed in sequence by second molars, first incisors, second incisors and premolars, third molars, third and fourth premolars. There is considerable variation in this sequence (Grigson, 1982a) and there are recognisable early- and late-developing breeds. Eruption of the third molars continues for some years and root formation may not be completed until three or more years. The canines may not emerge until up to four years of age.

Andrews (1982) studied gingival emergence in detail for anterior teeth in many breeds of cattle, and the study was potentially very useful, because it distinguished between gingival emergence and arrival in the occlusal plane. Unfortunately, anterior teeth are usually missing from the jaw in archaeological specimens.

Higham (1967) devised a series of 18 eruption and attrition phases for cattle, and used them to discuss Neolithic and Bronze Age European animal husbandry. Grigson (1982a) outlined a shorter series of eruption and attrition stages for cattle jaws (Table 3.4). The Grant method (see sheep and goats above, and Appendix A) includes a system for scoring attrition in cattle teeth, which can again be used to calculate MWS scores. Crown height indices have also been used to estimate age in cattle (Ducos, 1969, 1968).

Pigs

Pig deciduous teeth erupt and come into wear during the first weeks and months after birth (Matschke, 1967; Silver, 1969). The permanent first molar crown starts to form at around birth and is completed at two to three months (McCance *et al.*, 1961). It erupts immediately behind the deciduous fourth premolar at 4–6 months and the roots are completed just afterwards (Table 3.5). This is followed by the first premolar, the crown of which starts to form in the first month after birth and is completed at four to five months. The first premolar erupts around the same time as the first molar, in front of the deciduous cheek tooth row. The crowns of the third

Table 3.5 *Gingival emergence timing in permanent teeth of pig*

	Early maturing domestic[a]	Middle maturing domestic[a]	Late maturing domestic[a]	Overall range of domestic pig[b]	Range for wild boar[b]
1st incisor	11	12	14	11–17	12–16
2nd incisor	14	16	18	14–20	18–27
3rd incisor	6	9	12	6–12	7–12
Canine	6	9	12	6–12	7–12
1st premolar	4	6	8	3.5–8	5–8
2nd premolar	12	14	16	12–16	14–17
3rd premolar	12	14	16	12–16	14–16
4th premolar	12	14	16	12–16	14–18
1st molar	4	6	8	4–8	4–6
2nd molar	7	10	13	7–13	12–14
3rd molar	16	18	20	16–22	21–33

Notes: [a] Habermehl (1975);
[b] Bull & Payne (1982).
Ages given in months after birth. Upper and lower dentitions included together

and fourth premolars and second molar all start to form at one to two months after birth, and are complete at six to eight months. The second molar erupts between seven and 14 months behind the permanent first molar, and its roots are completed around this time. The second premolar starts to form its crown later than the others, at three to four months, and completes it at nine to 10 months. All the premolars replace the deciduous premolars between 14 and 18 months of age. Finally, the third molar also starts to form its crown around three to four months and completes it around one year of age. The eruption time of the third molar is around two years, but is highly variable – the range is a good eight months either side. Lower third molars tend to be in advance of upper molars.

There is some variation between modern breeds in timing (Table 3.5) but, if earlier descriptions are discounted, the range is not too great (Bull & Payne, 1982). Wild pig dental eruption (Matschke, 1967) fits inside domestic ranges (towards the older end) up until about 14 months of age; after which wild pig eruption is relatively delayed. Overall, the variation is not large and it should be possible to use eruption for ageing pigs from birth up to two years or more. The Grant (1975c, 1982) method (above; Appendix A) includes a system for scoring eruption and attrition in the pig mandible. This produces some 47 mandibular wear stages.

Cervidae

Dental eruption and attrition have been studied in most deer. For red deer *Cervus elaphus* (elk or wapiti in North America), stages have been published by Mitchell (1967), Habermehl (1961) and Lowe (1967). The most detailed investigation has been by Brown & Chapman (1991a, 1991b), who used mandible X-rays to record dental development stages, and a scoring system to record wear (Table 3.6,

Table 3.6 *Dental development and wear scores in lower cheek teeth for* Cervus elaphus *and* Cervus (Dama) dama

	Range of development scores (*wear scores for molars*)						Summed scores for all teeth
Age (months)	3rd molar	2nd molar	1st molar	4th premolar	3rd premolar	2nd premolar	

Red deer (North American elk, wapiti) Cervus elaphus
1–11	0–2 (0)	1–6 (0)	4–9 (1–17)	0–1	0–1	0–1	5–19
13–19	3–6 (0)	7–8 (1–14)	9–10 (16–23)	2–7	1–7	1–8	23–45
26–33	7–9 (1–17)	8–10 (16–24)	9–10 (20–28)	8–10	8–10	8–10	50–99
38–42			Tooth development completed (*wear scores overlap*)				

Fallow deer Cervus (Dama) dama
2 days–11 months	0–1 (0)	1–6 (0)	2–9 (0)	0–2	0–2	0–1	2–22
18–22	6–7 (0)	8–10 (6–13)	8–9 (17–22)	7–8	6–8	6–8	42–49
26–35	8–9 (2–14)	9–10 (17–22)	9–10 (21–28)	9–10	9–10	9–10	56–59
38–43			Tooth development completed (*wear scores overlap*)				

Notes: from Brown & Chapman (1991a, 1991b). Wear scoring system given in Figure 3.4.
Development scores defined as follows:
1. Evidence of a crypt.
2. Evidence of mineralisation.
3. All cusps mineralising.
4. The infundibulum is formed.
5. Crown formation is complete.
6. Early root formation.
7. Half the root length formed.
8. Late root formation. More than half but less than full length.
9. Full root length formed with apex open.
10. Root apex closed.

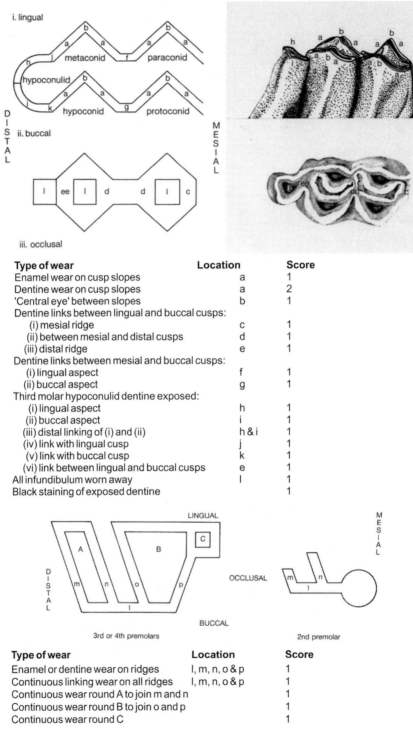

Type of wear	Location	Score
Enamel wear on cusp slopes	a	1
Dentine wear on cusp slopes	a	2
'Central eye' between slopes	b	1
Dentine links between lingual and buccal cusps:		
(i) mesial ridge	c	1
(ii) between mesial and distal cusps	d	1
(iii) distal ridge	e	1
Dentine links between mesial and buccal cusps:		
(i) lingual aspect	f	1
(ii) buccal aspect	g	1
Third molar hypoconulid dentine exposed:		
(i) lingual aspect	h	1
(ii) buccal aspect	i	1
(iii) distal linking of (i) and (ii)	h & i	1
(iv) link with lingual cusp	j	1
(v) link with buccal cusp	k	1
(vi) link between lingual and buccal cusps	e	1
All infundibulum worn away	l	1
Black staining of exposed dentine		1

Type of wear	Location	Score
Enamel or dentine wear on ridges	l, m, n, o & p	1
Continuous linking wear on all ridges	l, m, n, o & p	1
Continuous wear round A to join m and n		1
Continuous wear round B to join o and p		1
Continuous wear round C		1

Figure 3.4 Brown and Chapman (1990) wear recording scheme. Reproduced with permission from Cambridge University Press, from the *Journal of Zoology*, **221**, 668–9. The locations at which wear is scored on the occlusal surface are given as letters on the diagrams. The criteria and score system are given as tables. Scores are totalled for each tooth (see Table 3.16).

Figure 3.4). These were tabulated for each lower cheek tooth separately, and summed for all six teeth. The system was used by Carter (1998) to suggest, contrary to Legge & Rowley-Conwy (1988), that the British Mesolithic site of Star Carr was occupied by hunters during the winter, assuming that the deer were calved in May/June like living deer in north-temperate regions. A similar study of the Mesolithic site of Thatcham suggested that it was visited during the winter (Carter, 2000).

Uchiyama (1999) used eruption and wear of sika deer *Cervus nippon* to suggest the season of occupation of the Jomon site of Awashimadai in Japan. It appears that sika dental development progresses more rapidly than red deer. Aitken (1975) tabulated dental development in roe deer, *Capreolus capreolus*. Eruption is of relatively little help in ageing this species, because all teeth are erupted within the first year, but Carter (1997) was able to apply Brown & Chapman's methods (above) to a study of dental development in roe deer kids, and this also suggested a winter occupation of Star Carr. Miller (1972, 1974) studied dental development in Barren Ground caribou *Rangifer tarandus* and Spiess (1979) devised an age and season of death determination scheme for archaeological use (Table 3.7), using Miller's data. The stages of moose *Alces alces* eruption and attrition have been described by Passmore *et al.* (1955). The sequence for white-tailed deer *Odocoileus virginianus* has also been established (Severinghaus, 1949; Gilbert & Stolt, 1970) and is widely used for age estimation. Crown height measurements have also been used to estimate age in archaeological material (Walker, 2000).

Klein *et al.* (1981) used crown height measurements to study wear in red deer teeth from the Magdalenian cave site of El Juyo, Spain. Formulae were derived from studies of known-age animals for deciduous lower fourth premolar and lower second permanent molar. Age distributions constructed by these means were compared between different levels of the site. Legge & Rowley-Conwy (1988) used crown height measurements to try to determine the season of occupation for the site of Star Carr (above). They measured lower fourth premolars in red deer, in the hope that values of this measurement would show limited ranges for different seasons, as they had found in recent deer, but this was not the case. Kierdorf & Becher (1997) tried an alternative measurement of wear – the width of the band of exposed dentine on the mesiobuccal cusp of red deer first molars as a proportion of the buccolingual crown diameter (Chapter 4). This was modestly related to known age.

D'Errico & Vanhaeren (2002) established a scheme for recording wear in red deer canines, and showed the relationship of these scores to known age. This is useful because large numbers of deer canine ornaments have been found on European Upper Palaeolithic sites. It is also possible to use measurements to determine sex from the same specimens, so that details of hunting strategies can be examined.

Other Artiodactyla
Eruption timing for chamois *Rupicapra rupicapra* is given by Habermehl (1961) and age groups for eruption and wear have been devised by Pérez-Barbería (1994).

Table 3.7 *Dental development groups for sorting caribou/reindeer mandibles*

Group	Permanent tooth features	Deciduous tooth features
1	Permanent 1st molar unerupted	Deciduous premolars worn but still showing high relief and sharp cusps
2	1st molar quarter to three quarters erupted	
3	1st molar fully erupted, crypt of 2nd molar not yet appeared in bone of jaw	Deciduous premolars moderately worn, with rounded cusps
4	2nd molar quarter to three quarters erupted	
5	2nd molar fully erupted, crypt of 3rd molar not yet appeared in bone of jaw	
6	3rd molar quarter to three quarters erupted	Deciduous premolars very worn, roots being resorbed or permanent premolars erupting
7	3rd molar fully erupted	
8	2 years old. Wear light on all molars except 1st molar. Buccal cusps still high and sharp. Wear of permanent premolars very light, with exposed dentine line narrower than the investing cement layers.	
9	3–5 years old. Molars and premolars, buccal cusps worn, lingual cusps noticeably higher and still sharp. Premolars with exposed dentine line wider than investing cement.	
10	6–9 years old. Molars, buccal and lingual cusps worn to even height, but occlusal surfaces still undulating from mesial to distal. Exposed dentine broad quarter or half-moon shaped, rather than thin crescent or teardrop shape. Premolars, lingual cusps blunted.	
11	More than 10 years. All teeth very worn, occlusal surfaces flat buccolingually and mesiodistally. Cement lining of infundibulum reduced to small oval or circle. Very old individuals may have the teeth worn down to the roots. Maximum recorded age is 18 years.	

Source: Spiess (1979).

Robinette & Archer (1971) defined eruption/wear classes for Thomson's gazelle *Gazella thomsoni* from Tanzania, and Davis (1980) gave mandibular eruption data for *G. gazella* and *G. dorcas* from Israel. Davis (1983) also used the height of *G. gazella* deciduous fourth premolar and permanent third molar crowns to compare season of occupation for Mousterian Kebara Cave (50 000–40 000 BC), Aurignacian Hayonim Cave (*c.* 24 000 BC) and the Natufian Hayonim Terrace outside the cave (*c.* 10 000–9000 BC), all in Israel. Today, most *Gazella* in the area are born in the spring. The restricted range of worn crown heights (estimated to represent 8–12 months of age) at Kebara and Hayonim Caves suggested seasonal occupation, possibly during winter, whereas the wider range of heights at Hayonim Terrace suggested year-round occupation.

Lubinski (2001; Lubinski & O'Brien, 2001) constructed dental development and attrition recording schemes for pronghorn *Antilocapra* like Brown & Chapman's for deer (above). The development scores were based on the visible eruption of the teeth through the jaw, instead of X-rays, and the wear scores were simplified. Nonetheless, the scores showed a reasonably good match with independently

Table 3.8 *Gingival emergence times for camel lower jaws*

	Emergence time
Deciduous teeth	
1st incisor	4–6 weeks
2nd incisor	3–4 months
3rd incisor	8–9 months
Canine	10–12 months
Premolars	4–6 months
Permanent teeth	
1st incisor	4 years
2nd incisor	5 years
3rd incisor	6 years
Canine	6.5 years
Premolars	4–5 years[a]
1st molar	1 year
2nd molar	3 years
3rd molar	5 years

Source: Silver (1969).
Note: [a] If lower 3rd premolar erupted, usually lost by 6–7 years.

known age. Dental eruption in the tahr *Hemitragus jemlahicus* has been studied by Caughley (1965). Laws devised a system for determining age from eruption and attrition in hippopotamus (Laws, 1968; Kingdon, 1979).

In the High Plains of the USA are a whole series of bison kill sites, dating throughout the prehistoric period, where men on foot stampeded the animals down drive lines, over a drop or into a trap which was reused over many years. Very large accumulations of bison bones have been subjected to careful study, and one of the central questions is the season of the hunt. Frison & Reher (1970) established dental eruption and wear sequences for prehistoric bison in the Glenrock Buffalo Jump site, central Wyoming, and used a modern commercial herd to attach ages to these sequences. For most similar sites, mainly females and young seem to have been driven, and eruption/attrition sort the dentitions into five clear age categories, with few intermediates, separated by one year (0.6, 1.6, 2.6, 3.6 and 4.6 years, see Reher, 1974). Assuming birth of calves in the spring, it was suggested that this represented a hunt in the fall. Heights of worn tooth crowns in the older age groups, where all teeth were erupted, showed multiple modes, again suggesting periodic hunts. This idea has become an accepted interpretation for similar sites, but Whittaker & Enloe (2000) have pointed out that there is little supporting evidence from known-age bison, despite the efforts of Gifford-Gonzalez (1992).

Silver (1969) listed gingival emergence ages for camel (Table 3.8). The Payne sheep/goat attrition scoring system has been adapted by Wheeler (1982) in an ageing system for other camelids – llama and alpaca.

Table 3.9 *Eruption through bone (and first signs of wear) in the horse*

	Eruption time
Deciduous dentition	
1st incisor	5 days–2 months
2nd incisor	1–2 months
3rd incisor	5–10 months
Premolars	0–1 months
Permanent dentition	
1st incisor	2–3 years
2nd incisor	3–4 years
3rd incisor	4–4.5 years
Canine	4–5.5 years
1st premolar	–
2nd premolar	2.5–3.5 years
3rd premolar	2.5–3.5 years
4th premolar	c. 3–5 years
1st molar	7–12 months
2nd molar	16–24 months
3rd molar	2.5–5 years

Source: Levine (1982). Upper and lower dentitions combined.

Perissodactyla

Horses

The traditional method of ageing horses is from attrition in the anterior teeth. Versions of this method are given by Silver (1969), Getty (1975) and Habermehl (1975). The outline of the single infundibulum in horse incisors (Chapter 1) is oval early on in wear, but gradually changes to a circle and eventually is lost. The pulp cavity fills with secondary dentine, seen as a dark mark on the occlusal surface – the 'dentine star'. These changes form the basis of the method used widely by farriers and horse dealers, but it is less useful in archaeology, where cheek teeth are much more common than incisors (and isolated incisors are difficult to tell from one another). If a reasonably complete set of incisors survives in a jaw, this method is useful up to eight years of age (Levine, 1982).

In the horse (Table 3.9), the deciduous first incisors and premolars are present at birth, or appear within a fortnight afterwards. The second and third incisors follow and all deciduous teeth emerge into the mouth by nine months after birth. The first permanent tooth to emerge is the first premolar, or 'wolf tooth', at five to six months. This is not always present. The first molar usually emerges by the end of the first year and is followed in order by second molars, second, third and fourth premolars, and third molar. The anterior teeth appear one after another, from first incisors to canines, between two and five and a half years. By this time, the whole permanent dentition is present in the mouth.

Levine (1982) defined age ranges for several different stages of eruption and attrition, which are useful until the last tooth erupts and comes into full wear at six years of age. After this, an alternative is needed that can be used on isolated teeth, as well as cheek tooth rows. Levine developed a system (above) for determining age from crown height measurements of cheek teeth. This approach is attractive in horse teeth because of the very tall, regular crown, but one major problem is variation in original crown size and form – before any wear actually takes place. This was partly addressed in Levine's method by finding a reference group of recent horses of similar size to fossil horses. Levine's technique was used by Fernandez & Legendre (2003) to derive an age distribution for horses from the Mousterian levels of Bau de l'Aubesier, Vaucluse, France. In the earlier parts of the site, adult horses were preferentially hunted whereas, in the later levels, both juveniles and adults were included. Brown & Anthony (1998) studied the relationship between lower second premolar crown height and known age in a recent horse collection, as part of their experimental examination of bit wear damage. The identification of bit wear is important in discussion of human exploitation and domestication of horses, particularly as there is little evidence for morphological change with domestication.

Rhinoceros

Some reference information is available on dental eruption, particularly for *Diceros*, and it is possible to use this to reconstruct age profiles for extinct forms of rhinoceros (Tong, 2001).

Carnivores

African lions *Panthera leo* (Smuts *et al.*, 1978) are born without any teeth present in their mouth. In one to two weeks the deciduous first incisors start to erupt, to be followed over the next month by deciduous second and third incisors, and canines. Deciduous cheek teeth start to erupt between one and three months of age. Incisors are the first permanent teeth to erupt, starting at eight months. Permanent carnassials and canines erupt between 11 and 15 months. These are followed over the ensuing few months by the remaining premolars and upper molars. Smuts *et al.* defined wear stages up to 10–14+ years of age. Eruption and wear stages have also been defined for leopard *P. pardus* (Stander, 1997). Domestic cats follow a similar pattern (Silver, 1969; Habermehl, 1975). Deciduous incisors and canines erupt within the first month of postnatal life; deciduous premolars within the second. Permanent anterior and cheek teeth emerge over a short period from three to six months (Table 3.10).

In domestic dogs *Canis familiaris*, deciduous teeth emerge within the first two months after birth. They are replaced by permanent teeth between three and seven months. First to erupt are permanent incisors, followed by molars (first, second then third), premolars and canines (Table 3.11). Dog incisors come into wear by the first year of age. Their three-lobed crown form (Chapter 1) is lost completely

Table 3.10 *Gingival emergence stages for domestic cat*

Stage	Definition	Suggested age
1	Deciduous incisors and canines emerging	1–4 weeks after birth
2	Deciduous premolars emerge	5–6 weeks
3	Full deciduous dentition present	6 weeks–3.5 months
4	Permanent incisors emerge, permanent premolars may start to emerge at 4 months, permanent molars at 4.5 or 5 months	3.5–5.5 months
5	Permanent canines emerge, permanent premolars continuing eruption to 6 months, molars to 5.5 or 6 months	5.5–6.5 months

Source: Silver (1969) and Habermehl (1961).

Table 3.11 *Gingival emergence stages for domestic dog*

Stage	Definition	Suggested age
1	Deciduous teeth emerge, incisors and canines first, then premolars	1–2 months
2	Permanent incisors emerge, permanent 1st molars and 1st premolars emerge later in this stage	3–5 months
3	Permanent 2nd molars, 2nd, 3rd and 4th premolars emerge. Permanent canines may emerge in this stage or the next	5–6 months
4	Permanent 3rd molars emerge. Permanent canines may only emerge at this time	6–7 months

Source: Silver (1969) and Habermehl (1961).

between one and two years. Habermehl (1975) described criteria for determining age from incisor and canine wear in domestic dogs, and Gipson *et al.* (2000) tested a scoring system for attrition in wolf, finding that it performed reasonably well in known-age animals. Red fox *Vulpes vulpes* teeth erupt in a similar sequence and timing to dog (Habermehl, 1961) between one and two months of age. Van Bree *et al.* (1974) gave four stages of wear for upper permanent first molar, for ages up to three years.

A certain amount of information on eruption and attrition is available for mustelids. In *Mustela*, by around 28 days the deciduous canines and carnassials are emerging through the gums (Mazak, 1963). Permanent teeth replace them over a period lasting from seven to eight weeks to about 10–13 weeks. Permanent incisors emerge first, followed by molars and then canines and premolars, but there is some variation in this order (Slaughter *et al.*, 1974). *Martes* seems to follow a similar pattern, with permanent teeth emerging from 7–8 weeks to 16 weeks (Mazak, 1963). A system for scoring wear has been devised for the badger *Meles meles* (van Bree *et al.*, 1974), with six stages up to three years. In the wolverine *Gulo gulo*, emergence of permanent teeth starts with incisors at about four and half months and this is followed by eruption of canines and cheek teeth. Zeiler (1988) devised age criteria based on wear in Danish otter, calibrated by cement layer ages (below).

Garniewicz (2000) developed a series of dental wear age stages for the raccoon *Procyon*.

Rodents

According to Osborn & Helmy (1980) *Mus* and *Rattus* are mature when the cusps of the molars are coalesced by attrition to form laminae, *Acomys* are mature when there is some wear on molar cusps, and the high-crowned molars of *Nesokia* show maturity when their laminae are broadly crescent shaped. Brothwell (1981b) adapted an attrition grading system for house mouse *Mus musculus* lower molars and showed that the mouse population from Çatalhöyük consisted predominantly of young individuals. Armitage (in Morales & Rodríguez, 1997) developed a scheme for scoring attrition in black rat *Rattus rattus* upper molars.

In the Mongolian gerbil *Meriones unguiculatus* (Wasserman *et al.*, 1970) the incisors emerge at 12 days, first and second molars at 18–21 days and third molars at 30 days of age. In *Meriones* and the similar genus *Psammomys* cheek tooth crowns have infoldings or 'pleats' in their buccal and lingual sides, which smooth out towards the cervix of the crowns. Osborn & Helmy (1980) regarded individuals as immature while these infoldings still extended below the alveolar margin. Individuals in which eruption had exposed the smooth cervix were labelled mature. Other gerbils *Sekeetamys* and *Dipodillus* were regarded as mature when the mesial cusps of the first molars coalesced due to attrition.

In most voles, the cheek teeth grow continuously and remain unrooted throughout life. The bank vole *Clethrionomys glareolus*, however, starts to form roots at about two to six months of age and Lowe (1971) devised an age determination method based on measurements of root length in the first molar. Tupikova *et al.* (1968) devised similar methods for bank vole and northern red-backed vole *C. rutilis*.

Shorten (1954) defined six cheek tooth wear stages for grey squirrel *Sciurus vulgaris*, aged up to 8–9 months. Munson (1984) established a dental development chronology for woodchuck *Marmota monax* and used this to suggest the season of occupation of the Late Woodland/Early Mississippian Haag site, Indiana, USA. In the beaver *Castor fiber* the deciduous molars emerge at around one month after birth, and permanent molars (first, second then third) by six months. The premolars erupt last and come into wear by 10–12 months (Mayhew, 1978). Van Nostrand & Stephenson (1964) devised a scheme of stages for root development and cement layering (below).

The variation with age of dentine exposure patterns in rodent teeth is illustrated in many texts (e.g. Ognev, 1948). For sites with large rodent collections, it should prove possible to develop scoring systems for comparisons of age distributions.

Insectivores

Crowcroft (1957) was able to differentiate between sexually mature and juvenile common shrews *Sorex araneus* using measurements of their lower canines. He demonstrated that summer populations had two generations, whilst winter populations had only one. This type of analysis could prove very useful in archaeology for

Table 3.12 *Elephants and mammoths, loph numbers and attrition ages for lower cheek teeth*

Cheek tooth	Laws[a] tooth number	Number of lophs (plates, lamellae) present[b] in *Loxodonta*	Number of lophs (plates, lamellae) present[b] in *Elephas*	Number of lophs (plates, lamellae) present[b] in *Mammuthus columbi*	Number of lophs (plates, lamellae) present[b] in *Mammuthus primigenius*	Age range when in wear (years)[c]
Deciduous 2nd premolar	M1	2–4	5	4	4–6	0.1–2.0
Deciduous 3rd premolar	M2	5–7	6–9	6–8?	7–10	0.3–4.1
Deciduous 4th premolar	M3	7–10	11–14	9–12	8–15	2.3–13.0
Permanent 1st molar	M4	6–10	14–17	11–15	10–17	5.5–21.0
Permanent 2nd molar	M5	8–12	16–21	11–16	12–21	14.5–40.0
Permanent 3rd molar	M6	9–14	21–29	15–23	20–27	27.0–61.0

Notes: [a] from Laws (1966).
[b] from Haynes (1991).
[c] summary of overall range of several studies cited in Haynes (1991).

Teeth and age 245

determining the season of site occupation. Funmilayo (1976) developed a five-stage attrition scale for upper second molars of the mole *Talpa europaea*, and Hartman (1995) devised a similar scale for *Scalopus aquaticus*.

Dugongs, elephants, mammoths and mastodons
Age determination from dugong teeth has been studied in some detail (Marsh, 1980). The six simple conical cheek teeth (Chapter 1) erupt in sequence, starting with the second premolar and ending with the third molar. Premolars are present in the mouth at birth. First molars erupt between two and three years of age (sometimes earlier), second at three to four years and third molars at six to nine years. The tusks erupt in males between 12 and 15 years of age. Marsh found that counts of dentine layers (below) were a reliable index of age.

Age estimation in Proboscidea is summarised in detail by Haynes (1991). See also Laws (1966), Sikes (1966, 1968) and Roth (1988). The deciduous to permanent dental progression lasts much longer in elephants than in most other mammals. Six cheek teeth in each quadrant of the jaw (Chapter 1) follow one another in succession – at most ages, one or two teeth only are in wear per quadrant at the same time but around 2 years of age there are three (Table 3.12). They are, in fact, a series of deciduous premolars and permanent molars, but they are often numbered M1 to M6 (Laws, 1966). The first requirement in age estimation is to identify the teeth present, by the number of lophs, or plates, present. This varies within genera and populations, and it is generally easiest to distinguish M1 and M2, but most difficult to distinguish M4 and M5. Most is known about the lower dentition, and one full quadrant is used to estimate age from the teeth in wear at the time of death (Table 3.13). Extinct *Mammuthus* and *Mammut* have a similar pattern of dental eruption and wear, and bone development, to living *Loxodonta* and *Elephas*, so it is assumed that they had similar age ranges (Haynes, 1991).

Circum-annual layering in cement and dentine
The counting of layers in cement and dentine is now widely recognised as a technique for age and season of death estimation in mammals for wildlife management purposes (Klevezal & Kleinenberg, 1967; Klevezal, 1996; Morris, 1972, 1978; Casteel, 1976; Perrin & Myrick, 1980; Scheffer & Myrick, 1980). It has been widely considered in archaeology, but is much less used in general practice.

Marine mammals
The technique has its origins in the work of Scheffer (1950), who noted concentric rings on the root surfaces of permanent canines from the Alaskan fur seal, *Callorhinus ursinus*. Although the crown of the canines in this species is practically complete by birth, the root continues to grow for up to 20 years or so. Dentine is deposited as a series of thick layers (Scheffer & Myrick, 1980), the edges of which bulge out to form the ridges around the root surface, seen even under a coating of cement. Scheffer discovered that, in a group of tagged, known-age animals, the

Table 3.13 *Lower jaw tooth wear stages in elephants*

Stage	Teeth in wear[a]	Age range (years)[b]
Animals sexually mature, growing rapidly		
A	None in wear, or M1 only	0.0–0.5
B	M1 and M2	0.5–2.0
C	M1, M2 and M3	2.0 (+ 3–4 months)
D	M2 and M3	2.5–4.0
E	M3	4.0–5.5
F	M3 and M4	6.0–13.0
Animals sexually mature, some limb-bone epiphyses fused or fusing		
G	M4	13.0–14.5
H	M4 and M5	14.5–22.0
I	M5	22.0–27.0
Animals reaching end of growth, or fully grown and in prime		
J	M5 and M6	27.0–40.0
Animals passing prime or in old age		
K	M6	40.0–60.0

Notes: [a] see Table 3.13.
[b] from Haynes (1991).

number of main ridges ('growth rings') corresponded to the age of younger animals although, in individuals older than four years, the rings were less closely related to age. In winter and spring, fur seals live at sea, feeding on abundant resources, but in summer they stay on shore, the males not feeding and the females nursing young. Each ridge seemed to represent winter growth and the troughs indicated the summer. In their first year fur seals do not return to land and have a less pronounced trough and ridge for that year. This type of life cycle, together with long-growing canine roots, is seen in other Otariidae (Chapter 1) and Scheffer found similar rings in other genera. He also noted them in a phocid, the elephant seal *Mirounga leonina* (Figure 3.5). In a histological study of South Atlantic elephant seal canine sections Laws (1952, 1953a, 1953b) found alternating layers of translucent and opaque dentine. It was possible to recognise a neonatal line (Chapter 2) under the crown, followed by a series of variable zones which appeared to represent the period of growth to sexual maturity at the end of the third year in males and second year in females. After this, four layers were deposited each year in both sexes – a translucent layer representing the spring (September/November in the South Atlantic) breeding season, an opaque layer representing early summer feeding at sea, another translucent layer during the late summer and autumn moult, and another opaque layer during the winter feeding at sea. The total sequence for one year was around 1–1.5 mm thick. Later work by Carrick and Ingham (1962) on tagged animals suggested the layering was more complex, but the principles were nevertheless established.

Teeth and age 247

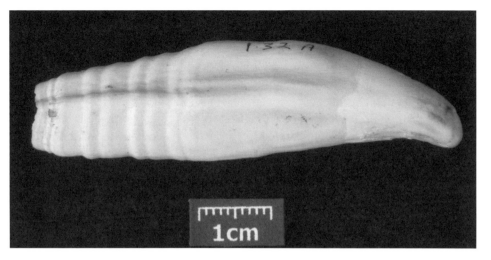

Figure 3.5 Canine tooth from an elephant seal *Mirounga leonina* in the Institute of Archaeology, UCL teaching collection. The crown is to the right, and the growing apex of the root to the left. Ridges on the root mark out seasonal fluctuations in diet and physiology. In the central part of the tooth, the ridges have been covered by cement.

Since this early work, similar methods have been tried on most marine mammals (International Whaling Commission, 1969; Jonsgard, 1969; Perrin & Myrick, 1980; Scheffer & Myrick, 1980). Standard terminology has been developed (Report of the Workshop, 1980). *Incremental growth layers* (also called laminae, bands, zones and lines) are defined as:

> distinct layers parallel with the formative surface of a hard tissue (dentine, bone, cement and their subtypes) which contrast with adjacent layers. The contrasting nature may be:
>
> (a) translucency and opacity of a section examined by transmitted light,
> (b) dark and light (correspondingly less light-scattering and more light-scattering) layers seen on a surface with incident illumination (reflected light),
> (c) high or low relief of the surface following the application of an etching technique, the surface being examined by incident light (microscopy) or scanning electron microscopy,
> (d) more or less intense staining of a decalcified cut surface, where haematoxylin, a basic stain, is the most commonly used stain.
>
> (Report of the Workshop, 1980:2)

Often, regular groupings of layers can be recognised. These are called *growth layer groups* or 'GLGs' (Report of the Workshop, 1980) and are recognised as cyclic repetitions of incremental growth layer sequences. It is these GLGs that form the basis for methods of age determination.

These thick incremental growth layers in dentine clearly represent a much longer periodicity of growth than the Andresen and von Ebner lines (Chapter 2), and the spacing of each GLG in large marine mammals is reasonably compatible with one year's accumulation – about 1.5 mm might be expected from the growth rate (2–4 μm per day) estimated for other mammals. Most sections show finer layerings (accessory layers) within each incremental growth layer, and these presumably represent the more prominent Andresen lines, although this has not been investigated. Some marine mammal GLGs are highlighted by bands of interglobular dentine representing deficient mineralisation, but the main feature seems to be an alternation in the balance between ground substance and collagen, which shows in the haematoxylin stain reaction (Boyde, 1980). Ground substance is basophilic; that is, it stains intensely with haematoxylin, whereas collagen is acidophilic and stains much less heavily. It might be expected from studies of bone development that ground-substance-rich layers would be deposited just before dentine development stops for a period, and again when it restarted. Boyde (1980) was able to demonstrate specimens in which organic dentine matrix formation had ceased and the developing surface had become fully mineralised during this quiescent phase (actively growing dentine always has a surface layer of unmineralised predentine).

Similar seasonal and annual layering has also been recognised in the thick cement of marine mammals (Report of the Workshop, 1980). Laws *et al.* (2002) found a good match between counts of dentine and cement lines in seals, up to the age of 5 years, reasonable agreement up to 10 years and a tendency for the dentine counts to underestimate above that age. This dentine fall-off is presumably due to the completion of the roots and filling-in of the pulp chamber. Cement is more like bone as a tissue than dentine, so it is more logical to equate the layers with resting lines in bone, which represent a pause in the formation of new tissue (Boyde, 1980; Boyde & Jones, 1995). Resting lines are also marked by strong staining with haematoxylin and therefore presumably represent deposition of ground-substance-rich matrix. On this principle, the more strongly staining lines of cement should represent the seasons of slow growth in the life of marine mammals. It is possible to check if the outermost layers of cement represent a resting phase, or active deposition, by looking at the mineralisation of extrinsic fibres seen by SEM examination of the root surface (p. 194).

The sperm whale *Physeter* has well-marked, conical layers of dentine filling up the pulp chamber of its simple teeth (Chapter 1). A neonatal line is recognisable, with poorly laminated dentine laid down during the prenatal period. In transmitted light microscopy of intact sections, the structure of postnatal dentine shows up as alternating translucent and opaque layers. A GLG consists of two layers, one of each type, but accessory layers may be apparent (Report of the Workshop, 1980) and cause different observers to arrive at different counts. Limited studies with tagged animals suggest that GLGs in *Physeter* represent one year's growth, but it is difficult to validate the method with such large, deep sea creatures, and age estimates are still seen as subjective (Evans *et al.*, 2002). A similar arrangement is found in the large Delphinidae and Monodontidae (including the dentine of the narwhal's tusk). The beaked whales (Ziphiidae) have a very different dental structure from other odontocetes. Their large, stout teeth (Chapter 1) have very thin enamel, commonly worn away. Dentine deposition is prolonged and the whole tooth is invested with thick layers of cement. Both tissues show GLGs.

The pulp chamber in large marine mammal teeth is often filled with masses of pulp stones (Chapter 2) held together with ordinary dentine (so-called *orthodentine*), or sometimes other types of dentine more commonly found in fishes (Osborn & Ten Cate, 1983). In *osteodentine*, pulp runs through tunnels, each of which has a dentine wall, a bit like the trabeculae of bone. *Vasodentine* includes small blood vessels, running through canals. These tissues are prominent in the infilled pulp chambers of ziphiid teeth (Boyde, 1980).

Amongst pinnipeds, similar layers to fur seal and elephant seal develop in the dentine of walrus tusks. These can be seen as ridges if the cement is chipped away from the surface (Scheffer & Myrick, 1980). From studies of known-age animals, the layers appear to be annual. Cement layering is used for age estimation in most phocids. Klevezal & Stewart (1994) found that cement layers were better age estimators than root ridges (above) in female elephant seal. Mansfield & Fisher (1960) described layers in the cement of the common seal *Phoca vitulina*. The GLG was a pair of layers, one translucent and thin, the other opaque and thick, and the count corresponded quite well with the known age of the individual. Mansfield & Fisher suggested that the thinner layer might represent spring/summer breeding and moult. Hewer (1964a, 1964b) defined similar layers in the cement of the grey seal *Halichoerus*.

Both male and female dugong show prominent layering in the dentine of their tusks, once again visible as ridges on the surface (Marsh, 1980). GLGs consist of a broad band (Zone A) which is more opaque in section, stains to an intermediate extent with haematoxylin and contains many faintly defined accessory layers, and a narrow band (Zone B) which is more translucent and includes one to three intensely staining lines. This strong layering is surprising in a herbivorous tropical/subtropical marine mammal, where strong seasonal differences would not be expected. It may perhaps represent some physiological rhythm. Nevertheless, the layering forms the basis of a reliable age-determination method.

Land mammals
In his study of the elephant seal, Laws (above) noted that land mammals showed similar layers. Sergeant & Pimlott (1959) followed this up with a transmitted light microscope study of cement in thin sections of moose *Alces alces* incisors. They noted thin, translucent zones, alternating with thick and opaque zones (which had more cementocyte lacunae; Chapter 2). Moose first incisors erupt at seven to eight months, and animals this age had a single layer, about 120 μm thick, consisting of an opaque zone followed by a thin translucent one. By the end of the first year, two layers were complete, each consisting of two zones. In animals over one year of age, further layers were present. Counts of cement layers corresponded well with independently known age, although this relationship was less strong in older

animals. Mitchell (1963, 1967) studied similar cement layers in known season-of-death red deer *Cervus elaphus* from Scotland, using reflected light microscopy of cut, polished sections of molars. A high proportion of the animals killed between January and March showed zones of the translucent type (darker in reflected light) outermost, and opaque (brighter in reflected light) in animals killed between May and December. About one quarter of the stags killed during October instead showed thin translucent zones outermost – interpreted as a 'rutting layer'. Out of 22 deer of known age, the count of cement layers (each consisting of one main translucent and one opaque zone) came within one year of true age in all but four cases. Similar results were produced by Low & Cowan (1963) for black-tailed deer *Odocoileus hemionus*. Not all studies yielded such encouraging results, however. Lowe (1967) used methods similar to Mitchell (above) on 34 red deer *Cervus elaphus* from the Island of Rhum, Scotland. All had been marked when only a few hours old and their age when shot was known to the month. Out of 28 animals older than one year, only half the cement layer ages were correct, compared with 89% of the ages derived from dental eruption and attrition. Similarly Gasaway *et al.* (1978), also using intact thin sections, found that for 45 moose of known age two years or older, about half the cement-layer age estimates were correct. Out of 23 yearling animals, however, 21 were correctly aged.

Cement has the advantage that it continues to be formed throughout the life of the animal, even when the dentine of the root has been completed. Layering in cement is therefore widely used in wildlife management to estimate age and season of death. It is offered as a service by commercial laboratories (e.g. Matson's Laboratory, Milltown, Montana), and has been studied for many years by Galina Klevezal at the Russian Academy of Sciences in Moscow (Klevezal, 1996). It is important to emphasise that the technique uses demineralised sections (10–30 μm thick), stained with haematoxylin. These are very different from 'thin' (usually 150 μm or more thick) sections of intact teeth, reflected light microscopy or SEM-BSE examination of polished surfaces. From the illustrations in Klevezal (1996), it is apparent that the structures being counted are clear, narrow lines of darker staining. The cementocytes are also sharply defined, but do not show much change in density relative to the lines. In addition, because polarising microscopy is not being used, the lines cannot represent changes in extrinsic fibre orientation and, because the inorganic component has been removed, they cannot directly represent changes in mineralisation (although the organic component of cement is intimately linked with this). In sections with distinct layers ('readability' is reckoned on a 5-point scale), there are narrow, darkly stained incremental lines, with wider intervening poorly stained bands. Klevezal called the narrow dark lines the *principal element* of the growth layer, and the broad intervening bands were the *intermediate element*. In some sections, there were also finer and less distinct dark lines within the intervening bands, and these were called *additional elements*. The whole combination was termed a *growth layer* (avoiding use of the word 'annual' because this is not always the case). Klevezal felt that this was a better terminology for terrestrial mammals than GLG (above) or *lignes d'arrêt de croissance* (LAC, see Castanet, 1981) – also known as *lines of arrested growth* or LAG (Burke & Castanet, 1995), because the counted unit in cement was not a group of lines, but a single line, and a whole variety of factors could be involved in its formation. Many studies of known-age animals (by tagging or keeping in captivity) showed that readability of cement sections varied between species, between populations, and between individuals within a population. Some study groups had a large proportion of individuals with readable layers, but some had a very small proportion. The distinctness of the principal elements varied, and the additional elements sometimes became so prominent that it was difficult to distinguish between the two. In general, within the Holarctic region, the principal elements of the growth layers corresponded to the winter and spring, and their prominence increased with the contrast between the seasons, as it impinged on the animals. There were therefore differences between populations from different regions. Domestic animals, particularly those kept indoors, usually had less distinct principal elements and more distinct additional elements than wild animals of the same species. Hibernators might be expected to show more distinct layering, but this varied. With experience, however, it was usually possible using known-age specimens of a whole variety of mammals to identify growth layers that represented an annual rhythm in at least a proportion of animals from most study groups. The conversion of growth layer counts to age at death depended on identification of the first principal element formed, during the first winter of the animal's life. In order to establish this reliably, it was necessary to examine young animals. Season of death could be determined by identifying the last formed element, although once again care was needed in recognising the very thin principal element when it was close to the surface.

Cement layer counting is particularly well established as an age estimation method in bears, where the small second and third premolars can easily be extracted even in living animals, and attrition does not provide a good alternative. Harshyne *et al.* (1998) found only 9% of a large number of age estimations were in error by one year or more. Ungulates, by contrast, are routinely aged by eruption and attrition. Hamlin *et al.* (2000) found that cement growth layer counts performed better as age estimators, relative to eruption and attrition, in two groups of independently aged deer. It should be pointed out, however, that the basis of the eruption/attrition estimates was not closely defined in this study. Moffitt (1998) found that 41% of 42 molars from known-age bison were aged by cement to the correct year, and a further 33% were only 1 year astray. When several teeth were sectioned per individual bison, 93% of 28 animals were aged to the correct year. Where known-age animals are not available, comparisons have sometimes been made between cement layer counts and age estimated from attrition (O'Brien, 2000), but care needs to be taken because cement deposition may well be directly related to continuous eruption and attrition, so they may represent two aspects of the same phenomenon.

It might be expected that animals living in the tropics would have less well defined cement layering, because there are no real seasons. Layers can, however, still be seen although they are indistinct and the accessory elements are relatively strongly developed (Klevezal, 1996). In some cases, the principal elements are double (Spinage, 1976). There must be some physiological fluctuations through the year, in order for the layers to form. Similarly, humans should be the most buffered of all mammals from seasonal changes, but cement growth layers are seen here too. Some studies of thin sections have suggested annual layering (Stott et al., 1980, 1982; Naylor et al., 1985) but this is controversial (Lipsinic et al., 1986; Miller et al., 1988). More recently, a large study based on thin sections of cement has reported a close relationship between layer counts and age (Wittwer-Backofen et al., 2004). With demineralised, stained sections, the preferred technique in most mammals, Charles & Condon (1986, 1986, 1989) found a moderately good relationship between cement layer count and age at extraction of teeth, as did Klevezal (1996). Human cement, like domestic animals, shows a complex pattern of layering in which the additional elements are prominent and make counting difficult.

Archaeological applications

Archaeological studies would be best integrated with work on living populations, but the central problem is that the method of choice, demineralisation, microtome sectioning and staining, usually does not work with archaeological specimens. They just do not have a sufficiently well-preserved organic component. Instead, most studies have used ordinary transmitted light microscopy, or polarising microscopy, of thin sections. These rarely show such distinct layers (above), and they are hard to relate to the stained specimens. One guiding principle of microscopy is that sections should be thinner than the spacing of the structures being imaged (Boyde, 1980) and, for most mammals, thin sections of intact teeth are too thick. This problem might be addressed by confocal microscopy, which focuses on a very thin plane within the section, or by SEM-BSE imaging of flat, polished surfaces (Chapter 2). The features seen in these images, however, have still not been matched to the structures seen in stained preparations. In addition, cement is often badly affected by diagenetic change, because it is a relatively thin layer on the outside of the root. All this can make it very difficult to see layers distinctly, and image analysis is unlikely to improve matters. This is not to say that archaeological applications are impossible, but care is needed, particularly where the cement layers are thin.

The pioneer study for archaeology is Saxon & Higham (1968) who demonstrated cement layers in thin sections of Iron Age sheep molars. Kay (1974) used cement layers in thin sections to estimate age and season of death in white-tailed deer *Odocoileus virginianus* from the Middle Woodland Mellor site, Missouri, USA. Spiess (1976) had some success with demineralised sections in seal *Phoca* spp. teeth from protohistoric to historic sites in Okak Bay, Labrador. He distinguished summer sites from those occupied in winter by the cement layer lying outermost. Bourque et al. (1978) demonstrated cement layering in reflected light microscopy of intact polished surfaces from a variety of animals at American sites. Mayhew (1978) used cement layer counts in thin sections of third molars to estimate age in subfossil beavers from the Cambridge Fens, England. Coy *et al.* (1982) were able to use decalcified, stained preparations to demonstrate cement layering in about half a group of Anglo-Saxon cattle molars. Thick slabs were cut from the teeth, including the desired plane of section, demineralised in nitric acid and stained with toluidine blue. They illustrated a section with clear layering in at least part of the cement thickness, and this suggests more attempts should be made to try demineralisation if the low success rate can be accepted.

Lieberman and colleagues (Lieberman *et al.*, 1990; Lieberman & Meadow, 1992; Lieberman, 1993a, 1993b, 1994) used thin sections of gazelle teeth at the side of the root near the crown, rather than the apex or the 'molar pad' between the roots because, although the layers were thinner, they felt they were more evenly deposited. The sections were examined with polarising microscopy and Lieberman argued that the bright and dark layering represented changes in the orientation of extrinsic fibres which could result from contrasts in fodder texture between seasons. Estimation of the season at death from the outermost cement layer made it possible to suggest a shift in the pattern of mobility of hunters in the Natufian of Israel. Landon (1993) used similar techniques to examine cement layering in cattle, pig and sheep teeth from historic sites in Massachusetts, comparing urban and rural sites. This suggested that cattle were slaughtered from late summer to winter, pigs in fall and winter and sheep/goats at a peak in late summer/fall.

Other age-related histological changes
Secondary dentine deposition
Regular secondary dentine is deposited inside the pulp chamber through life (Chapter 2). It covers all walls of the chamber and continues regardless of other changes to the tooth (Philippas & Applebaum, 1966) but, in addition, patches of irregular secondary dentine are deposited, to seal the ends of dentinal tubules exposed by attrition or caries. The pulp chamber is progressively filled in and, in humans, this has been suggested as the basis of an age estimation method. The proportion filled by secondary dentine can be scored by visual examination of a section (Gustafson, 1950; Johanson, 1971), or the decreasing size of the pulp chamber can be measured (Solheim, 1992). Only modest correlations with known age are achieved, and other methods of age determination seem to be better in humans. One advantage, however, is that the size of the pulp chamber can be measured non-destructively on radiographs (Kvaal *et al.*, 1995), giving reasonable age results (Willems *et al.*, 2002) and a useful extra age estimate in archaeological material that cannot be damaged (Kvaal & During, 1999). Secondary dentine also fills in the pulp chambers of teeth in many other mammals, and could potentially be used for age estimation. This has been done, for example, in *Mustela* (King, 1991) and wolf (Landon *et al.*, 1998).

Root dentine sclerosis

In human teeth, as an adult becomes older, the primary dentine seen in root sections becomes progressively translucent, due to sclerosis of the dentinal tubules (p. 187) (Gustafson, 1950; Miles, 1963b, 1978; Bang & Ramm, 1970; Vasiliadis *et al.*, 1983a, 1983b). The apex of the root changes first and the affected zone spreads up towards the cervix. It has an irregular boundary, with a complex three-dimensional form inside the root. Thick intact sections of the tooth, examined in transmitted light, show the zone most clearly, but it can be seen in both recently extracted and many archaeological teeth simply by shining a bright light behind the unsectioned root. In the first large study, Bang & Ramm (1970) measured the length of the sclerotic zone in both unsectioned and sectioned roots for 1000 teeth extracted from 265 living patients. They found moderate to high correlations for most teeth between zone lengths and the age at which the tooth had been extracted. From this sample they devised an age estimation method, which they tested on 24 individuals not included in the original study. If the results of this test are plotted, they suggest that the method worked reasonably well (± 5 years) up until the age of 60 years, but then seriously underestimated. An independent test of Bang & Ramm's intact tooth method suggested (Willems *et al.*, 2002) that their regression equations worked well on another European sample.

Vasiliadis *et al.* (1983a, 1983b) made serial transverse sections of the roots, in which they digitised the transparent area, so that they could establish the volume occupied by sclerotic dentine as a proportion of the whole root volume. This showed a strong correlation with age at tooth extraction that suggests this would be the best method to use. Ages estimated from the regression differed from the actual ages by ± 3.5 years (although the sample did not include many teeth from people over 50 years). These results are better than most alternative ageing methods for adult humans (Hillson, 1996).

Even in recently extracted teeth, a small proportion cannot be used, because the area of transparency is unevenly distributed or the junction between opaque and transparent dentine is unclear. Archaeological teeth may show diagenetic changes that make it difficult to see the transparent area. Sengupta *et al.* (1999) compared sections in a group of recent extracted teeth with known-age archaeological teeth from Christ Church, Spitalfields, in London. They found moderate correlations between the size of the translucent area as a proportion of the root size in the recent teeth, but much lower correlations in the Spitalfields specimens. This seemed largely to be due to diagenetic changes. Other studies of archaeological material, however, have produced more consistent results (Kvaal & During, 1999) and it is clear that dentine preservation varies greatly.

Root dentine sclerosis seems to have considerable potential as an age estimation method in adult humans, which are otherwise difficult to age. It has not yet been demonstrated in the teeth of other mammals, but there is no reason why it should not occur. It might well be used for age estimation in some of the other long-lived mammals.

Gustafson's method

Age estimation is always a problem in forensic examination of human adults, and Gustafson (1950, 1966) devised a method combining six factors seen in sections of teeth:

1. attrition
2. secondary dentine deposition
3. periodontosis (recession of gingivae)
4. cement apposition
5. root resorption (at the apex)
6. root dentine transparency (RDT, root dentine sclerosis).

Each section was assessed by eye and given a score in points according to a series of criteria. For each tooth, the points were then summed and applied to a simple linear regression of summed scores against age. Correlation of scores versus age was very high and the method has been applied widely in forensic work. Independent tests (Nalbandian & Soggnaes, 1960) demonstrated a strong relationship between estimated and true age. Unfortunately, Maples & Rice (1979) found statistical errors in the method and, even with their revisions (Burns & Maples, 1976), the method performed much less well than originally suggested. Problems were later also found with Maples & Rice's approach (Lucy & Pollard, 1995; Lucy *et al.*, 1995, 1996, 2002), and with the multiple regression alternatives which have been developed. These were due to Johanson (1971), who further subdivided the scores for the criteria, and calculated a multiple regression. Similar methods were used by Maples *et al.* (Maples, 1978; Rice & Maples, 1979). An examination of the regression coefficients makes it clear that most of the criteria contribute little to the age estimation, and root dentine transparency actually does rather better on its own. 'Two criteria' formulae, including just secondary dentine and RDT (Maples, 1978), or periodontosis and RDT (Lamendin *et al.*, 1992; Foti *et al.*, 2001), have been developed but, aside from the fact that periodontosis cannot be determined in archaeological material, there seems little advantage to including even one additional criterion.

Conclusion

Most of the more complex age-estimation techniques described above have only been applied to humans. In zooarchaeology, most material comes from animals that are still developing dentally and in which attrition is so rapid that there is a clear succession of stages which were still unfolding at the age they were hunted, or culled from the herd. The difficulties of ageing in a long adult period, during which dental changes are slow and variable, are therefore less relevant. Instead, the difficulties are more to do with the butchery of carcasses and the isolated, fragmentary nature of a lot of dental material. In addition, there is variation between different populations (breeds in more recent times) of domestic animals, and even in hunted populations, where there has been strong selection particularly for the rate and pattern of development. This has made it doubly necessary to find robust seriation methods for dental development and attrition which allow comparison of dental ages, rather than chronological age. The main zooarchaeological interest in chronological age is often more related to identification of season of site occupation. By contrast, in human remains, particularly in forensic archaeology, the pressure has been to establish a reasonable estimate of chronological age. This is very difficult to do, particularly in older adults, because, as people age, all the age-related changes in their dentition and skeleton become more and more variable. It leads to difficulties in forensic identification, and for wider archaeology it is not currently possible to provide the reliable adult age estimates necessary for life tables in demographic studies. What can zooarchaeology learn from biological anthropology and vice versa? The main procedure, used routinely in studies of human remains, that might help age estimation in most non-human mammals would be X-rays of the developing dentition. In addition, most dental histology has been done using human teeth, and this shows the importance of establishing the biological basis of the structures in dentine and cement which are routinely used for ageing in wildlife management. Many human remains specialists, in turn, could learn from zooarchaeologists the important lesson that developmental stages are independent of chronological age. It is important to start from a clearly argued seriation of all developmental changes (not just dental ones), and use well-defined points in the series for comparison between different bone/tooth assemblages. The comparison is therefore in the state of development at death. Assignment of chronological age, or season, at death is hypothetical; an interpretation that must be discussed and argued separately. In particular, more attention needs to be paid to statistical methods used to infer age. There have been fundamental misunderstandings in the application of statistical techniques to age calibration which have had far-reaching effects on widely believed interpretations of archaeological development series (Lucy *et al.*, 2002).

4

SIZE AND SHAPE

Size, shape and populations
Variation in size and shape of teeth and jaws distinguishes between species (Chapter 1), but also occurs within species. Each species can usually be divided into several *populations*, or groups of males and females who tend to breed together rather than with members of other populations. They may be isolated by geography, or by behaviour that acts as a barrier. Some features or characters of tooth shape may be more common in some populations than in others, or the average of a measurable feature may differ. Within any population, there is also variation between individuals so, over a range, a feature may take on different forms or its measurable size may vary. Some individuals may have a form or size of this feature that fits within the range found in other populations, but the population as a whole can be distinguished by the frequency of different forms, or distribution of sizes. In most mammals, males and females differ in tooth and jaw size and shape – so-called sexual dimorphism – within each population. This dimorphism is largely in size, and males of most mammal species tend to have larger teeth than females. The level of dimorphism often varies between populations of the same species, and this needs to be taken into account when populations are compared. With archaeological material, it is difficult to distinguish males from females in some species, which often complicates interpretation.

Genetics, development and morphology
The mature form of any anatomical structure (the *phenotype*) is determined by a complex interaction of genotype and environment. The *genotype*, an individual's particular collection of genes, sets limits on size and shape and on the body's reaction to the environment in which it develops. The genotype represents a potential which the individual may achieve, depending on environmental factors such as diet, disease and psychosocial stress. Teeth are useful in archaeology, because each is formed once only, in childhood, when it takes the form it will have throughout life (subject to wear and disease). Unlike bones, teeth of immature individuals may therefore be compared with those of adults. In addition, environment seems to play a lesser role in dental development than in the skeleton, so it may be easier to distinguish the role of the genotype and there has been much recent work on the activities of different genes during dental development (a comprehensive catalogue is kept at http://www.bite-it.helsinki.fi). Most characters of size and shape in teeth and jaws are thought to be related to genes in several (perhaps many) loci

on the chromosomes. This is described as non-Mendelian, or multifactorial inheritance. Together with the varying environment in which growth took place, these multiple loci may combine to produce *continuous variation* in size and shape, where a measurement can take any value, within a range. This is in contrast to Mendelian inheritance, where characters are present or absent depending upon genes at one single locus. In humans, about 5000 Mendelian characters are known (see OMIM database: http://www3.ncbi.nlm.nih.gov/Omim) including some characters that involve teeth. Non-Mendelian characters, by contrast, involve many loci, when they are called polygenic, or just a few, when they are oligogenic. For continuous characters, each individual polygene has a relatively small effect on the character, but discontinuous characters are more complex. A *discontinuous character* is one which may be present or absent but such dental characters, once present, often show themselves in a continuously variable way. Many features of dental anatomy are probably inherited like this. One example is congenital absence of teeth (below), with continuous variation in tooth size but a lower limit, or threshold, below which the tooth is not formed at all. The same may be true of cusps and other features on the crown. Inheritance may be polygenic, or a few loci (even one single locus) may have a larger effect than others – major susceptibility loci. It is also possible that, if a character tends to run in families, its expression in different families may be controlled by different major loci. In addition, it is possible that the common environment shared by family members has an effect on the development of some characters.

Because members of a population tend to breed together, they share a similar collection, or pool, of genes. A reproductive barrier isolates, to a greater or lesser extent, one population's gene pool from the pools of others. Within the limits of the pool, genotypes of individuals vary, but the population as a whole is characterised by the overall frequencies of particular genes within its gene pool. If the physiology and morphology of the phenotype is linked to the genotype, then populations should similarly be distinguishable in the distribution of phenotypic characters. This is the assumption behind many interpretations in archaeology and palaeontology. For example, dental crowding and reduction of teeth are associated with what are thought to be the earliest stages of dog domestication. It might be assumed that there was deliberate selection in what, in effect, would be a carefully isolated incipient dog population, for some mix of inherited characters that included a shorter muzzle. The interpretation therefore assumes the isolation of a separate gene pool, in which selection causes a change in the frequencies of genes. Similarly, the form of teeth varies characteristically in humans between different groups of indigenous people in the world today. A few forms are very distinctive so, for some people, it is possible to say with a good deal of confidence, simply by looking at their teeth, approximately where their family comes from. If it is assumed that these differences represent frequencies of genes in the gene pools of populations separated by geography, then the distribution of dental characters can be used to reconstruct the migrations and origins of those populations. One of the difficulties with this approach is that those parts of the genome which are expressed in the phenotype are subject to selection – that is, the frequencies of these genes are determined, not only by origins, but also by evolutionary adaptation to the environment in which the population finds itself. To reconstruct population history, geneticists deliberately use parts of the chromosome (in fact the bulk of the sequence) which are not expressed as genes. The use of phenotypic characters to reconstruct population history is therefore problematic. For example, most anthropologists postulate a selection for smaller tooth size during human evolution (p. 270) so, if tooth sizes distinguish between recent human populations, it is difficult to argue that the pattern of differences results from migrations of people, rather than from their dental adaptations to their environment.

Archaeological assemblages and living populations
A population is a living entity. An archaeological collection of bones and teeth is by contrast the remnant of a death assemblage. Those individuals that die each year are a particular subset of a population, which is characterised by particular age cohorts, balance between males and females, social position and so on. In farmed and hunted animals, deliberate human selection defines the subset very closely. The central assumption of any attempt to use morphology of teeth and jaws to characterise ancient populations, and their gene pools, is that the collection recovered from an archaeological site is, in some way, representative of that ancient population. Furthermore, if statistical techniques are to be used, it is assumed that the collection is a random sample of the population – that is, any individual in the population has an equal chance of being selected. This is almost certainly untrue in most cases. Not only is initial inclusion in the death assemblage often non-random, but inclusion and preservation on the site, not to mention excavation and recovery, may favour one size over another. The difficulty is to decide how severe the problem is. One possible approach is to look for consistency. For example, do different archaeological collections behave in the same way? Does the pattern of relationships suggested by males match that of females? Do different age cohorts show the same thing? Do the measurement distributions in males and females match what might be expected in a real population (archaeological distributions are notoriously irregular, with gaps, multiple modes, and outliers)? When using statistical tests to summarise and compare results, it is necessary to be careful in the use of words which have a particular meaning in statistics. It is better not to describe an archaeological bone and tooth collection as a 'population' or 'sample' because, to a statistician, a sample is a random selection of individuals from the total population. This is a problem in many statistical applications, for example in the design of opinion surveys, but the situation in archaeology is not only far from ideal, but untestable. 'Significant' is another loaded word, implying that a test has been carried out to show that there is an acceptably small probability that the

260   Teeth

observed difference could have arisen by random variation during selection of the sample, and depends on such things as the amount of difference and the size of the sample. In effect, it is an expression of the strength of the evidence. Such tests have a number of assumptions and it seems likely that archaeological collections often violate them. In any case, it is difficult to check and it is critically important to be aware what assumptions are being made, and the nature of the data being investigated.

Measurable variation

It is usual to distinguish between *metrical* and *non-metrical* variation in tooth form. 'Non-metrical' does not imply that the feature cannot be measured; just that it is difficult to define measurements that can be taken consistently (see p. 274).

Measurement of teeth

Teeth are difficult to measure – complex, variable and rounded in form, without flat surfaces or right angles and with few easily definable reference points. Reference points for measurement need careful definition, because different observers may interpret these points slightly differently. Care is needed, not only in taking measurements, but also in interpreting published studies. Measurement of teeth, or *odontometry*, has been applied to primates for many years. Owen, in 1845, measured the teeth of chimpanzees and Muhlreiter carried out the first study of human teeth in 1874 (see Lunt, 1969). Commonly used measurements for non-Primates are given in von den Driesch (1976) which includes both measurements of the whole cheek tooth row and some individual teeth.

The main measurements taken for all mammal teeth are the maximum dimensions of the crown, as seen in occlusal view (Figure 4.1). In effect, this is the size of the box into which the tooth could be fitted. For most measurements, the mesial–distal axis is the foundation on which they are defined (although this is not always so and not necessarily explicitly stated). The *mesiodistal crown diameter*, or *length*, is therefore its dimension along a mesial to distal line. This line is mostly defined in relation to the contact points between teeth. The *buccolingual diameter* or *breadth* is usually assumed to be at right angles to the mesiodistal diameter, but this often has to be varied. The measurements are taken with simple calipers, so a perpendicular relationship can only be guessed and it is much more practical to define the measurement in terms of consistent points on the buccal and lingual crown sides. This definition of points is one of the fundamental issues in biometry. If comparisons are to be made between species, or populations, or sexes, then the measurement points used must define structures that have the same function in all the animals being compared – that is, they are homologous. The rationale for using occlusal crown diameters is that they define the main axes of occlusal, or chewing area, which is presumed to be functionally the same. In fact, if occlusal area is the important aspect, there are better ways to record it using digital photography (below), but caliper-measured diameters remain the accepted standard.

Size and shape

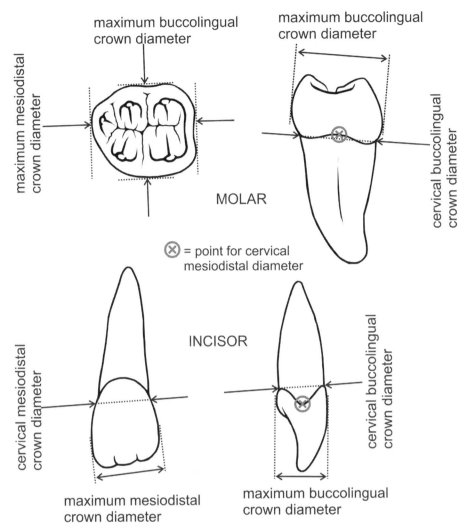

Figure 4.1 Positions of mesiodistal and buccolingual crown diameters in human teeth, at the maximum bulge of the crown sides, and cervical margin.

Definitions of measurements

Maximum mesiodistal crown diameter (length): The distance between two parallel planes, tangential to the most mesial and most distal points of the crown side. This measurement often follows an axis nearly parallel to the occlusal plane and nearly perpendicular to the buccolingual plane of the tooth, although the calipers are not deliberately held in this position. The maximum bulge of the mesial and distal crown sides, at which these measurement landmarks lie, may correspond to the contact points in normal occlusal relationship, but in some dentitions contact occurs between other parts of the teeth, or there may be malocclusions. In such cases, the line of the tooth in the dental arcade, in normal occlusion (below), is used to define the most mesial and most distal points. Where teeth are measured in the jaw, this usually requires calipers with sharp points.

Maximum buccolingual crown diameter (breadth): The maximum distance between two parallel planes, one tangential to the lingual crown side and the other tangential to the buccal crown side. For simple crowns (many incisors or canines), the buccal and lingual points are on the maximum bulge of each crown side. For more complex teeth, there may be several bulges so the measurement may be between the largest of these, or one jaw of the caliper may be laid along two bulges or ridges on one side, and the maximum taken from this point. Sometimes, several breadths are taken between different bulges. The axis of the measurement may not be far from perpendicular to the axis of the mesiodistal crown diameter (including parallel to the occlusal surface), but this is not the defining character for the measurement and, for some animals, the calipers are moved about to find a maximum. If the teeth are large enough, it is often better to have calipers with flat jaws rather than narrow points for this measurement.

Definitions are given for cheek teeth in horse, dog and bear, and for lower third molar only in cattle, pig and camel, in von den Driesch (1976). Swindler (2002) has published definitions and measurements for most Primates. Several publications deal with human measurements (Kieser, 1990; Hillson, 1996). The definitions above come largely from the discussion of Tobias (1967).

If the mesiodistal diameter is at the contact points of the crown, it is strongly affected by approximal attrition. This can readily be checked by plotting values of the measurement against attrition scores (Payne & Bull, 1988). Wear takes place at both ends of the tooth and the more convex the surface, the more rapidly the diameter changes. Most mammals, even humans, in archaeological assemblages show heavy wear so this is a serious problem. Where one site shows larger mesiodistal diameters than another, this may simply reflect a different age structure. The measurement has been proposed as a means of monitoring wear (Pike-Tay *et al.*, 2000), but there are real difficulties in using it as a measure of morphology. Buccolingual diameters are much less affected by attrition, because they are at the cingulum of the crown, near the cervix of the tooth, so occlusal attrition has to be greatly advanced to involve them. One other way to include worn teeth in a study of morphology is to take crown diameters at the cervix of the tooth, along the margin where the enamel meets the root. This has been done, for example, for cattle molars (Beasley *et al.*, 1993), deer (Pike-Tay *et al.*, 2000), and humans (Colby, 1996; Hillson *et al.*, in press). If teeth are being measured in the jaws, then strongly pointed calipers are needed to fit between teeth for the cervical mesiodistal diameters. For cervical buccolingual diameters, a special design of jaws may be needed to fit around the bulge of the cingulum (Figure 4.1). The cervix has a very different form from the contact points and the cingulum, so the measurements need to be defined differently. Often, there is a concavity at the mesial and/or distal midline of the root, and the enamel margin of the crown frequently has a curve to occlusal at this point. It is difficult to take a maximum cervical diameter because, to do that, very thin, parallel caliper points would need to be introduced between the teeth, and be made stiff enough not to bend as the calipers closed. In practice, more consistent results are obtained by using caliper points angled towards one another and taking the cervical mesiodistal diameter as a minimum (Hillson et al., in press) between the mesial and distal concavities.

Dental measurements are taken to 0.1 mm (as compared with 1 mm for most bone measurements). Teeth are small, so the precision is needed, and they yield sufficiently consistent results. It is good practice to test measurements by repeatedly measuring the same collection of teeth – by both the principal observer and different observers. These are known as intra- and inter-observer error studies and should be the first step of any project. The simplest way to present the results is to plot, for each tooth, the measurements of different observers against the mean of all observers, but several statistical tests can be used (Kieser, 1990). Hunter & Priest (1960) and Lunt (1969) found a maximum of 0.1 mm difference between observers. Barrett et al. (1963) found that the standard deviation for several observations of the same measurement was between 0.09 mm and 0.13 mm. The mean absolute error of Kieser and Groeneveld (1991) varied from 0.03 mm to 0.3 mm for an intra-observer study, and 0.13 mm to 0.47 mm for their inter-observer study. Hillson et al. (in press) found mean absolute errors of less than 0.1 mm for most measurements of most teeth in an inter-observer error study of human teeth. On this basis, the normal unit of measurement is 0.1 mm, to take into account the majority of observer error.

For humans, the definitions of crown diameters have been discussed many times (Goose, 1963; Moorrees & Reed, 1964; Tobias, 1967; Kieser, 1990). One difficulty is that definitions vary, perhaps by as much as 0.9 mm for one measurement (Goose, 1963). It is therefore important for researchers to make the definition used clear, and for care in comparing studies. Measurements may be taken using casts, from living or fossil jaws, instead of measuring the teeth themselves. The dimensions of anterior teeth measured in this way are very little different, but diameters of cheek teeth casts are often about 0.1 mm larger than the originals (Hunter & Priest, 1960). Both left and right teeth can be measured. A degree of asymmetry is common, but does not usually occur in a predictable or systematic way and does not prevent the left from one individual being compared with the right of another. Lunt (1969) measured only the right-hand side of the dentition, bringing in the left side only when teeth were missing.

These uncertainties might make tooth measurement appear somewhat futile, but it depends on what the measurements are to be used for. Differences between wild and domestic forms in the mean values of tooth size may be 5 mm or more (below). This is much greater than the variation discussed above. Even in humans, the differences in mean tooth diameter between the largest and smallest toothed populations may be as much as 3 mm (tables in Kieser, 1990). Differences between fossil and living hominoids are much larger. In monkeys and apes, sexual dimorphism (below) in canine size may be several millimetres, whereas in humans it may be just 0.5 mm, and carnivore canines, perhaps unexpectedly, often show modest dimorphism.

Crown outline may be summarised using indices. The most frequently used is the *crown area* (mesiodistal diameter × buccolingual diameter). It is common in anthropology to summarise the whole occlusal surface in the cheek row as a summed crown area. Another possibility is the *crown index* (100 × buccolingual diameter/mesiodistal diameter). Neither of these, however, gives more than an approximate idea of the crown shape. An alternative is to use shape factors which relate the occlusal area to the length of its perimeter (Barnosky, 1990). These measurements are taken from digital images. In effect, shape factors measure how strongly the outline undulates, but several other approaches are possible (O'Higgins & Johnson, 1988).

Inheritance of tooth size
Studies of laboratory mice have indicated that a considerable proportion of variability in tooth size is determined genetically (Bader, 1965; Bader & Lehmann, 1965). Studies of human families can be used to determine heritability, by comparing the measurements of (usually) fathers and their children. These have recorded high heritabilities for crown diameters (Goose & Lee, 1971; Townsend & Brown, 1978, 1979; Townsend, 1980), varying between different teeth in the dentition (Alvesalo & Tigerstedt, 1974). Twin studies – comparing monozygotic (identical) twins with dizygotic twins – have also suggested a strong inherited component (Lundström, 1948; Horowitz *et al.*, 1958; Osborne *et al.*, 1958; Osborne, 1963; Dempsey *et al.*, 1995; Hughes *et al.*, 2000). Inheritance appears to be multifactorial (Potter *et al.*, 1968; Goose, 1971; Potter & Nance, 1976; Goose & Roberts, 1982) and it appears that genes affecting crown size are present on both X and Y chromosomes (Alvesalo, 1971; Townsend & Alvesalo, 1985) as well as on other chromosomes (Potter & Nance, 1976; Potter *et al.*, 1976). Environmental factors are clearly involved in crown size variation (Garn *et al.*, 1979b; Potter *et al.*, 1983), but one of the advantages of teeth is that environmentally induced growth disruptions register as dental enamel defects (Chapter 2) so, potentially, these effects can be recognised (Hillson, 1998).

Relationships between teeth
Tooth measurements may be related in various ways:

- deciduous with permanent teeth
- interrelationships of different classes of teeth in one dentition
- maxillary (upper) dentition with mandibular (lower)
- left half of dentition with right.

Human crown diameters show a consistent pattern of positive, significant intercorrelations when one population is considered on its own. Diameters of different permanent teeth are correlated in this way (Harris & Bailit, 1988; Harris &

Rathbun, 1991). Deciduous crown diameters show low, but significant, correlations with crown diameters in permanent teeth that follow them (Moorrees & Reed, 1964; Arya et al., 1974). Maxillary crown diameters are quite highly correlated with mandibular diameters (Arya et al., 1974), and left with right (Moorrees & Reed, 1964). Size of one tooth is therefore part of a web of relationships with sizes of other teeth, and it does not make biological sense to consider a single tooth in isolation (Harris & Rathbun, 1991; Harris, 1998b). Variation in human teeth can be reduced to a few underlying axes of variation, expressed statistically as principal components (below). By far the largest source of variation (Harris, 2003) in dental measurements is tooth class (incisor, canine, premolar, molar) and it seems likely that this is also true in other mammals. Within each class, there is far less variation between the different teeth, or between upper and lower teeth. Variation between tooth classes far outstrips differences between sexes or between populations. This suggests that it is important in any study to have at least a representative tooth from each class. The difficulty for archaeology is that dentitions are rarely complete, and comparisons often have to be made using single teeth, or just a few.

Within modern human populations, tooth size does not seem to be strongly correlated with body size (Garn et al., 1968; Henderson & Corruccini, 1976; Perzigian, 1981), or with head size, but it is related to the size of tooth supporting structures (Harris, 1998b). In the Primates as a whole, there is a strong relationship between mean tooth size and body size for different species, and between dental and body size sexual dimorphism (Leutenegger & Kelly, 1977; Gingerich, 1977). The same general relationship is shown in a number of mammal groups (Gould, 1975), although there are variations.

Field and clone theory
Principal components are a statistical technique in which the variation shown in a complex original set of measurements can be expressed by a reduced number of calculated variables. If this approach is applied to a large sample of crown diameter measurements from a single human population, four main principal components are found (Harris & Rathbun, 1991; Harris, 1998b). One expresses variation in the mesiodistal diameters of incisors and canines, the second buccolingual diameters of these teeth, the third both diameters of premolars, and the fourth both diameters of molars. The same is true of maxillary and mandibular dentitions. Differences between the main tooth types are by far the greatest source of variation in the human dentition (Harris, 2003). This analysis is consistent with the *field theory* of dental development. Butler (1939, 1982) suggested that there were three developmental fields – incisor, canine and molar – each being controlled by a morphogen which diffused in a gradient along the dental lamina. The concentration of this hypothetical morphogen would be highest at the most mesial tooth in each field (the 'polar tooth') and fall off to distal. Polar teeth would show more stability in size than teeth away from the poles. Dahlberg (1945) suggested that there were four fields in humans, by adding the premolars. The field theory is still regarded as a useful concept, although

the details of its regulation have not been established. From more recent studies (above), however, it appears that canines should be included in an anterior tooth field with incisors (Harris, 1998b). In addition, within different tooth types, there is little evidence that the most mesial pole tooth is less variable than the others (Harris, 2003). More recently, doubt has been cast upon the role of morphogens (Meikle, 2002). They are certainly produced in the internal enamel epithelium, but if tooth germs are isolated at the cap stage, the teeth develop the expected cusp pattern in the absence of these morphogens.

Another possibility is that the developing tooth germs are prepatterned. This was first proposed as the *clone theory* (Osborn, 1978), which suggested that separate populations of cells (clones) define the different tooth classes. So a molar clone might start with deciduous first premolars then, by further cell division, second, third and fourth premolars, followed by the permanent molars, first, second and finally third. As cell division accumulated along the series, variability would accumulate. To this has been added the idea that the infolding of the internal enamel epithelium (Figure 3.1) and thus the form of the tooth, is controlled by the combined action of various homeobox genes in the late bud stage tooth germ, in the enamel knot (Meikle, 2002) which is associated with position of each cusp. If the action of genes in patterning the crown form is local to each tooth germ, then it is less likely that morphogens have a dominant controlling role.

Patterning of crown size in dental identification
It is possible to use measurements to distinguish between different teeth in a series. Herbivore cheek tooth rows often contain several teeth of very similar crown form, so correct identification of an isolated tooth may be difficult. Klein *et al.* (1981) used crown measurements to distinguish between lower first and second molars in red deer *Cervus elaphus*. A similar approach has been taken for cattle (Beasley *et al.*, 1993). Cervical crown diameters are used, so that the effects of different wear states can be discounted.

Dental asymmetry
Although crown diameters of teeth are correlated with those of their *antimeres* (left or right pair), there is some asymmetry in the dentition. Three types of asymmetry are recognised (van Valen, 1962; Kieser, 1990; Harris, 1992). *Directional asymmetry* is when the mean value of a measurement is larger on one side than the other – that is, one side is consistently larger as shown by a difference in means. *Antisymmetry* is similar, but applies to dichotomous features such as handedness. *Fluctuating asymmetry* is random, with the larger side varying between individuals. Statistically, it is the variation of left/right differences about the mean difference between the sides, expressed as a standard deviation.

The level of asymmetry in human dental measurements is usually quite small (<0.1 mm), relative to the precision of measurement. From the point of view of comparing species, populations and sexes, fluctuating asymmetry can be seen as random noise, which will not affect comparisons. More serious is directional asymmetry, because it implies that both sides of the dentition should be included. There is also the possibility that such asymmetry results from the observer's handedness in taking the measurements with calipers (Greene, 1984). These can be investigated when setting up a study, but most published work makes no distinction between left and right. There are, however, often patterns in directional symmetry. Where, say, in upper first incisors, the mean diameter on the left is consistently larger than the right, the lower mean tends to be larger on the right than the left. This pattern varies between teeth, and between studies (Garn *et al.*, 1966b; Townsend & Brown, 1980; Harris, 1992).

It can be assumed that genotype and environment of development would differ little on the left and right sides of the jaw, so fluctuating asymmetry is seen as evidence of reduced developmental stability (Sciulli, 2003). Experiments with laboratory rats (Siegel & Doyle, 1975; Siegel *et al.*, 1977; Sciulli *et al.*, 1979) have shown that heat, cold or noise, applied to the pregnant mother, were associated with greater fluctuating dental asymmetry in the offspring. It is suggested that such environmental stresses might decrease the ability of the organism to buffer against 'developmental noise' (Sciulli, 2003). This is related to the idea of a Selyean response to environmental stress. It is however difficult to link fluctuating asymmetry to particular prenatal stresses (Smith *et al.*, 1982). There is evidence that children of parents who smoked, or drank alcohol, during the pregnancy show more fluctuating dental asymmetry than children of non-smokers (Kieser & Groeneveld, 1998). Another study, however, showed no relationship between a number of assumed indicators of prenatal stress and fluctuating dental asymmetry in living children (DiBennardo & Bailit, 1978). High asymmetry in Neanderthals (Suarez, 1974) matched similar levels among the Inuit (Mayhall & Saunders, 1986), implying the Neanderthals were environmentally stressed, but whereas some studies of dental enamel defects support this, others do not (Chapter 2).

Sexual dimorphism
Size differences between males and females of one species are measured as a percentage (mean difference in size/mean female size × 100). The level of dimorphism varies between mammals and depends upon the aspect of size which is used. Generally speaking, it is greater in large mammals than in small, but this is not always so. Walrus *Odobenus* sexual dimorphism is 20% by body length, 50% by weight and the canine tusk dimorphism is probably greater than that (figures from King, 1964). For narwhal *Monodon*, body length dimorphism is 18%, body weight 78% and the tusk length dimorphism is around 1000%. Sperm whale *Physeter* maximum body weight dimorphism is probably the largest for any mammal, at around 200%. By contrast striped hyaena males are very similar in size to females, and spotted hyaena *Crocuta* females are something like 10% heavier than males (figures from Nowak & Paradiso, 1983). Between these extremes, most mammal families show sexual dimorphism in at least some of their genera.

Dental sexual dimorphism is most marked in tusks. This includes narwhals, as mentioned above, walrus, dugong, elephant, hippopotamus and pig. Fragmentary archaeological tusks are often difficult to measure, but the difference can be so large that males and females can be recognised by eye (e.g. pig, see Chapter 1). Carnivores show dental dimorphism, usually demonstrated in the canines and carnassials (Gittleman & van Valkenburgh, 1997), but the level of dimorphism varies widely. It does not seem to be associated with size of the teeth but, when canines and carnassials are plotted together, can be seen to have some family patterning. Canidae, Felidae and Ursidae are generally in the middle. The Mustelidae overlap them but, as with their dental morphology, this family shows a great deal of variation in sexual dimorphism and includes both the highest and some of the lowest levels. Hyaenidae and Viverridae have in general the lowest levels of dental dimorphism amongst the carnivores. Kurtén (1955, 1976) used canine diameters in fossil cave bears *Ursus spelaea* to investigate sexual dimorphism. Bishop (1982) did the same for fossil *Ursus denigeri* and *Mustela*, as did Turner (1984) for fossil cave lion *Panthera leo* canines and carnassials. Dental dimorphism has been extensively studied in the Primates, particularly in the canines and the lower third premolar which, in the males of many genera, acts like a hone against the upper canine (Garn *et al*., 1967b; Plavcan, 2001). The developmental mechanism behind Primate canine dimorphism has been investigated by Schwartz & Dean (2001). This 'canine-premolar complex' dimorphism is particularly strongly developed in the cercopithecoid monkeys and is at a maximum in baboon *Papio*, with 69% dimorphism in upper canines and 75% in lower third premolars (calculated from Swindler, 2002). *Macaca* is not far behind, with 48% upper canine and 49% lower canine dimorphism. Apes generally have somewhat less strong dimorphism, with 40% upper canine and 14% lower premolar in *Gorilla*. Humans are also dimorphic, but this is greatly reduced to around 5% for upper canines, up to 7.5% for lower canines, and practically no dimorphism in lower premolars (which clearly do not have the same honing function). Despite this greatly reduced level, sexual dimorphism has been more studied in humans than in any other mammal. It is found in both the deciduous and permanent dentitions, although it is less in the deciduous teeth (Harris, 2001). Like other Primates, human dimorphism centres on the canines (Garn *et al*., 1967b) and, combined with the rest of the teeth in a discriminant analysis, can be used to assign a sex correctly to a large proportion of individuals (Garn *et al*., 1967a, 1979a; Ditch & Rose, 1972; Garn *et al*., 1977; Black, 1978; Rösing, 1983; De Vito & Saunders, 1990). This is useful in archaeology, because the teeth are full size from the start and can be used to sex children's remains as well as those of adults. It is possible to use a baseline collection of known sex material as a reference group against which to classify fossil individuals (Bermudez de Castro *et al*., 1993) although, in a large archaeological assemblage, a better approach might be to develop a dental discriminant function from males and females within the assemblage which have been sexed by pelvis or skull. The difficulty with archaeological material is the fragmentary and incomplete nature of many dentitions, which makes it difficult to use multivariate methods.

Molleson (1993) used male and female ranges of canine diameters alone to classify children's remains.

Variation in tooth size between species and populations
Tooth measurements are sometimes crucial in identification, for example in fossil mustelids (Bishop, 1982), cats (Davis, 1987), or living rodents (Osborn & Helmy, 1980). They also show variation between members of one sex, within and between populations. Absolute size variation of individual teeth is often less marked than relative size variation between different parts of the dentition, so it may be important to use several tooth classes at the same time. In living and recent humans, for example (Harris & Bailit, 1988; Harris & Rathbun, 1991; Harris, 1998b, 2003), the sizes of any one tooth in the dentition overlap extensively between geographically defined populations, with the exception of indigenous Melanesians and Australians who tend, on average, to have larger teeth than any other recent people. If, however, the effects of size are removed, other differences are apparent. One is in the balance between short (mesiodistally), broad (buccolingually) teeth which are found in Europeans and Asians and long, narrow teeth which are found amongst sub-Saharan Africans and native Americans. Another is in the balance between anterior teeth and premolars in size. Europeans and native Americans tend to have relatively small premolars, at the expense of their anterior teeth, and Africans have particularly large premolars.

Dental reduction in human evolution
One of the main trends in the evolution of the Hominidae (the family of Primates that includes living humans and our extinct ancestors) has been one of reduction in jaw and tooth size. *Australopithecus* had relatively much larger cheek teeth than those of modern humans and, in *Paranthropus* (the robust forms), this was even more marked because the cheek teeth were greatly expanded and the anterior teeth reduced. Early forms of *Homo* in Africa had somewhat smaller cheek teeth than *Australopithecus* and the forms of *Homo* which spread outside Africa had smaller teeth still. Neanderthals were distinguished by having smaller cheek teeth, but large and robustly built anterior teeth, which were accommodated in a much broader jaw and palate. The most rapid reduction, however, and the one which has stimulated most discussion, occurred in *Homo* of Upper Palaeolithic and Epipalaeolithic/Mesolithic contexts at the end of the Pleistocene and early part of the Holocene. Studies have been carried out, in particular, using assemblages from Europe (Frayer, 1978, 1980, 1984; Hillson & Fitzgerald, 2003), Israel (Smith, 1972b, 1982; Smith *et al.*, 1984, 1986) and Egypt/Nubia (Calcagno, 1989). Specimens are often fragmentary, and few in number, there are problems in distinguishing males from females (although this is not likely to make a large difference – above), and the teeth are often worn but, if Upper Palaeolithic, Mesolithic and Neolithic, and in some studies later assemblages, are compared, there is a consistent trend for reduction in most crown diameters. This seems to be paralleled by a rise in occlusal anomalies and a

reduction in the robustness of the skull and facial skeleton. The reduction takes place over a short period, in evolutionary terms, and there are four main hypotheses to explain it:

1. The Probable Mutation Effect (PME), which suggests that a reduction in selective pressure for larger teeth, through more sophisticated food preparation and cooking, allowed teeth to reduce in size as a result of the accumulation of random mutations (Brace, 1964; Brace & Mahler, 1971; Brace & Ryan, 1980; Brace et al., 1984, 1987, 1991).
2. Selection for smaller, more energy efficient males, with correspondingly smaller teeth (but note poor correlation with body size; above). In Upper Palaeolithic Europe, males were affected more than females, and the appearance of the bow and arrow (rather than spear) at about this time might have changed the balance of selective advantage away from large, robust males (Frayer, 1980, 1984).
3. Selection of smaller teeth to match jaw reduction due to lower functional stimulation. In Central Europe, reduction was associated with the adoption of agriculture and, if a softer diet resulted in smaller jaws, this might have led to malocclusions and increased dental disease (y'Edynak, 1978, 1989, 1992; y'Edynak & Fleisch, 1983).
4. Selection for a smaller face, jaws and dentition, as one functional unit. Development of teeth is clearly related to the development of the face, and dental and face measurements are correlated (Carlson & van Gerven, 1977).

The idea of selective neutrality at the heart of the PME has been criticised (Calcagno, 1989). The last hypothesis has an appeal because, not only does it makes developmental sense not to consider the teeth in isolation, but it is difficult to see how the other explanations would constitute a strong selective advantage. Put crudely, why would having smaller teeth confer a larger than average share of offspring in the next generation?

Domestication
One of the most important archaeological discussions relating to tooth size variation concerns the early domestication of dogs. Archaeological canid skeletons which are reported as dog, rather than wolf, have been dated to 9000–14 000 years ago from western Asia (Dayan, 1994; Tchernov & Valla, 1997), northern Europe (Degerbøl, 1961), central Europe (Benecke, 1987), and north America (Grayson, 1984). There is, however, debate about both the identifications and dates (Olsen, 1985; summarised in Dayan, 1994). At the centre of the debate are the size and proportions of the teeth and the jaws. It is widely held that the earliest human manipulation of dogs is marked by a shortening of the snout (Dayan, 1994) without reduction in the teeth, and therefore also crowding and overlapping of the cheek teeth. According to this interpretation, the teeth reduced in size only later. The fossil record suggests that this imbalance would be expected in rapid evolutionary reduction in size (Gould, 1975). The specimens are, however, few in number and often somewhat fragmentary, and it is not always clear what factors other than domestication should be considered. One approach to resolving these issues is to search for independent evidence of human intervention, such as the finding of dog skeletons in association with human skeletons in burials. The earliest example of this

is from Bonn-Overkassel in Germany, dated to about 14 000 years ago (Benecke, 1987). Larger numbers have been found at Natufian sites in Israel, dating to around 11 000 years ago. Also in Israel are Neolithic and Bronze Age dogs or wolves from Jericho, dating perhaps between 10 000 and 5000 BP. This volume of evidence has therefore made Israel a central point for the debate, but there are complicating factors. Today, there are two populations of wolves in Israel: a larger Mediterranean pallipes form which is found in higher rainfall areas and a smaller desert pallipes found in dryer areas. Middle Palaeolithic wolves, foxes (possibly jackal too), wild boar and cattle of Israel were much larger than the Holocene forms (Davis, 1981; Dayan, 1994). This reduction in size took place around 10 000–9000 years ago, and is interpreted as a response to the global climatic change which occurred at the end of the Pleistocene. The dogs of the Natufian, however, were small even with respect to either modern form, as judged by the size of their limb bones. Their snouts were short compared with living wolves, although only in the anterior part, in front of the carnassials, and they were not narrower (Tchernov & Valla, 1997). The canines were similar in size to those of living wolves, but the lower and upper carnassials were markedly smaller, both in mesiodistal diameter (length) and buccolingual diameter (breadth). The more anterior premolars were slightly reduced in size so, in spite of the shortening of the anterior muzzle, there was no crowding of the teeth. Neither was there any crowding in the posterior jaw because the carnassials were small and jaw length was not reduced. The Natufian dogs, therefore, did not follow the expected pattern of changes with domestication, with reduction in snout length before reduction in teeth, but Tchernov & Valla (1997) have suggested that such changes occurred later, during the Neolithic, and represented a later stage of domestication in which deliberate selection took place. This analysis of Natufian dogs is under discussion, and Quintero & Köhler-Rollefson (1997) have instead argued that many of the dogs buried with humans might instead be wolves kept as pets. There are examples from Neolithic sites in Israel of other animals (such as gazelle, wild cattle and boar) being buried with humans in a similar way.

Skeletal and dental size differences are used to distinguish between wild and domestic forms of a number of animals, including wild boar/pigs, sheep and aurochs/cattle. This is striking, for example, in northern Europe (Higham, 1968; Grigson, 1982b), where the earliest Neolithic domestic cattle were smaller than most modern breeds, and markedly smaller than the wild aurochs of the first half of the Holocene, which were huge animals. Sexual dimorphism needs to be taken into account, because this was strongly marked in aurochs, giving a bimodal distribution for many measurements, but the size difference is so great that there is only a little overlap between the Neolithic cattle and the smaller, female mode of auroch measurements. Dentally, the long (mesiodistally) permanent lower third molar is used to demonstrate this difference. In western Asia, aurochs seem to have shown the same size decrease as wolf and fox at the Pleistocene/Holocene boundary (above; Davis, 1981). Aurochs/cattle bones from Pre-Pottery Neolithic levels at Jericho are thus smaller than Pleistocene bones, but the Pottery Neolithic cattle bones are much smaller, as are Chalcolithic, Bronze and Iron Age (Grigson, 1989). This second reduction is taken to be a result of human manipulation and domestication. A similar pattern of changes is apparent in pigs, although the amount of difference between wild boar and domestic pigs varies (Payne & Bull, 1988). Once again, the large molars, particularly the permanent lower third, are used as one of the indices of size, and one of the confounding issues is sexual dimorphism (although this is relatively modest in pig cheek teeth – as opposed to the massively dimorphic tusks). As with dog and cattle, wild boar in Israel seem to have undergone a size reduction at the Pleistocene/Holocene transition (Davis, 1981), and then a further size reduction with domestication.

There has been considerable discussion of this size reduction. Some, for example, have suggested that it was a result of selection for more easily controlled animals (Boessneck & von den Driesch, 1978), whilst others have hypothesised that it might have been due to poor nutrition and management in early flocks and herds (Higham, 1968). In the latter case, it might be expected that such growth-disrupting factors might show as developmental defects in dental enamel (Chapter 2). One of the interesting things about the size reduction, both at the Pleistocene/Holocene boundary, and with domestication in western Asia, is that there was a human dental reduction at the same time (above). That is not to say that they share the same causes, but it is a notable coincidence nonetheless.

Non-metrical variation
Inheritance of non-metrical variation
Non-metrical dental variants include the presence and size of cusps, the form of fissures in molar occlusal surfaces, presence of pits, form of ridges, presence or absence of teeth. Some aspects of these can now be measured, for example the proportion of the occlusal area occupied by a cusp, but they are normally scored as present, or absent, or graded into categories defined by a set of rules.

One of the most studied phenomena of this kind is congenital absence of teeth that would normally be expected in the dental formula. The teeth fail completely to develop, and there is no sign of them even in an X-ray. This is known particularly from the experimental work of Hans Grüneberg, who used to tell the story of arriving in Britain with his precious laboratory mice hidden in his hat (Chris Dean, pers. comm.). Some strains of laboratory mice are prone to congenital absence of the third molar whereas in other strains this almost never occurs. Grüneberg (1963) concluded:

1. Absence of third molar is linked with tooth size. Strains which frequently have absent teeth also have smaller and more variable teeth. Less affected strains have larger teeth, less variable in size.
2. The unit of inheritance is size of tooth. Absence is merely an extreme expression of this. Once size variation falls below some critical point, tooth formation is arrested altogether.

3. Disposition to third molar absence is inherited, apparently through many genes acting together. It also depends partly on lactation in the mother during tooth development, so the genotypes of both mother and offspring are involved, and environmental factors such as diet have an effect.

Grüneberg described this as *quasi-continuous variation*. The expression of the underlying variation (genetically and environmentally mediated) is controlled by a threshold; above this the characteristic is present and variable, below it the characteristic is absent altogether. This is the same as non-Mendelian discontinuous characters described above. Many similar non-metrical dental variants are known in the mouse (Grüneberg, 1965), including the appearance of extra cusps or different crown forms. Sofaer (1969) investigated a supernumerary cusp of the lower first molar and concluded that it was also best explained as variation of the quasi-continuous type.

Most work on the inheritance of non-metrical traits has been carried out on humans. Dental morphology can quickly be recorded in living people by taking dental impressions, so very large collections of casts have been accumulated. Twin and family studies have suggested a strong inherited component for at least some non-metrical features of the crown (Sofaer *et al.*, 1972; Biggerstaff, 1973, 1975, 1976; Berry, 1978; Sharma, 1992). Nichol (1989) proposed from a family study of dental characters that a range of patterns of inheritance was involved, from simple Mendelian, to oligogenic and polygenic. In baboons, two cusps arising from the cingulum, the interconulus on the lingual side of upper molars and the interconulid on the buccal side of lower molars, have been shown to have a high heritability (Hlusko & Mahaney, 2003). Genetic correlations were very high between left and right sides of the dentition, almost as high between first and second, and second and third molars, and rather lower between upper and lower teeth. The same gene or group of genes therefore controls development of the cusps on both sides, and within the molars on each side. There are, however, some differences in the group of genes controlling development of the cusps in upper and lower dentitions. The development and position of a cusp is determined mainly by the point in the internal enamel epithelium at which ameloblast precursors (Chapter 2) stop dividing and differentiate into matrix-secreting ameloblasts (Thesleff & Åberg, 1999). Developing mouse teeth have proved a good model in which to study growth, and a great deal of detail is being accumulated about the activities of different genes during dental development (Peters & Balling, 1999; Sharpe, 2000; Jernvall & Thesleff, 2000a, 2000b). It seems quite likely that in the not too distant future, some aspects of dental morphology might be linked to particular genes.

The lack of a strong directional asymmetry in non-metrical dental characters suggests that it does not much matter which side is recorded, and the usual rule is to record the strongest development of the character, whichever side it is (Turner *et al.*, 1991). There is also little evidence of strong sexual dimorphism (Portin & Alvesalo, 1974; Aas, 1983), although there is some (Harris &

Bailit, 1980). Most studies in humans have considered the sexes together. It is not clear whether or not there is a link between non-metrical characters and crown size. Numbers of the main cusps in human lower molars increase with crown diameters (Dahlberg, 1961; Garn et al., 1966a), but the fissure pattern shows no relationship with the diameters. There are conflicting results for Carabelli's trait in human upper molars (Garn et al., 1966c; Reid et al., 1991, 1992) and the interconulus–interconule of baboon teeth shows no relationship with crown size (Hlusko & Mahaney, 2003).

Recording non-metrical dental variation
Non-metrical variants in the teeth can be recorded in casts made from dental impressions taken from living mouths, and compared directly with teeth in fossil jaws. For humans, large collections of casts have been accumulated, for example in North America and Australia. A scheme has been developed for recording an array of features – the Arizona State Dental Anthropology System (ASUDAS). This was initiated by the work of Al Dahlberg who, in the 1940s (1945, 1947, 1963), started to make sets of casts (called plaques) marking different stages of development of several different dental variants. These have since been added to by Christy Turner, his students and colleagues into 30 crown traits, 6 root traits, and 2 occlusal traits. Most of these have an associated plaque and all are defined in Turner et al. (1991). The plaques may be obtained from the Department of Anthropology, Arizona State University, Tempe, AZ 85287-2402, USA. Variation for each trait is arbitrarily divided into several stages (usually five or more) which are defined simply for consistency of scoring between different observers. They summarise variation from the slightest hint up to the strongest development, and are not intended to imply that the trait is actually divided into a series of discrete stages, with no intermediates. There is therefore a certain subjective element for scoring, although this is improved by practice with the plaques and definitions, and by intra- and inter-observer error studies, before work starts on a project.

Scoring with the ASUDAS is rapid and it is the usual way in which non-metrical variants are recorded, but there are in fact several ways in which they can be measured. One possibility is to measure with calipers or a microscope eyepiece graticule between landmarks on the occlusal surface defined by intersections of fissures with one another, or with ridges (Corruccini, 1977a, 1977b, 1978). The equivalent in the ever-growing molars of voles, measurements of the prismatic elements and re-entrants, is well established (Barnosky, 1990; Nadachowski, 1991), as it is in the persistently growing cheek teeth of horses (Payne, 1991). An alternative to caliper measurements is to take photographs or digital images of the occlusal surface, and derive measurements from these. In fossil hominids, this has been used to measure the area occupied by the base of each cusp, bounded by the outline of the crown and the occlusal fissures (Wood et al., 1983; Wood & Abbott, 1983; Wood & Uytterschaut, 1987; Wood & Engelman, 1988; Suwa et al., 1994). Standard image analysis programs can be used to digitise the outlines simply and measure areas, but approximal wear causes changes which require some estimation of the outline. Heights of cusps, or three-dimensional volumes, can be determined by photogrammetry of stereo-pairs of photographs (Hartman, 1989), or by Moiré fringe contourography (Mayhall & Kanazawa, 1989; Mayhall, 1992; Mayhall & Alvesalo, 1992), in which a grating is interposed between the camera and the tooth, arranged to superimpose contours at 200 μm intervals on the tooth image. The contour outlines are digitised, so that areas, heights and volumes can be calculated. Similar three-dimensional models can be made by scanning the surface of a high-resolution tooth crown replica (Chapter 2) using a laser scanner (Ungar & Williamson, 2000). In this case, the resolution can be just a few micrometres, and it is possible to treat the 3D model like topography, measuring slopes, aspect, surface areas, volumes and many other features. Currently, this equipment is expensive and the procedure is more time-consuming than the traditional scoring of traits, but it does allow proper measurement.

Morphological variation of teeth in the Hominoidea
Close study of the teeth of any family of mammals reveals a great variety of different forms and, over 200 years of study, many papers have been written on minute variations in human and great ape tooth form. A fuller description is given in Turner *et al.* (1991), Hillson (1996), Scott & Turner (1997). Only a few variants are widely recognised and they include:

Shovel-shaped incisors: Upper incisors are normally somewhat shovel-like in hominoids, with a concave lingual surface bound by mesial and distal ridges. In some individuals, particularly people from north-east Asia and the Americas, the ridges are more strongly developed. The labial/buccal surface of the incisors may also show mesial and distal ridges – double shovelling. Some upper second incisors have greatly developed lingual ridges that join together to make a 'barrel shape incisor'.

Lower premolar lingual cusps: The buccal cusp dominates the crown in lower premolars, particularly in the sectorial third premolar of apes (Chapter 1). There are, however smaller cusps on the lingual side and these are variably developed. For recent humans, ASUDAS plaques record variation in lingual cusp number and relative size. In the extinct hominid *Paranthropus* the premolars were 'molarised', with a single lingual cusp nearly as large as the buccal, and additional small cusps along the distal margin of the crown.

Molar main cusps: Upper molars normally have four main cusps, with the hypocone being the smallest (Figure 4.2). This cusp usually reduces along the tooth row to distal, but the extent of this reduction varies (it may be lost altogether). There may also be a fifth small cusp on the distal marginal ridge of the crown. The lower

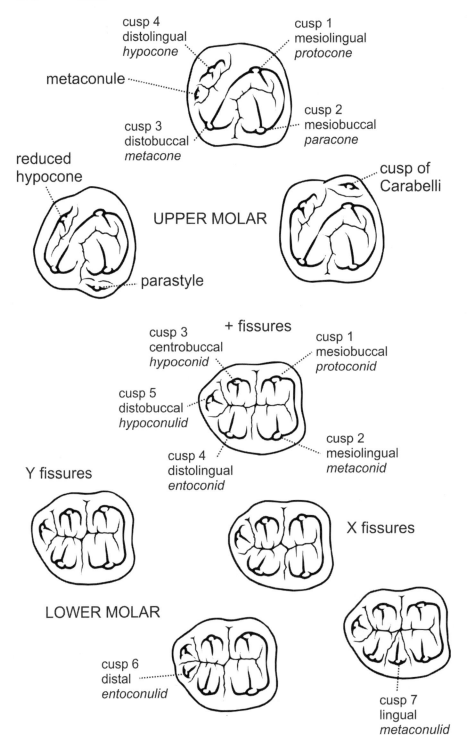

Figure 4.2 Human molar cusps and fissures.

Size and shape 277

second molar in humans usually has four regular cusps (protoconid, metaconid, entoconid and hypoconulid), the first molar five cusps (adding hypoconid), and the third molar an irregular five. In apes, all three molars generally have five cusps. Sometimes, however, they may be reduced to four (Swindler, 2002). In addition, in both humans and apes, there may be a sixth cusp (entoconulid) at the distal end of the crown and, particularly in people from Africa, a seventh cusp (metaconulid) on the lingual side. The extinct hominids *Australopithecus* and *Paranthropus* generally had five or six cusps.

Lower molar fissure patterns: In apes, the fissures of the lower molar usually meet in the middle of the occlusal surface in the so-called 'Y' configuration, where the metaconid cusp meets the hypoconid, but the protoconid and entoconid do not meet (Figure 4.2). Sometimes, all the fissures meet at one point, and this is called the '+' configuration. Both Y and + are found in living human molars, although Y is more common in first molars, and + in second and third. *Australopithecus* and *Paranthropus* lower molars were usually Y. Another form found in modern humans is the 'X' configuration, where the protoconid and entoconid meet at the centre.

Additional cusps arising from the cingulum of molars: The cingulum is the broad thickened band along the base of the crown in some mammals. It is not strongly developed in hominoids, but a number of cusps arise from this position of the crown. In upper molars, they include the cusp of Carabelli (Figures 4.2–4.3), which varies from a small rippling of the lingual crown side, up to a full-sized cusp, and the rarer parastyle, which arises in a similar way on the buccal side. In lower molars, the buccal side equivalent is the protostylid.

Root numbers: Upper premolars normally have one or two roots, lower premolars one root, upper molars three and lower molars two. These roots may show varying degrees of fusion, and occasionally there are additional roots.

These are some of the dental characters which are defined by the ASUDAS. Once they have been scored, a 'cut-off' point is usually decided for each character, a score above which it is regarded as present and below which as absent. These are then used to calculate percentage presence for each character in different populations. The percentages for several characters can be combined into a distance statistic (usually the Mean Measure of Divergence or MMD) which expresses the degree of similarity between populations, in terms of those characters. If this is done, the 'average' tooth form is found in people from South-east Asia, an area comprising Sumatra, Java, Borneo, the Malay peninsula and Indo-China. This condition was called Sundadont by Turner (1990). Least divergent from this group are people from the Pacific islands and Australia. Next in divergence are people from central and northern Asia. People from Europe, western Asia and North Africa are generally similar

Figure 4.3 Pronounced cusps of Carabelli in a deciduous upper fourth premolar and permanent first molar. Dental model of 7-year-old child.

to one another in tooth form, and diverge more markedly still from Sundadonts. Most divergent, on the one hand, are people from north-east Asia and the Americas (Sinodonts) and, on the other hand, people from sub-Saharan Africa (Irish, 1997). Dental characters are therefore able to distinguish between broad population groups of living or recent people in a similar way to skull measurements (Howells, 1973, 1989, 1995; Lahr, 1996). Genetic studies also show this pattern of geographical groupings, which is usually interpreted as evidence of an African origin for all modern human populations (Stringer & McKie, 1996; Mountain, 1998). Turner instead argued (Turner, 1987; Haeussler & Turner, 1992) that the Sundadonts might be the centre of dispersal for modern humans, but a reanalysis of the dental characters suggested that they could also be interpreted in terms of an African origin (Stringer et al., 1997). Attempts have been made to use dental characters on a finer level, to look at possible family relationships in archaeological burials. Corruccini & Shimada (2002) investigated Middle Sícan shaft tombs, dated about 1000 years ago, from Huaca Loro in Peru. In the West Tomb, one male burial was in the central chamber, with two 'sacrificial' women in niches and, at a higher level, there were two groups of nine women (South and North groups). In the East Tomb, a central male burial was accompanied by two children and two women. The dental characters suggested that the individuals in central burials of the West and East Tombs were more closely related to one another than to any of the other burials. The South group of women appeared to be closely related to one another, and to the southern 'sacrificial' woman, whereas the North group were less closely related to each other, and to any other burials. DNA was also extracted from these teeth, and mitochondrial DNA sequences were amplified for the majority of burials (Corruccini *et al.*, 2002). The results implied that the principal occupants of the West and East Tombs were indeed maternally related (the mitochondrial genome is transmitted only down the maternal line). They also implied a pattern of relationships within the South group of women, and the North group, but no relationship between them. The results gave 'some confidence' for the use of dental characters as indicators of genetic relationships.

Morphological variation of teeth in other mammals
The classic work is by Sir Frank Colyer (Colyer, 1936; Miles & Grigson, 1990), who assembled the world's greatest collection of anomalous teeth at the Odontological Museum of the Royal College of Surgeons in London. Some of the variations described are quite extraordinary, but most involve congenitally missing and additional teeth, which are described in a separate section (below). Nevertheless, similar variants to those described in humans occur in many other mammals. This includes additional roots and cusps which translate, in hypsodont teeth, into extra columns and folds down the sides which complicate the worn occlusal surface. Such variation is used, in any case, to distinguish between species (Miller, 1912; Ognev, 1948; Osborn & Helmy, 1980; Hall, 1981), but this is only one extreme of a variation which is present within species.

Premolars to mesial of the carnassials commonly vary in size and shape within species of carnivores, and may have additional cusps or roots. The carnassial teeth themselves vary too, for example in the presence or absence of a metaconid cusp in felids. The complex, bunodont molars of bears quite often have additional roots. This is also a particular feature of pigs, where the four main roots are often supplemented by additional small roots between them. Kratochvil (1981) devised a scheme to describe the complex variation in third molar cusps of pigs from the Czech site of Mikulčice. Ten or more cusps were present in many teeth. Some of these arose from the cingulum in a similar way to the interconulus–interconulid of baboons (p. 273). Examples of third molar variation are shown in Figure 4.4. The lower third molar of Bovidae and Cervidae normally has an additional element, the hypoconulid, at its distal end (Chapter 1). This varies in size (Grant, 1975a, 1975b, 1977), or may be absent altogether (Miles & Grigson, 1990; Wilson, 1974). Mouse populations can be defined by the cusp morphology of their molars, allowing the history of migration between islands and other separated populations to be reconstructed (Berry, 1968, 1969, 1975; Berry & Jakobson, 1975; Berry *et al.*, 1977; Sikorski, 1982)

The pattern of folds exposed in the occlusal surface of persistently growing equid cheek teeth varies. This is one of the ways in which different species of horse and ass are distinguished (Payne, 1991; Zeder, 1991), but there is also variation within species. The ever-growing molars of voles show consistent patterns of

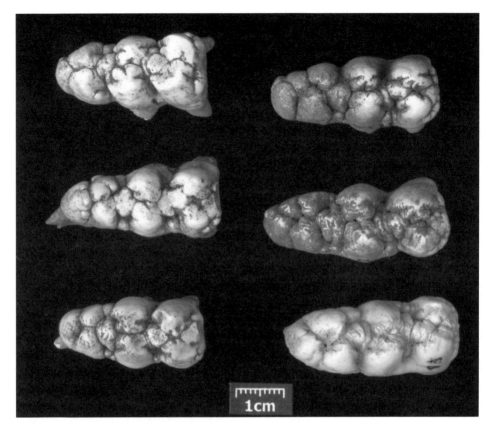

Figure 4.4 Pig lower third molar variation in cusps.

subspecific variation in the depth of re-entrant folds and form of prismatic elements (Berry & Rose, 1975; Corbet, 1975; Sikorski & Bernshtein, 1984; Barnosky, 1990; Nadachowski, 1991). A study of permanent lower third premolars in brown hare *Lepus europaeus* defined 29 variants of infoldings (Suchentrunk *et al.*, 1994). These were analysed using the MMD (above) in a similar way to studies of human morphology, showing a patterning of dental variation in Austrian hare populations that matched their geographical distribution.

More marked anomalies include *connate teeth* – partly joined, or partly divided, depending on the degree of separation (Miles & Grigson, 1990). These are seen, for example, in the incisors of Primates and carnivores, which are often joined side by side at the crown and a variable amount of root. Other teeth and other mammals may also be affected. Whales may show such teeth, sometimes joined by cement rather than by dentine. Similarly, elephant cheek teeth are occasionally joined by cement. Elephant cheek teeth may show a range of distortions to their lamellae (Roth, 1989) and, rarely, the normal curve of the tooth may be accentuated so the occlusal surface forms a 'U' shape.

Variations in tooth number

Any mammal may diverge from the numbers of teeth in its normal dental formula, either by *congenital absence* of teeth, or the presence of *supernumerary* teeth. Proper investigation requires the use of X-rays, because teeth which may apparently be absent at the jaw surface may in reality not have erupted, but still be present inside (impaction). Similarly, supernumerary teeth often do not erupt. Absence of just a few teeth is called *hypodontia*, but rarely all teeth in the dentition fail to develop (anodontia). Presence of supernumerary teeth is known as *polydontia*, or polygenesis.

In humans, at least one third molar may be congenitally absent in up to one third of the population, although the figure varies around the world (Brothwell *et al.*, 1963; Lavelle & Moore, 1973). After that, the upper second incisors and second premolars are most commonly missing, but only in a few per cent of people. The high frequency of missing human third molars is quite rare amongst mammals, although fox populations, for example, may have a congenital absence rate of 10% for this tooth (van Bree & Sinkeldam, 1969). Missing teeth are uncommon in most canids, but may be somewhat more common in domestic dogs than in their wolf progenitors. In bovids and cervids, the small permanent lower second premolar (Chapter 1) is quite frequently missing. The tooth is absent in any case in *Saiga* and *Pantholops* and, although it would normally be present in the other genera, it is quite often absent (Wilson, 1974; Andrews & Noddle, 1975; Miles & Grigson, 1990). The small, globular upper canine is usually absent in bovids, but more often present in deer – although they are quite often exfoliated at a young age. Both upper and lower first premolars are often absent in pig. Lower first premolars are almost never found in horses, but the upper first premolar 'wolf tooth' is characteristic of stallions, and may be found in some mares (Miles & Grigson, 1990).

Supernumerary teeth are much rarer than missing teeth in humans, with just a few per cent of individuals affected (Lavelle & Moore, 1973). They usually do not erupt and often have an anomalous form (a simple conical crown, or a complex form unlike any normal tooth), although they may take on the character of neighbouring teeth. Supernumeraries are more common in the apes, particularly an extra molar at the distal end of the tooth row (sometimes called a fourth molar). In other mammals too, polydontia is more common than hypodontia. Extra teeth are not uncommon amongst carnivores. Amongst canids, they are more common in domestic dogs than wolves and additional incisors are especially frequent in dogs with shortened muzzles (Miles & Grigson, 1990). Variations in tooth number generally are less common amongst ungulates but, for example, additional incisors may be present in horses and double (or even multiple) tusks in elephants.

Occlusion and malocclusion

The term *occlusion* is used to describe the way in which teeth fit together. This involves the relationship between teeth in the same jaw, and between lower teeth against upper. The latter relationship is often difficult to observe in archaeology, because upper and lower dentitions are often found in isolation and, even if it is certain they fitted together in life, most skulls and mandibles are distorted in the ground to some extent.

To dentists, there is a concept of normal occlusion, and failure to meet it is known as *malocclusion*. Cats and particularly dogs receive orthodontic treatment as well as humans and so also have a defined normal occlusion. The normal relationship is for upper incisors to overlap the lower incisors just to buccal/labial. The spatulate canines of humans do the same, but in carnivores, the canines should interlock so that the lower is to mesial of the upper. Human lower cheek teeth fit with their buccal row of cusps in the fossae of the upper molar occlusal surfaces, and the lingual row of upper molar cusps in the fossae of the lower molars. In dogs with a normal length muzzle (not a short-faced breed), the points of upper and lower premolars to mesial of the carnassials should lie alternately in between one another (Wiggs & Lobprise, 1997). The upper carnassial must overlap to buccal of the lower carnassial if it is to work properly with a shearing action. The tolerances are quite fine in the fit of a carnivore dentition, and the tall points tend to interlock with one another from an early stage of eruption, keeping them in occlusion. Malocclusions are relatively uncommon in wolves, but quite common in dogs, and seem to have appeared early in the progress of domestication (above). In short-muzzled breeds of dogs, a relationship that would be considered a malocclusion in a wolf is the normal condition. Amongst humans, much effort and money is spent in correcting occlusion but, whatever the appearance of the smile, the bunodont cheek teeth and form of the temporomandibular joint make it possible to adapt to a large range of malocclusions without great impairment of function.

Orthodontic work is also commonly carried out on horses, but here, normal occlusion is defined by the regularity of the wear facets as they meet in the occlusal plane (Allen, 2003). It is expected that the wear facets of the cheek teeth will all lie along a smooth, slightly curved occlusal plane, with the teeth all uniform in height. In the same way, the incisor occlusal plane should lie along a horizontal line, with the mesial plane between upper first incisors matching that of the lower first incisors. In the horse, departures from this even wear pattern may result in feeding difficulties and the animals are valuable enough to make it desirable to have orthodontic work to correct it.

The same terms are used for occlusions in both human and veterinary dentistry. Within one jaw, teeth may deviate from the expected line of the arcade, either to buccal or to lingual of its neighbours. This is called an *irregularity*. In addition to this, or whilst keeping in line, the tooth may be *rotated*. Sometimes a tooth erupts in the wrong position, *transposed* with its neighbour. For carnivores, it is normal for there to be *spacing* of canines and premolars, with a *diastema* (gap) between each, but this is not normal for humans. The opposite condition of *crowding* may occur too, with teeth overlapping one another. When the upper and lower dentitions are closed together, the mesial division between the upper first incisors normally lines up with the same division between the lower first incisors, at the median sagittal plane. A departure from this is called a *midline deviation*. In the cheek teeth, the lower cusp tips may be to anterior or posterior of their expected position, and this is called anteroposterior relation. This is traditionally summarised in human dentistry by Angle's classification. Angle Class 2 is where the lower teeth are displaced to distal (posterior in cheek teeth) and Angle Class 3 is where they are displaced to mesial (anterior). *Buccal crossbite* is when the lower teeth are to buccal of their normal position, whereas *lingual crossbite* is the opposite condition. In anterior teeth, *overjet* is often used to mean lingual crossbite, and *underjet* is buccal/labial crossbite. Finally, *openbite* is when the teeth fail come together at the occlusal plane. Anterior teeth are normally in slight overjet and *overbite* – that is, their incisal edges overlap. It is common for dogs with long narrow muzzles to show marked overjet, with the upper incisors projecting far forward relative to the lower incisors. Short muzzled dogs, by contrast, show pronounced underjet, and this is regarded as normal for the breed (Wiggs & Lobprise, 1997).

Malocclusions appear to be more common in modern humans than in people of the Pleistocene. This has been linked with the phenomenon of dental reduction (above), because smaller teeth are often found in association with anomalous dentitions. For modern humans, there is a Fédération Dentaire Internationale (FDI) system for recording malocclusions, which can be adapted for use with archaeological material (Baume *et al.*, 1970; Hillson, 1996). Malocclusions of all kinds are common amongst Primates in general, and in the apes almost as much so as in humans (Miles & Grigson, 1990). In wild carnivores, they are much less common, and usually involve crowding of the lower incisors, or rotations of the premolars to mesial of the carnassials. They seem to be more common in captive animals, and are pronounced in domestic dog. The lower jaw is often of normal length and occlusion in short-muzzled breeds, but the upper jaw is both shorter and wider so the two sides of the dentition come together at a pronounced angle. The only way in which the crowded premolars can fit is through overlapping and rotation. Underjet can be so pronounced that the upper incisors fit well inside the lower dental arcade. Generally, malocclusions are not common in high-crowned ungulates. In any case, the heavy wear would be likely to reduce malocclusions as the teeth continue to erupt through life to make up for the dental tissue lost at the occlusal surface (Chapter 3). Irregularities in the incisors are, however, common in horses and pronounced overjet turns the upper incisors into a beak-like protrusion known as 'parrot mouth' (Wiggs & Lobprise, 1997). Some breeds of domestic sheep commonly display underjet of the lower incisors/canines, so that they do not meet the horny pad on the upper jaw. This is commonly called undershot jaw. In pigs, there are some short-faced domestic breeds, often with irregularities in the cheek tooth row. In animals with persistently growing curved canine tusks, such as pigs and hippopotamus, or incisors, such as rodents, the teeth may fail to meet in occlusion, so that the teeth continue to grow unchecked and enter the skull or lower jaw again. It is often considered that this condition might be lethal, but van der Merwe (1997) has observed in African cane rats that the overgrowing tooth may break before it can enter the skull and then return once more into a more normal occlusion.

The differences between domestic dog breeds imply a strong inherited component to occlusion, but there is little supporting evidence from the many studies of humans. Twin studies do not produce high heritabilities for occlusal characters (Corruccini & Potter, 1980; Corruccini *et al.*, 1986; Sharma & Corruccini, 1986; Potter *et al.*, 1981; Corruccini, 1991) and comparisons of isolated villages on Bougainville Island showed so much occlusal variation within villages that it swamped the differences between them (Smith & Bailit, 1977; Smith *et al.*, 1978). One factor that must be important is functional stimulation of the jaw. Bone remodelling is known to react to the forces that are regularly acting upon it, so it is possible that heavy forces might lead to a larger jaw, with more space for the teeth (Corruccini, 1991). It is just as clear that attrition must modify the occlusion (Kaifu *et al.*, 2003). Teeth in most mammals cannot function properly until attrition has exposed ridges of the enamel in their occlusal surfaces. Approximal attrition may also proceed rapidly. Parts of the teeth that are preventing them from meeting normally are frequently worn away. The alveolar process remodels to make the adjustments required for continuous eruption (Chapter 3), and this is also likely to reduce malocclusions. It is therefore likely that older animals would have fewer malocclusions evident than younger. All these factors might also explain why captive or domesticated animals tend to have more malocclusions than wild animals. In humans, it is possible to follow these trends for living people who have recently settled from a hunter-gatherer existence. A long project (Corruccini *et al.*, 1990) with Aboriginal people from Central Australia has taken a series of dental impressions, and showed that the more recent generations have more malocclusions than the older generations who had spent part of their life eating a much tougher diet, which was prepared largely with the teeth.

Conclusion

Teeth have a complex form and, when a large collection of dentitions from one species is seen, show a surprisingly large degree of variation in form. This variation seems to be relatively little affected by the environment in which growth took place and, in any case, disruptions to growth can be recognised in affected individuals as developmental defects (hypoplasia) of the enamel and dentine. In addition, teeth are formed full-size from the start. Teeth of juveniles can be compared directly with adults, in strong contrast with the morphology of the skull, for example. Not only that, but teeth are visible in the mouths of living animals and their morphology can be studied relatively easily by taking dental impressions. By comparison, bone morphology in the living is much more difficult to study, but dental impressions of living animals can be compared directly with archaeological dentitions, and even the teeth of extinct fossil creatures. As a bonus, the teeth are one of the most commonly preserved parts, and some fossil mammal species are defined solely from teeth or jaws. Last of all, but most exciting, is the rapid progress being made in understanding the way in which tooth morphology is controlled during the growth of the jaws. The dentition makes a convenient model for studying the activity of genes during development. It is becoming one of the best known parts of the body from this point of view. This gives a real potential to studies of dental variation in archaeological and fossil mammal remains.

5

DENTAL DISEASE

Disease is an abnormality in the structure or function of the body. The definition of 'abnormal' is difficult. On the one hand, all animals show extensive variation in size, shape and physiology. What is normal? On the other hand, some diseases are so common that almost all individuals show some sign of them. Do these still qualify as diseases? Allowance has therefore to be made for a normal range of variation that does not lead to impairment of function but, once a structure or process cannot operate efficiently, it is considered to be in a state of disease. The process through which a disease is caused is known as its *aetiology*, and the site at which an abnormality occurs is known as a *lesion*. This may only be a chemical change. It may be confined to soft tissue, or it may involve widespread hard tissue destruction and repair. Usually, the bone and dental tissues that are preserved in archaeology represent only one part of the lesion. The rest has to be reconstructed and this makes diagnosis difficult. This chapter includes dental caries, periodontal disease, periapical inflammation and injuries. It does not include the defects of enamel hypoplasia even though, if they are pronounced, they may predispose to caries or tooth fracture, because they are dealt with in Chapter 2.

Disease is part of ecology. It represents the impact of the environment and the body's reaction to it. This makes disease a very useful source of information in archaeology. 'Environment' is used here in its widest sense. It may involve physical factors such as temperature and humidity, but it also involves other animals and plants. A major way in which the environment impinges is in the form of food. Diseases of the teeth reflect much of what is in the diet. Teeth are in direct contact with all the foodstuffs entering the mouth, but most of the dental diseases are related to interactions between diet and the micro-organisms that live in the mouth. The biology of much dental disease is therefore very largely the biology of the dental microbial flora.

Dental plaque
The mouth is home to a wide variety of microscopic organisms – bacteria, viruses, yeasts, protozoa – and provides many different habitats for colonisation (Hardie & Bowden, 1974). Some sites are on soft tissues (cheeks, lips, tongue and gums) but these are all impermanent from the point of view of the micro-organisms because the surface cells are constantly shed. The teeth provide non-shedding sites which allow large communities to build up, known as *dental plaque*. Different parts of the teeth create different habitats: fissures and pits; smooth surfaces of the cusps and

sides of the crown; the approximal area in between neighbouring teeth; within the gingival crevice or sulcus around the neck of the tooth. When diseases such as caries or periodontitis occur, new and quite different sites are offered as lesions develop (below). Each plaque site imposes its own conditions on its microscopic inhabitants. These include the degree of protection against the cleaning actions of saliva, lips, other teeth and tongue, and the influence of the body's defence mechanisms. They also include gradients of temperature, pH, and availability of oxygen and nutrients within the plaque deposit. Deeper parts of a plaque deposit have a different chemistry from superficial layers. Other micro-organisms occupying the site may modify the habitat. Even in sites of similar type, these conditions vary throughout the day, over periods of years, in different parts of the mouth, between individuals and between species. Disease processes complicate matters still further.

Animals are normally born with no micro-organisms in their mouths but, within a few hours, bacteria start to colonise the soft tissues. The main source of infection is presumably the mother (Marsh & Martin, 1992). When teeth appear, specialist bacteria colonise the enamel surfaces, apparently again from other members of the family group (Rogers, 1981). At each site there is a gradual build-up of species, until a stable plaque climax community is established.

Bacteria are classified by shape and by their reaction to physiological tests. They are split into two main groups through the way in which they take up Gram's stain. Gram-positive bacteria predominate in most plaque. The most common genera are *Streptococcus* (small, spherical bacteria) and *Actinomyces* (rod or filament-like micro-organisms, found only in the mouths of animals). *Lactobacillus* is another genus of Gram-positive rods. The Gram-negative group includes two genera of spherical forms (cocci) – *Neisseria* and *Veillonella* – and many rod or filament forms, notably *Haemophilus*, *Bacteroides* and *Fusobacterium*. Also important at some sites are the corkscrew-shaped, swimming, spirochaetes.

Bacteria adhere to the tooth surface, and to one another, by factors on their cell walls and by adhesives which many species produce. *Streptococcus*, *Actinomyces*, *Lactobacillus* and *Neisseria* all include species that produce polysaccharide adhesive from dietary sugar in humans. The bacterial community on the tooth surface is therefore embedded in a matrix, partly manufactured by the bacteria themselves and partly derived from proteins in the saliva and gingival crevice fluid. Matrix and community build up into dental plaque deposits with a well-organised architecture, based on the relationships between different colonies of bacteria. *Streptococcus* and *Actinomyces* are consistent inhabitants of plaque in a wide variety of animals (Dent & Marsh, 1981). *Veillonella*, *Neisseria*, *Fusobacterium* and some *Bacteroides* also occur regularly. Of the streptococci, it is the species and strains that do not produce polysaccharide adhesives that are found most widely in mammals. Streptococci that do synthesise such adhesives are common only in humans, apes, monkeys and some bats – animals that may eat fruit or other foods containing sugars.

The flora of plaque also varies between different sites. Plaque deposits in the fissures of human molars include mostly *Streptococcus*, with relatively few other genera. This restricted flora may be because fewer nutrients make their way into the fissures. Approximal plaque has fewer streptococci and more *Actinomyces*, with Gram-negative forms, many of which do not need oxygen to live. This implies that the approximal sites are less aerated than the fissure sites. For similar reasons, plaque in the gingival crevice has a much more diverse flora, including even more species which do not require oxygen.

The quantity and type of extracellular material in plaque varies. Bacteria may be widely separated by copious matrix, or may be closely packed (Frank & Brendel, 1966). Partly, the matrix consists of the polymers produced by various microorganisms to bind the plaque together. The polysaccharides synthesised from sucrose seem to be particularly important in plaque growth in humans. Plaque grows faster when sucrose is added to the diet than when other sugars, such as fructose or glucose, are added. Another component of the plaque matrix is material derived from the saliva or the gingival crevice fluid. This may also be involved in bacterial adhesion. Bathing the bacteria and matrix is plaque fluid, the chemistry of which is important in determining a variety of dental diseases. Nutrients are present as proteins and glycoproteins in the saliva and crevicular fluid. They also come into the mouth with food. Food debris is not often incorporated into plaque at most sites (Hardie & Bowden, 1974) although such particles are common in fissures (Theilade *et al.*, 1976). The nutrients diffuse into the plaque fluid through the surface membrane. Sugar molecules are small enough to enter directly, but starches are long molecules and need to be broken down into their component sugars first. Enzymes present in the saliva, and produced by some plaque bacteria, accomplish this. Similarly, proteins and fat molecules are too long to diffuse into plaque directly and a battery of bacterial enzymes allows proteins to be broken down into their constituent amino acids, which then enter the plaque. The role of fats in plaque physiology is not clear. Once in the plaque fluid, sugars and amino acids are metabolised by the bacteria. Sugar metabolism in the plaque produces mostly acid waste products and amino acid metabolism produces alkaline. The pH of the plaque fluid therefore reflects at least partly a balance between nutrients available to the bacteria. 'Resting' pH of plaque is near neutral, like that of the saliva. Changes from this state occur on a timescale of minutes, and can be monitored in several ways, to produce time versus pH graphs known as Stephan curves (Mundorff *et al.*, 1990; Lingström *et al.*, 1993). In addition, the nutrients available control the flora of the plaque so that, for example, animals eating a sugar-rich diet have different species present from those that do not.

Dental calculus

The deepest layers of a plaque deposit, next to a tooth, often become mineralised (Ten Cate, 1989). The source of the mineral is the plaque fluid, although it ultimately derives from the saliva, which is saturated with calcium phosphate. The mechanism which triggers it is unclear, but the first parts of the plaque to be mineralised are the cell walls of the bacteria. These are followed by the matrix, and the centres of the

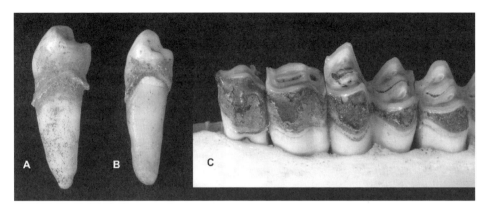

Figure 5.1 Calculus deposits. A: supragingival calculus in a human permanent premolar in which the root had been exposed during life (probably by periodontal disease, because the tooth is so little worn), and the calculus deposit has followed the line of the gingivae, down onto the root. B: a similar tooth, in which exposure of the root has not been so marked, and the supragingival calculus is present on both the crown and the root. C: sheep lower premolars and molars showing the dark, shiny calculus deposit which is often seen in ruminants.

cells are often left as voids. Mineralised plaque deposits are commonly seen on the surface of teeth and they are known as *dental calculus*, or *tartar*. The bacteria whose outlines are fossilised in such deposits are dead but in the living mouth, the calculus is covered with a layer of living plaque. In dry skulls and archaeological specimens, the appearance of the calculus varies between animals. Humans often have thick, clay-like deposits (Figure 5.1) with an uneven surface, fixed to the crown running along a line representing the edge of the gingivae in life. Fresh calculus deposits vary from pale brown or yellowish to darker brown or black but in ancient material, the colour may be affected by diagenetic changes and most are a uniform pale brown. Properly speaking, such deposits are called supragingival calculus (p. 305). Where the deposit is large, it may overhang the gingivae and, where the roots have been exposed by eruption or periodontal disease, it may spread onto the roots. Different individuals show wide variation in the extent of calculus deposits, but the mixture of inherited, dietary and other factors responsible is not well understood. The distribution through the mouth is strongly related to the position of the main ducts for the salivary glands, because the ultimate source of the mineral is the saliva. The ducts of the sublingual and submaxillary salivary glands emerge under the tongue, and those of the parotid gland emerge inside the cheeks by the upper molars. In humans, calculus therefore builds up particularly on the lingual surfaces of incisors and canines, and buccal surfaces of upper molars.

Domestic cats may have large calculus deposits like those of humans (Figure 5.4 below), particularly around the carnassials and canines (Jubb & Kennedy, 1963). Calculus is rare in wild cats and, because it is only domestic cats fed chopped meat (rather than larger pieces they have to use their teeth on) that seem to develop the deposits, the cause is thought to be the texture rather than the content of the diet (Miles & Grigson, 1990). Domestic dogs may develop similar large calculus deposits. Other domestic animals, and captive wild animals kept in parks and zoos, may show them as well. Many other animals, both wild and domestic, develop darkly stained deposits on their teeth which are presumably also calculus, but little is known about them. Usually, they are thin and adhere strongly to the tooth surface. In deer and sheep, they often have a metallic lustre (Figure 5.1). The origin of the colour may be dietary, but it is also possible that some plaque bacteria produce a dark stain.

The best technique to examine the microstructure of calculus is SEM-BSE (Chapter 2), where it shows as moderately heavily mineralised – more so than dentine and cement, but less than enamel. It often shows patches of regular layerings of different density, interspersed with irregular voids and areas of lower density. The positions of bacteria are represented by small (about 2 μm diameter) rounded voids representing coccal forms and tubules representing filamentous forms. Such 'fossil bacteria' have been demonstrated in Medieval human calculus (Dobney & Brothwell, 1986). The mineral component of human calculus is mostly calcium phosphate, but there are whitlockite, brushite and octacalcium phosphate as well as apatite. In other animals carbonate may be a major component in addition to calcium phosphates (Unmack & Rowles, 1963; Weaver, 1964).

In humans, large food components are rarely incorporated into calculus but fibrous plant material may be wedged between teeth and into the infundibulum, particularly in horses and rhinoceros, but also cattle. Dobney and Brothwell (1986) found a range of food debris in archaeological calculus: pollen grains, chaff and animal hairs, and phytoliths. Phytoliths are microscopic silica bodies found in many plants. They are extracted from calculus by dissolving it in hydrochloric acid, centrifuging the resulting suspension and using heavy-liquid flotation to separate the phytoliths (Gobetz & Bozarth, 2001). Phytoliths have been extracted in this way from cattle (Armitage, 1975), deer, moose, tapir and mastodon calculus (Gobetz & Bozarth, 2001). Also incorporated into calculus are shed cells from the oral mucosa lining the mouth. These cells, gathered by rubbing a swab inside the cheeks, are the usual source of DNA in genetic studies of living people. Calculus therefore acts as a repository of DNA in forensic and archaeological material. It has been shown that sex can be determined from the DNA extracted from small calculus samples removed from human teeth (Kawano *et al.*, 1995).

Dental caries

The disease dental caries, commonly called dental decay, is a progressive demineralisation of the enamel, dentine and cement. It is common particularly in living humans, but is also seen in domestic animals and, in the wild, amongst apes and monkeys. The cavities which are the most obvious sign of the disease are, in fact, a late stage in the development of the lesion. A great deal of demineralisation takes place underneath the surface, before it becomes visible. In most cases, the disease is slowly progressive and shows a strong relationship with age. The demineralisation of dental caries is caused by organic acids which are formed during the fermentation of carbohydrates by the plaque bacteria. The strongest association is with the proportion of sugars in the diet. This is known from the effect of sugar rationing during the Second World War in a number of countries, and other studies which have shown that there is a clear relationship between the weight of refined sugar consumed per person per annum and the number of teeth affected by caries (Sheiham, 1983; Rugg-Gunn, 1993; Navia, 1994). The relationship of other carbohydrates, such as starch, with caries is less clear but they do seem to have a role. Proteins and fats in the diet do not seem to be involved, although there is some evidence that the protein casein, which is present in milk products, has a protective effect (Bowen & Pearson, 1993; Mundorff-Shrestha et al., 1994).

Coronal caries
Carious lesions in living humans are most commonly initiated in the enamel of little-worn tooth crowns (Thylstrup & Fejerskov, 1994; Hillson, 2001). The two main sites are the fissures of the occlusal surface, and just below the contact points between neighbouring teeth (the approximal areas of their crowns). These are often distinguished as *occlusal caries* and *contact area caries*. Plaque in these locations is protected from the cleaning actions of lips, cheeks and tongue. In approximal caries sites, the lesion is initiated under the smooth enamel surface. At first the changes are microscopic, and can be seen only in sections (Chapter 2). The first clear sign at the surface is the appearance of a tiny opaque white or brown spot in the otherwise translucent enamel. This gradually grows in size and the enamel surface becomes slightly roughened to the touch. It is only at about this stage that the lesion becomes visible in an X-ray as a radiolucency. The surface of the enamel eventually breaks down and the cavity forms (Figure 5.2) but, even then, the lesion usually progresses slowly, and it may stabilise. The cavity gradually increases in size until it reaches the EDJ, at which point the lesion often becomes wider as the walls of the enamel cavity are undermined. Once it reaches the dentine, the character of the lesion changes and the dentinal tubules are invaded by a specialised bacterial flora. From the first involvement of the dentine, the pulp protects itself by deposition of a reparatory patch of secondary dentine. Eventually, however, these defensive measures may be breached and the pulp is exposed to infection. The pulp becomes inflamed (below) and may die, resulting in an inflammation of the periodontal tissues around the root apex. This can be very painful and, for most of human history, the only sure remedy has been extraction of the tooth. For at least the past two centuries this has been the most common cause of ante-mortem tooth loss.

Occlusal carious lesions are usually initiated deep within the fissures of molars in living humans, so that most of the early changes cannot be seen from the surface. Often, there are several lesions in one fissure system. Even X-rays do not provide much evidence, because the overlapping folds of enamel in the occlusal part of the tooth crown obscure what is going on. Once a cavity is formed, this usually

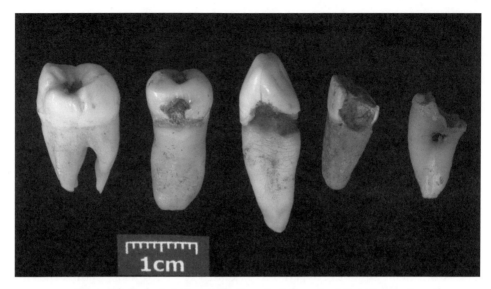

Figure 5.2 Caries in human teeth. From the left: occlusal fissure caries in an upper third molar; contact area caries in a lower fourth premolar, root surface caries in a lower canine; gross approximal caries with the lesion bridging the root surface, crown side and worn occlusal surface; gross gross caries in which the lesion has removed the whole tooth except for the apical region of the root.

becomes apparent at the surface by a slight widening of the fissures (Figure 5.2), or perhaps a dark stain (Penning *et al.*, 1992; van Amerongen *et al.*, 1992; Tveit *et al.*, 1994; Ekstrand *et al.*, 1995). The lesion usually rapidly involves the dentine, and undermines the enamel. Pressure on the occlusal surface then causes the enamel to collapse and this exposes the cavity. X-rays may be useful once the dentine is sufficiently demineralised to show as a radiolucency below the pronounced radio-opacity made by the occlusal enamel (Espelid *et al.*, 1994; Lussi, 1996). Progression through the dentine, and pulp exposure, follow as described above. In domestic horses, a similar lesion is initiated in the deep infundibulum which is analogous to the fissures of human molars. The infundibulum is mostly filled with cement, but there is a narrow space down the middle within which plaque accumulates and the lesion progresses rapidly deep inside the tooth crown (Miles & Grigson, 1990).

Root surface caries
In humans, where the cement–enamel junction (CEJ) and root surface are exposed above the gingival cuff by continuous eruption, or periodontal disease (below), plaque accumulates and carious lesions may be initiated (Thylstrup & Fejerskov, 1994). They are first apparent as a dark stain running around the root just above the line of the gingivae (Figure 5.2). Frequently, the stain runs along just below the CEJ. The lesion usually progresses very slowly, with an increasing area of stain and, in living people, softening of the root surface so that it feels leathery when pressed

with a probe. Eventually, a broad, shallower cavity forms. In humans, the layer of cement is very thin, so the dentine is quickly involved in the lesion. It is relatively rare, however, for the pulp to be exposed in the same way as for coronal lesions. Root surface caries is very much a disease of middle-aged adulthood, because it is necessary for the root to be exposed first. In archaeological specimens it can be difficult to decide if this has taken place. For statistics on root surface caries to mean anything, it is not only necessary to count the lesions present, but also the number of exposed root surfaces which could potentially have developed lesions. The simplest approach is to measure the CEJ-AC distance (p. 312) which, in the absence of periodontal disease or before substantial continuous eruption, should be less than 2 mm. This can be done rapidly with a graduated periodontal probe (Hillson, 2000, 2001). If the measurement is more than 2 mm, a potential site for a carious lesion is counted as present. Care is needed in the case of postmortem damage to the alveolar process.

Caries of worn tooth surfaces
Most of the dental literature deals with the relatively unworn tooth crowns of living people. The great majority of archaeological dentitions, however, show a great deal of wear which exposes dentine in the attrition facets (Chapter 3). Hillson (2000, 2001) found that carious lesions were initiated in both occlusal and approximal wear facets, or may perhaps have remained from a previously existing lesion in the crown or root surface. Very little is known about the aetiology of such lesions, but they are very common, for example, in European Medieval dentitions. They seem particularly to develop along the EDJ where it is exposed in wear facets but, often, the cavity is too large for the site of initiation to be apparent. In much archaeological material, amongst recent people with a heavy tooth wear rate, and in wild apes, dental caries is particularly a disease of the worn teeth of older individuals. This may be related to dentine exposure, or the trapping of food between teeth, or the fracturing away of enamel from the attrition facet edges, to create traps for plaque to accumulate. In any case, the lower prevalence of caries in younger individuals whose teeth were already worn enough to expose dentine implies that such lesions are actually initiated in the wear facet. There have been some suggestions that heavy wear reduces dental caries by removing the vulnerable occlusal fissures. Whilst this may be true in younger individuals, wear seems positively to be related to caries in older adults.

Gross caries
A carious lesion is described as gross when it is so large that it is no longer possible to determine the exact site at which it was initiated (Moore & Corbett, 1971; Hillson, 2001). There are various categories of gross caries (Figure 5.2). It may, for example, be possible to see that the lesion was initiated either in an occlusal fissure or at the contact point of the crown. This would be gross occlusal–approximal coronal caries. Alternatively, the lesion may involve both the root surface and the approximal crown

side, so that it is not possible to determine whether it was initiated on the crown or the root. This would be simply a gross approximal lesion. Sometimes, a lesion may be so large that it is only possible to say it was initiated in that particular tooth. This is a gross gross lesion. Gross lesions cause difficulties in interpretation because the original lesion at its site of initiation may have developed in a variety of ways, which cannot be assumed.

The histology of caries

The progress of a carious lesion in enamel can be followed by ordinary bright field transmitted light microscopy of a thin section mounted in Canada balsam or quinoline (Darling, 1959, 1963; Silverstone *et al.*, 1981; Thylstrup & Fejerskov, 1994). At the earliest stage of development, the lesion appears as a small triangular translucent zone which underlies a thin layer of unaltered enamel at the crown surface. The enamel is translucent at this point because demineralisation has increased the pore space to the point at which it can imbibe mounting medium, which reduces RI contrast between pore space and mineral in the enamel (Chapter 2). As the lesion grows, this translucent zone increases in size, and a dark zone appears at its centre. The dark zone is also partially demineralised, but the pore spaces are still sufficiently small that the mounting medium cannot enter them, so that air is trapped. RI contrasts are increased, light is scattered, and the zone is dark in transmitted light microscopy. As the lesion grows further in size, another translucent area appears within the dark zone. This is known as the body of the lesion. Once again, it is demineralised to the extent that mounting medium can be imbibed. Only after this point is the lesion large enough to be seen as an opaque spot at the surface, and the thin surface layer is still intact (although SEM studies show that it is penetrated by tiny channels, see Haikel *et al.*, 1983). After this, the surface starts to break down, takes on a chalky appearance and can be scratched with a sharp probe. As enamel is lost, a cavity forms. Examination by compositional SEM-BSE confirms that there is a zone of deficient mineralisation underlying a thin intact surface layer in a carious lesion (Jones & Boyde, 1987). It also shows that the advancing margin of the lesion is irregular and exploits the prism boundaries, brown striae of Retzius and prism cross striations.

Thin sections in ordinary light microscopy also show the effects of caries on the dentine (Thylstrup & Fejerskov, 1994). Even before the lesion reaches the EDJ, the odontoblasts with processes in the dentinal tubules immediately underlying the lesion are affected by its chemical products. A patch of reparative secondary dentine is deposited over the pulpar ends of the tubules. The mounting medium is then unable to penetrate into the affected tubules, which remain filled with air and are dark in transmitted light. This is known as a 'dead tract'. In SEM-BSE examination, the secondary dentine can be identified by its more irregular pattern of dentinal tubules. When the lesion moves into the dentine, it is marked by four zones. Closest to the pulp is the zone of sclerosis, in which large crystals fill the lumen of the dentinal tubules (Schüpbach *et al.*, 1992). They show in compositional SEM-BSE as dense material within the peritubular dentine (Jones & Boyde, 1987). Next is the zone of demineralisation, in which the organic component of the dentine remains relatively intact, but the mineral component is removed. In ordinary light microscopy, the structure looks normal but in SEM-BSE, the peritubular dentine can be seen to be missing. Further into the core of the lesion is the zone of bacterial invasion, where bacteria are present in the tubules. Where there are clusters of bacteria, they break down small areas of organic matrix and distend the soft demineralised dentine structure, with other tubules compressed in between. Finally, at the centre of the lesion is the zone of destruction, where masses of bacteria occupy larger areas in which the matrix has been removed.

Root surface lesions quickly involve both cement and dentine (Nyvad & Fejerkov, 1982; Schüpbach et al., 1989, 1990, 1992). In the earliest stages, there is a thin fully mineralised surface layer, with a demineralised zone underneath. The demineralisation exploits the layering within the cement and when it enters the dentine it also follows the incremental structure. Lesions of this type spread more readily sideways, making broad shallow cavities, and do not penetrate towards the pulp chamber so rapidly as coronal lesions.

Counting carious lesions

Epidemiological studies of living people use DMF scores (Decayed Missing Filled) to quantify carious lesions. DMF-T scores count the number of teeth in each individual which show a carious lesion, are missing, or have a filling. The unwritten assumption is that all normal people have 32 permanent teeth and that any missing tooth has been lost due to caries. Such scores also fail to account for the possibility that some teeth have more than one lesion. For this reason, some studies use DMF-S scores in which each of the five crown surfaces (occlusal, mesial, distal, buccal, lingual) is counted separately. The assumption here is that $5 \times 32 = 160$ surfaces were present and that all five surfaces were carious in the case of missing teeth. No DMF scores distinguish between teeth, types of carious lesion or age groups, and modifications need to be used for more detailed studies even in the epidemiology of the living (Manji et al., 1989, 1991; Thylstrup & Fejerskov, 1994; Hillson, 2001). The assumptions of DMF scores are not appropriate for archaeological assemblages, in which teeth may be missing due to trauma or periodontal disease during life, postmortem tooth loss, or loss of whole segments of the jaw. The crown may also be missing due to heavy attrition and therefore incapable of showing whether or not it was carious. Instead, many archaeological researchers have found it most practical to treat each tooth as an isolated case, and express teeth with carious lesions as a percentage of all teeth present, whichever individual they belong to. This, however, makes no distinction between different tooth types, lesion types or age groups, all of which are important information in understanding the changing nature of dental caries in archaeological assemblages (below). It is necessary instead to count each potential lesion site on each tooth separately (Hillson, 2001). These were called 'sites at risk' by Moore and Corbett (1971). Lesions are

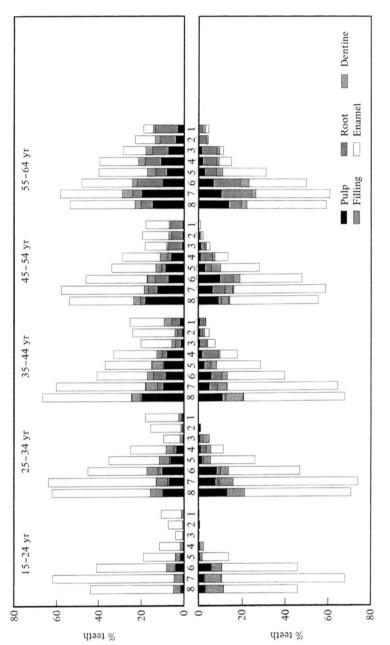

Figure 5.3 Distribution of dental caries in different tooth classes (numbered from 8 for third molar to 1 for first molar), in upper and lower jaws, for different age groups of recent rural Kenyan people. The different types of shading represent different types of caries. Reproduced from Figure 4.7, page 86 in Manji *et al.* (1991), with permission from Cambridge University Press.

then expressed as separate percentages for each site at risk. This makes for complex tables which can be summarised most easily by stacked bar charts (Figure 5.3 above).

Another problem is the diagnosis of carious lesions. If the early 'white spot' or 'brown spot' lesions are included, there may be variation between observers and potentially confusion with diagenetic changes. By contrast, if they are excluded, then a large amount of evidence for caries is lost. One answer is to record both spot lesions and cavities, and make it explicitly clear which is which in the scoring system (Table 5.1). It is also necessary to be clear what type of gross lesion is present, because of the inferences that can, or cannot, be made.

The distribution of caries
In human permanent teeth the main home of caries is the molars. Upper molars are affected more than lower molars. Incisors and canines are much less affected by caries, although they become more involved in older age groups. Upper incisors and canines tend to show more caries than lower. The left dentition mirrors the right so, if time is pressing, it is possible to study just one side. Ideally, separate statistics should be kept for upper and lower teeth, although the difference is not large, but the most important thing is to distinguish between different tooth types. On archaeological sites, incisors and canines are much more frequently lost post-mortem than molars. An apparent difference in caries rate between two sites may simply reflect differences in the preservation of incisors. Molars are the home of dental caries in deciduous teeth too, although these are much less studied in modern clinical literature. The great majority of studies involve the permanent teeth of children in Europe and North America. In industrialised urban populations over the past 150 years, caries has been a disease which starts in the crowns of children's teeth. There is a hierarchy of teeth involved (Sheiham, 1997). The primary site is the occlusal fissures of first and second molars, which are affected even in the lowest caries rate populations. Next are the approximal surfaces of the molars, followed by the occlusal surfaces of the premolars, and the upper incisors (usually approximal sites or the lingual pit which is found in some incisors), and finally the approximal surfaces of the premolars, and the upper canines. In general, dental health is poorer in less well-off socioeconomic groups, and amongst people who live in towns, rather than in a rural setting (Locker, 2000; Källestål & Wall, 2002). During the years leading up to the end of the twentieth century there was a marked decline in caries in Europe and North America, which seems to be related partly to fluoridation of water supply and toothpaste, and to a general improvement in the level of overall health.

There are far fewer studies dealing with all age groups, from children to elderly adults (Thylstrup & Fejerskov, 1994), but some of the most useful for archaeological purposes have been done in rural Kenya (Manji *et al.*, 1991) and the area of Beijing in China (Manji *et al.*, 1989). The Kenya study was carried out amongst smallholding farmers with very limited access to dentists. Few had fillings, although

Table 5.1 *Scoring for carious lesions*

Score	Crown caries (including occlusal, contact area and smooth surface)	Root surface caries
Missing value code	potential lesion site missing for any reason (attrition, tooth fracture, tooth loss), or fully obscured (calculus, concretion)	no part of root surface or cement–enamel junction (CEJ) preserved, or at least not visible if present
0	site present but enamel is translucent and with a smooth surface	root surface/CEJ present and visible, with no evidence of staining or cavitation
1	white or stained opaque area in enamel with smooth glossy or matte surface	area of darker staining along CEJ or on root surface
2	white or stained opaque area with associated roughening or slight surface destruction	Not used
3	small cavity where there is no clear evidence that it penetrates to the dentine	Not used
4	discoloration in exposed dentine of an attrition facet (only used for worn teeth)	Not used
5	larger cavity which clearly penetrates the dentine	shallow cavity (stained or unstained) following the line of the CEJ, or confined to the surface of the root
6	large cavity within the floor of which is the open pulp chamber, or open root canals	cavity involving the CEJ or root surface alone, within the floor of which is the open pulp chamber, or open root canals
7	gross coronal caries involving this particular lesion site and one or two more neighbouring coronal lesion site(s)	gross cavity, including the CEJ or root surface, which involves any neighbouring crown sites
8	gross coronal caries, within the floor of which is the open pulp chamber, or open root canals	gross cavity, as in Score 7, within the floor of which is the open pulp chamber, or open root canals

Notes: Adapted from Hillson (2000, 2001). Record separate scores for occlusal, mesial contact area, distal contact area, buccal crown side, lingual crown side, mesial CEJ, distal CEJ, buccal CEJ and lingual CEJ. Record separate scores for each tooth.

they had access to tooth extraction in the case of acute pain following exposure of the pulp. A total of 1131 people were studied, divided almost equally into males and females, and into five age cohorts from teenagers to elderly adults. In all age cohorts, molars (especially second molars) were more affected by caries than canines and incisors, with premolars in between (Figure 5.3). The number of molars with lesions rose until middle age, but stayed approximately constant thereafter. The number of premolars, canines and incisors with lesions just increased steadily with age.

- The 15–24 year age group was characterised by enamel lesions.
- The 21–34 year group had slightly fewer enamel lesions but more penetrating into the dentine, and some exposing the pulp.
- The 35–44 year group had similar rates of enamel and dentine lesions, but more of these exposed the pulp and more teeth were lost by extraction.
- The 45–54 year group showed fewer lesions involving only the enamel and dentine, more exposing the pulp, more extractions, and an increase in root surface caries.
- The 55–65 year group followed these trends further.

Overall, caries rates were similar in both males and females but, when the age cohorts were considered separately, females consistently had slightly higher rates of caries than males. Relatively few individuals had a large number of carious lesions. Far more people had just a few lesions, particularly in the younger age groups, and it was especially people with a moderate number of lesions who showed the largest rises with age.

The study of the Beijing area (Luan *et al.*, 1989) combined both urban and rural people with a greater access to dental care. They therefore showed more fillings, and a much stronger rise of caries involvement with age, in the molars as well as in the incisors and canines. In addition, carious lesions exposing the pulp rose strongly with age, until they predominated in the oldest age group. Otherwise, the general pattern of trends showed many similarities with the rural Kenyan study. Root surface caries was separately graphed in relation to the exposure of roots, and it was apparent that the degree of gingival exposure was not strongly related to the frequency of carious lesions on the exposed root surface, although both increased with age.

These studies have a number of important lessons for archaeology. One is that it is vital to split the assemblage studied into different age groups, if the nature of caries epidemiology is to be understood, and because varying preservation of different age groups between sites would make comparisons meaningless if these groups were not separated. Another point is that the different categories of lesions need to be separated because they show different trends with age. Few studies have done this, and the largest to date was carried out for British material, ranging from Iron Age, to Romano-British, Anglo-Saxon, later Medieval, seventeenth-century and nineteenth-century contexts (Moore & Corbett, 1971, 1973, 1975; Corbett & Moore, 1976). The first problem was to determine the age at death in a consistent way. Often, the jaws were preserved without the rest of the skeleton and it was most practical to use an age estimation method based on the teeth themselves. Moore and Corbett divided the individuals into five categories (Table 5.2) on the basis of tooth eruption and wear. These are neither independent of one another, nor independent of the pattern of caries, but no other method is available in these circumstances, although at some sites age may be estimated from another part of the

Table 5.2 *Dental caries in Anglo-Saxon British dentitions*

Age/tooth wear categories	1 Up to eruption of 3rd molar	2 Areas of dentine exposed on one or more molars/ premolars	3 Occlusal fissures obliterated on one or more teeth	4 Occlusal surfaces of all cheek teeth worn flat	5 Attrition removed crowns of cheek teeth
Caries sites (all teeth together)					
Occlusal gross caries	2% n = 92	9% n = 130	16% n = 114	14% n = 104	13% n = 39
Occlusal fissure caries	2% n = 92	16% n = 130	5% n = 114	4% n = 104	3% n = 39
Occlusal cuspal caries	0% n = 92	1% n = 130	1% n = 114	0% n = 104	0% n = 39
Approximal gross caries	0% n = 93	11% n = 130	23% n = 114	26% n = 106	18% n = 39
Approximal contact area caries	0% n = 93	10% n = 130	4% n = 114	1% n = 106	3% n = 39
Approximal CEJ caries	3% n = 93	11% n = 130	18% n = 114	32% n = 106	26% n = 39
Buccal gross caries	0% n = 93	7% n = 130	7% n = 114	10% n = 106	13% n = 39
Buccal fissure caries	2% n = 93	3% n = 130	2% n = 114	0% n = 106	8% n = 39
Buccal CEJ caries	0% n = 93	4% n = 130	4% n = 114	8% n = 106	8% n = 39
Teeth (all sites together)					
First incisor	0% n = 78	2% n = 108	0% n = 96	1% n = 77	0% n = 25
Second incisor	0% n = 72	2% n = 117	2% n = 104	1% n = 88	0% n = 30
Canine	0% n = 30	2% n = 122	0% n = 111	4% n = 95	3% n = 36
Third premolar	0% n = 63	2% n = 128	9% n = 112	5% n = 98	3% n = 36
Fourth premolar	2% n = 59	9% n = 129	6% n = 110	12% n = 100	9% n = 35
First molar	2% n = 92	12% n = 128	22% n = 113	25% n = 84	27% n = 33
Second molar	8% n = 61	14% n = 128	18% n = 113	32% n = 91	28% n = 32
Third molar	8% n = 12	2–2% n = 124	18% n = 103	25% n = 69	21% n = 28

Notes: Data from Moore & Corbett (1971). The table gives the numbers of teeth with caries 'sites at risk' (see text), and the percentage of these teeth with carious lesions.

skeleton, such as the pelvis. Most archaeological studies therefore examine, not age-related change, but the relationship between caries, dental development and wear. It is important to state explicitly how the different age categories were defined. Moore and Corbett gave most detailed tabulations for their Anglo-Saxon sample, although the Iron Age, Romano-British and Medieval material had a very similar pattern of occurrence for caries. In all the different age (tooth wear) categories, carious lesions were most common in molars and least common in incisors and canines. There were no significant differences between age/wear categories for incisors, canines, premolars or third molars, but the first and second molars showed a progressive increase in caries frequencies. In the youngest age/wear category, occlusal lesions and buccal pit lesions were most frequent, although never common. In all other tooth wear categories, lesions at or near the CEJ were the most frequent. The majority were approximal – they were also common on the buccal side, but rare on the lingual side. Most CEJ lesions were too large for it to be clear if they had been initiated in the cement or the enamel but, where they were small enough to see, the majority were in the cement. It is therefore likely that most of the CEJ lesions originated in root surface caries. Frequencies of these lesions increased with age/wear category. This pattern persisted into the seventeenth-century group, but caries frequencies in all teeth were higher in the first part of the nineteenth century and rose higher still in the second half of that century, with much more involvement of younger individuals. Caries rates thus became much more like the Kenya and Beijing studies above. Moore and Corbett attributed this change to increasing sugar consumption. Some form of carbohydrate would be needed for caries to occur with any frequency, so the implication is that the earlier pattern related to starch consumption rather than sugar. Wear may also be involved. Occlusal attrition must limit the survival of fissures in which coronal caries can develop (although it would have to be very rapid indeed to be the only factor involved). Similarly, approximal attrition must limit the development of contact area caries. Both are, however, related to the rate of continuous eruption which exposes the roots to caries and this, presumably, is one of the reasons why CEJ caries dominates in all the earlier material.

During the first half of the twentieth century, dentists visited people who still, to some extent, lived a life dominated by hunting, fishing and gathering of wild foods. Amongst the Greenland Inuit (Pedersen, 1938, 1947, 1949; Davies & Pedersen, 1955), people in the outlying villages ate only meat, and had almost no access to carbohydrates. Attrition was extremely rapid, with wear heaviest in anterior teeth and least in the third molars. Caries rates were very low indeed but, in the few cases where lesions were present, they were in the relatively little-worn fissures of the third molar occlusal surfaces. Another study of this kind (Campbell & Gray, 1936; Campbell, 1937) was done in central Australia where, in 1930, sheep and goat stations had only arrived some five years before. Most Aboriginal people still lived a nomadic hunting and gathering life, and had only limited contact with imported foodstuffs. They were able to gather large quantities of grass seed, which gave a much higher carbohydrate intake than was possible for the Greenland Inuit. Molars were the most frequently affected by caries, particularly the first molars, but all teeth showed at least some lesions in some individuals. The main feature of caries amongst the Aboriginal people, however, was a marked increase in caries frequency with age. Nearly 20 years later, in Arnhem Land, Moody (1960) described a 'dental break down' in the worn teeth of older people. The lesions were found particularly in approximal sites, and Moody concluded that the most common cause was fracturing away of the enamel rim of the occlusal attrition facet to create holes in which plaque could accumulate. Exposure of the pulp was very common, and many people had chronic periapical abscesses (below). Another factor was the heavy attrition, which had often proceeded so far that neighbouring teeth no longer met one another, making it possible for food to accumulate between them. In many cases, however, the teeth must have been lost quite rapidly because only a short segment of root was left and, with continued eruption, this would become so loose that it could be pulled out with the fingers.

A final archaeological example follows the effect of the adoption of agriculture. Along the Georgia and Florida coasts of North America a whole series of sites preserving human remains can be assigned to four periods. Precontact preagricultural (1000 BC to AD 1150) sites represent prehistoric hunters, fishers and gatherers who, from stable isotope evidence (Chapter 2), did not eat much in the way of C_4 plants (maize), and made substantial use of marine resources (Larsen *et al.*, 2001). Precontact agriculturalists (AD 1150–1550) had adopted maize agriculture, as can be seen from the carbon stable isotopes, but made a similar use of marine resources. The Early Contact group (AD 1607–1680) comprised the early Spanish missions established along the Georgia and Florida coasts and the Late Contact group (AD 1686–1702) the later missions. The burials from the Contact periods appear to represent not only Europeans, but also native Americans who had been brought into the mission way of life. Stable isotope evidence showed clearly that dietary concentration on maize intensified, and use of marine resources declined (Larsen *et al.*, 2001). These changes were reflected in the numbers of teeth with carious lesions (Larsen *et al.*, 1991). Frequencies of lesions were low for all classes of teeth in the Precontact preagricultural group. In the Precontact agricultural group the frequency of carious lesions increased most in the molars, followed by the premolars, canines and then, the least of all, the incisors. There were few differences in the Early Contact group, but a substantial rise in all tooth types (especially the molars) in the Late Contact group. Caries rates therefore showed a close link with the adoption of a more carbohydrate-rich diet.

Amongst non-human wild mammals, dental caries is only found consistently in Primates that eat a proportion of fruit in their diet (Lovell, 1990, 1991; Kilgore, 1995; Miles & Grigson, 1990). The disease is common in wild chimpanzees, showing a strong relationship with tooth wear. Lesions are almost all in approximal sites of worn incisors, of older individuals (one third of which may be affected). It is assumed that fruit provides the main source of carbohydrates for these high caries rates but, as most specimens would have been shot near human habitations, it is possible that these individuals raided human food stores. Caries is less common amongst wild orang-utans and, for the great apes, least common in gorillas. Once again, the lesions are all at approximal sites on worn incisors. The same pattern of caries is seen in wild Old World monkeys (Miles & Grigson, 1990), with figures in some species approaching those of chimpanzees. Many more specimens of Old World monkeys have been examined, and it is possible to compare captive with wild individuals. Caries not only is more common in the captive monkeys, but is also different in character, affecting particularly the occlusal fissures of molars. In New World monkeys, caries is less common overall. Caries is not reported in the other major group of fruit-eating mammals, Old and New World fruit bats. The family Phyllostomatidae (Chapter 1), which includes the fruit specialist genera of Central America, also has omnivorous, insectivorous and nectar-feeding genera. Most have very low caries rates. The spear nosed bat *Phyllostomus hastatus* is reported to be insectivorous (Nowak & Paradiso, 1983), but has the distinction of being the only bat to have a high caries rate, in the occlusal surfaces and buccal pits of molars (Miles & Grigson, 1990). The reasons for this are unknown.

Carious lesions are rarely reported in any other mammals in their wild state. They are, however, common in domestic animals which may be fed both starchy and sugary foods, including scraps of the same diet eaten by their human owners as well as specially prepared cake and feed pellets (Miles & Grigson, 1990). At the top of the list are domestic horses, in which the infundibulum of the cheek teeth is the vulnerable site for lesions. There is a cavity within the cement that fills each infundibulum, and a deep lesion akin to occlusal fissure caries in humans can develop. The number of individuals affected shows a strong increase with age. After the horse, cattle and sheep are said to be commonly affected, although figures are not available (Jubb & Kennedy, 1963). Caries is rare in carnivores generally, perhaps because the form of their tooth crowns reduces the number of caries susceptible lesion sites. Domestic dogs have the highest caries rates amongst carnivores, particularly in the occlusal surfaces of their bunodont distal molars (Miles & Grigson, 1990), but they never reach the levels found in horses.

Immunity and inflammation

Few bacteria in dental plaque invade normal tissues, but it is a large microbial community in close contact with the gingivae. In addition, caries and tooth fractures allow pathogenic bacteria directly to invade the dental pulp and the periodontal tissues. Bacteria produce a range of biochemical factors that are toxic to the cells of body, or that interfere with their metabolism. These are called *antigens*. The response of the body to this type of damage is known as immunity (Lehner, 1992; Marsh & Martin, 1992). It can be divided into two parts. *Innate immunity* includes the defences which are built into the body from birth. If these do not prove sufficient to remove the source of irritation and cell damage, *acquired immunity* becomes involved. This is a set of responses specific to particular antigens.

One essential part of innate immunity in the mouth is the saliva. Saliva aids and lubricates chewing, lip, cheek and tongue movements. Bacteria may be dislodged, or washed off mouth surfaces. Saliva also contains antibacterial agents. Another essential component of innate immunity is the epithelial covering of the gingivae, which protects the underlying tissues. The epithelium lining the gingival sulcus is, however, rather thin and does not form much of a barrier. Yet another component is the hard tissue of the tooth which seals away the pulp, and maintains its covering by depositing secondary dentine throughout life (Chapter 2). Innate immunity also includes the *phagocytes*; cells which move through the tissues, engulfing particles (including bacteria) and material in solution. When tissue cells are damaged, or when antigens are present, chemical factors are released that cause small capillaries in the vicinity to dilate and to become 'leaky'. Fluid exudes into the tissues, causing swelling (*oedema*) and phagocytes are able to pass out through the capillary walls, following the chemical signals, to arrive at the site of irritation. They remove antigens, bacteria and cell debris. This is the beginning of the process known as *inflammation*, and the site of inflammation is called a *lesion*. Plaque is present in most mouths and some parts of the gingivae usually show modest evidence of inflammatory response.

Acquired immunity involves different cells, *lymphocytes*, which are sensitised by exposure to a particular antigen. When exposed to it again, the lymphocytes react by two types of response. In humoral immunity, B-lymphocytes transform into plasma cells which produce *antibodies,* biochemical factors that bind to specific antigens and neutralise them, which are released into the blood, saliva and other body fluids. Cell-mediated immunity is the production of cells (T-lymphocytes) which move to the source of damage and react, either by transforming into killer cells which bind to foreign cells, or by releasing factors that change the action of the phagocytes. Lymphocytes are also present in a developing inflammatory lesion.

Inflammatory lesions may be *chronic* (slowly developing) or *acute* (rapid). They are characterised by exudation of fluid, the presence of lymphocytes and phagocytes, interference with the normal metabolism of cells in the tissues (both by bacterial action and the immune response), loss of normal tissue structure and its replacement by special inflammatory tissues. Some lesions (*abscesses*) accumulate pus, a mixture of fluid, dead lymphocytes and phagocytes, bacteria and tissue debris. In an acute inflammatory lesion, fluid generally exudes rapidly into the lesion. Dental tissues and bone are non-compressible so where, for example, the periodontal ligament is inflamed, the accumulation of fluid pushes the tooth up in its socket and causes sharp pain on biting. If the fluid accumulates inside the pulp, or the spaces of the bone, the venous return blood supply is choked off and areas of tissue die. Once the pulp has died, the tooth cannot react and the pulp chamber/root canal remain open. Bone, by contrast, can remodel, so the dead bone is removed and if the lesion is a chronic one, a chamber is formed around it. The rapid build-up of fluid (or pus) in an acute lesion is lost by diffusing through the spaces in the bone and if it resolves, no loss of bone takes place. Chronic inflammatory lesions of the gingivae and periodontal tissues that immediately underlie them do not cause loss of the blood supply but they interfere particularly with the turnover of collagen fibres that support the gingival cuff and form the periodontal ligament. It is this that causes most of the bone loss in periodontal disease (below). In many inflammatory conditions involving the jaws, there are alternations between chronic and acute phases over some years.

Periodontal disease
Most mouths, of any mammal, show at least some low-level, chronic inflammation of the gingivae, for much of the time, as normal soft tissue reaction to the presence

of large colonies of micro-organisms. The micro-organisms do not invade, so that the immune system is not able to eliminate them, and inflammation becomes a long-term condition. When confined to the gingivae, it is called *gingivitis*. In most adult humans, from time to time, there is an escalation of the immune response in one part of the mouth. This is not usually because of an injury, but due to a localised change in the balance between bacteria and immunity which tips it towards an inflammatory response. This type of 'overreaction', which does more damage to the tissues than to the bacteria, is called hypersensitivity. If the lesion starts to involve underlying periodontal tissues, it is called *periodontitis*. In the most common type (below) of chronic periodontitis, the development of the lesion can be divided into four phases (Schluger *et al.*, 1990; Lehner, 1992): initial lesion, early lesion, established lesion and advanced lesion. The first three phases are classed as gingivitis, and only the last as periodontitis. The initial lesion is low-level gingivitis, difficult to distinguish from the normal condition, in which fluid exudation causes the gingivae to swell slightly. Phagocytes in the exudate follow chemical signals from the plaque, out through the epithelium and into the gingival sulcus. Over the next few days, the movement of phagocytes may gradually increase until the epithelium is disrupted, and the underlying connective tissue becomes involved. This is the early lesion phase. There is no longer any barrier to the bacteria and their antigens, so the chemotactic gradient which led the phagocytes into the plaque no longer exists. Instead, they stay in the tissues and tackle the bacteria there. T-lymphocytes gather as well. The activities of inflammatory cells cause loss of collagen fibres in the gingivae, either because they compromise the ability of fibroblasts to synthesise them or, more likely, by producing collagenases that break down the collagen (Soames & Southam, 1993). Even at this stage, the lesion may resolve, or it may progress further to a concentration of B-lymphocytes and plasma cells, rising antibody levels, and increase in lesion size. It may become large enough to start affecting the collagen fibres at the top of the periodontal ligament, just around the neck of the tooth. This marks the beginning of the established lesion. Loss of periodontal ligament exposes the tooth root in a periodontal pocket which extends down from the sulcus. It is lined on the gingivae side by pathologically altered pocket epithelium, and contains subgingival plaque which has its own particular bacterial flora characterised by anaerobic forms. The subgingival plaque may mineralise to become a thin layer of subgingival calculus, which is only found in periodontal pockets. As with the other phases, the condition may remain an established lesion for months, or resolve – particularly if the plaque is removed. It may, however, develop into the advanced lesion through continued loss of the periodontal ligament at the neck of the root and extension of the pocket. Bone morphology is actively maintained by remodelling in response to the forces acting on it. Loss of attachment between the root and the alveolar bone, lining the tooth socket, causes the bone to be removed (in effect it is no longer needed). It is this involvement of the alveolar bone that distinguishes the advanced lesion as periodontitis. Even at this stage, the lesion may remain stable, or revert to one of the lesser phases. Bone loss usually proceeds gradually, through

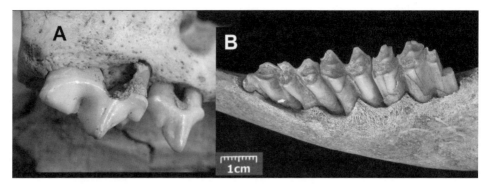

Figure 5.4 Bone loss due to periodontal disease. A: bone loss around the root of the upper carnassial of a domestic cat – the bone loss is trench-like (see text) showing that it is particularly the alveolar bone lining the socket that has been lost. The tooth also has substantial calculus deposits, indicating the presence of supragingival plaque.
B: bone loss around the cheek teeth in a sheep lower jaw – once more particularly the alveolar bone – to create a widening of the sockets. There is also some deposition of new bone at the crest of the alveolar process.

a series of escalations and diminutions of the level of inflammation, over many years. Eventually, so much support is lost that the tooth becomes loose, and falls out. Loosening of teeth is one of the key diagnostic signs of periodontal disease, which is otherwise mostly pain-free and progresses so slowly that many people do not notice. It is, however, one of the main causes of tooth loss in living people.

The pattern described above is by far the most common form in living humans, known as adult periodontitis. It is gradually age-progressive and the majority of adults in late middle age show at least some evidence of this type of bone loss. There are considerable individual differences in experience of the disease which presumably represent variation in plaque accumulation, in its flora and in the immune response, but there is very little evidence that differences in diet or physiology are responsible. In humans there are also other, more rapidly developing forms but they are considerably rarer (Soames & Southam, 1993). Some affect children, including prepubertal periodontitis and juvenile periodontitis, and others affect young adults, such as rapidly progressive periodontitis and acute necrotising gingivitis/periodontitis (ANUG/P). It is not clear how particular individuals come to be affected by these forms. Other mammals may be affected by human-like forms of periodontal inflammation. Wild chimpanzees, for example, may show it, although most of those studied may well have been animals who lived close to human habitation and supplemented their diet by raiding farms and refuse tips (Miles & Grigson, 1990). In effect, they might partly have been eating a human diet. Domestic cats (Figure 5.4), too, may develop periodontal inflammation like that of humans (Lane, 1981) and this has once again been linked to the texture and content of their food. Wolves in the wild seem to be free of it, but captive wolves and domestic dogs quite commonly show a pattern of disease not unlike that in humans (Page &

Schroeder, 1982). It is particularly prevalent in packs of hounds where, to keep down costs, meat may be mixed with large amounts of cereal. Some teeth in dogs have long roots, deeply embedded in the jaws. The progression of a periodontal pocket along these can cause a very large amount of bone loss. Domestic sheep have a particular type of periodontal disease in their incisors and canines, known as broken mouth (Spence *et al.*, 1980; Cutress & Schroeder, 1982; Page & Schroeder, 1982; Miles & Grigson, 1990). In some flocks, it is so common that it becomes a major economic problem for farmers. The lower incisors and canines in sheep are held by very flexible periodontal ligaments in shallow sockets which are rather wider than their roots. Many sheep show a chronic gingivitis around the incisors almost as soon as the teeth are erupted. At some farms, for reasons that are still not understood, this develops into periodontitis in which the periodontal pockets spread down the roots, and the incisors are gradually lost. This prevents the sheep from feeding and they usually have to be culled. The disease shows many similarities with human periodontitis (Duncan *et al.*, 2003).

Sometimes periodontal inflammation, at the side of the root, is also related to injury or disturbance of the gingivae. This is particularly common in domestic horses (Miles & Grigson, 1990) fed on chopped, tough fodder which becomes packed between teeth, pushes aside the interdental papilla (Figure 1.1; the extension of the gingival cuff that normally fills the space) and damages the periodontal ligament. This, in effect, creates a pocket without the intervening gradual stages described above. With inflammation and continued intrusion of the chopped fodder, the pocket extends rapidly. This type of condition may be the cause of most irregular bone loss seen in the jaws of herbivores so, for example, it is found in the cheek teeth of sheep (Figure 5.4; see below) as a separate condition from broken mouth of the incisors (above). The inflammation may become pus-forming (suppurative), in some areas. This is known as a *lateral periodontal abscess* (i.e. it is at the side of the root and is not associated with the apical foramen of the root, which distinguishes it from a periapical abscess, below). It sometimes arises in humans from some local escalation of inflammation in the periodontal pocket (above), due to the arrival of more virulent bacteria, some change in the immune response, blockage of the opening into the periodontal pocket, or impaction of food (Soames & Southam, 1993). The pus is usually released through the opening of the pocket but, if the abscess is deeper in the periodontal tissues, it may discharge through the gingivae. Often, such abscesses progress irregularly, with acute and chronic phases.

Periapical inflammation
The dental tissues are tough, and strongly resistant to damage, wear and caries but once the pulp chamber is penetrated, it provides a pathway for pathogenic bacteria which emerges deep in the bone of the jaw (Soames & Southam, 1993; Wood & Goaz, 1997). The entire blood supply to the pulp enters the tooth through the small apical foramen at the tip of the root, so the capability of the pulp to respond to infection is limited. Pulp inflammation is known as *pulpitis*. It is noticed by the

patient as pain, sharp or dull, often precipitated by hot or cold food against the tooth. As noted above, if exudate accumulates as part of an inflammatory lesion, it is likely to build pressure inside the pulp chamber and, in places, collapse the venous return blood supply. As a result, the pulp usually dies once it has been exposed to infection, although, in large multirooted teeth, it may not all die at once. The odontoblasts are lost at the same time, so the pulp chamber and root canal remain open. Bacteria, antigens, or the products of inflammation, emerge from the foramen at the apex of the root, where they initiate inflammation in the periodontal tissues around it. This is known as *periapical periodontitis*. Periapical inflammatory lesions have a completely different aetiology from the periodontal disease described above, which is centred in the gingivae at the neck of the tooth. There is, however, one confusing factor, which is the presence in some teeth of lateral root canals which branch from the main canal to emerge, not from the apex, but from a foramen in the side of the roots (Clarke, 1990; Clarke & Hirsch, 1991). These may sometimes be associated with the same type of inflammatory lesion as that seen more usually around the root apex and, because the lesion is nearer the cervix of the tooth, it may be confused with the bone loss of periodontal disease.

Periapical inflammation may be acute and, in many cases, this involves the periodontal ligament. As exudate accumulates, the ligament bulges and the tooth is pushed up a little proud in its socket. Any pressure on the tooth can cause great pain. In some cases, pus accumulates rapidly. This is called an acute periapical abscess and, although it is painful, it does not usually cause loss of bone around the apex, because the pus migrates through spaces in the bone. After this it may resolve. Bone changes are much more frequently associated with chronic periapical periodontitis. In humans, zones of periapical bone loss are monitored in X-rays as round zones of lower density – radiolucencies. The great majority of these are small (under 15 mm in diameter) and almost all turn out to be the chronic type of lesion known as a *granuloma* (Figure 5.5). This is an accumulation of granulation tissue (a special tissue formed to contain the lesion) which surrounds the root apex, inside a bony chamber with a smooth compact bone wall. They are usually painless (although there may be an occasional bout of acute periodontitis) and the patient is unaware of their presence until they are detected by X-ray. Larger chambers are usually classified by radiologists as *radicular cysts*. These are fluid-filled cavities which, although they usually have their origins as granulomata, are not themselves inflammatory lesions. In humans, they are found most frequently in upper anterior teeth, and have a tendency to grow slowly. The alveolar process remodels around them and they sometimes grow outside it (Soames & Southam, 1993) to bulge up the periosteum (the fibrous membrane which covers bones). Bone formation is continued under the membrane to make a thin shell which, in living people, exhibits 'eggshell crackling' as the bulge is pressed. In archaeological specimens the shell is almost always lost, leaving a large, smooth-walled cavity. It should be borne in mind that size is not completely diagnostic of the difference between granulomata and cysts, which may also remain relatively small.

Dental disease 309

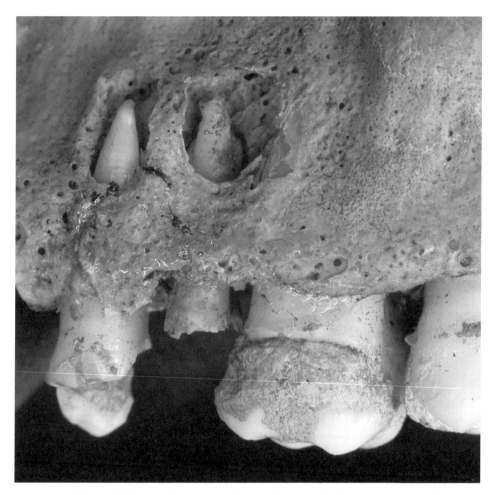

Figure 5.5 Periapical bone loss. A human jaw in which the permanent premolars have gross carious lesions, which have opened the pulp chamber. Periapical periodontitis has caused the loss of bone around the apices of both tooth roots, to make chambers which are close enough to the surface of the alveolar process to become exposed. The edges of the openings are wafer thin, which is not the typical appearance of a sinus opening, so it seems likely that these lesions were granulomata, rather than abscesses (see text) at the time of death.

Chronic periapical abscesses may form by the slow accumulation of pus, but they are rarely associated with evidence of bone loss in X-rays (2% of cases, see Wood & Goaz, 1997). The pus may drain through the open root canal and pulp chamber, track up through the periodontal ligament, or escape through a narrow canal – a sinus – passing through the alveolar process. This usually emerges through the oral mucosa on the buccal side, but may go to lingual, into the maxillary sinus, into the nose, or into the soft tissue of the cheeks or chin. Where it emerges from the bone into the soft tissue, the pus often bulges up the periosteum to form another abscess, so there may be bony signs of inflammation on the surface of the alveolar process as well. Although abscesses are not usually associated with radiolucencies, on the rare occasions that they are, they have most probably developed from a pre-existing granuloma (Dias & Tayles, 1997).

The most common cause of pulp exposure in living humans is dental caries. This is also often true of domestic animals, such as horse, cat and dog (Miles & Grigson, 1990). Thinning of dentine, and opening of 'dead tracts' (above), expose the pulp to bacteria and antigens some time before the carious cavity penetrates completely to the pulp chamber. Another likely cause is cracking, or fracture, of the tooth. Even a fine crack may allow bacteria to enter the pulp. Tooth wear, which is very rapid indeed in some ungulates, is unlikely on its own to expose the pulp chamber because secondary dentine (Chapter 2) is laid down just as rapidly in the roof and sides of the chamber. X-rays of teeth in older adult humans with heavy tooth wear still show fine threads of living pulp remaining in an almost solid dentine tooth (Elvery et al., 1998). Sometimes in heavily worn human teeth the floor of a pulp chamber is exposed in the dentine of the attrition facet. Anthropological literature (Hillson, 2000) often suggests that the pulp was opened to infection by wear, but this is unlikely because the pulp chamber shows little evidence of infilling by secondary dentine. Instead, it is more likely that a fracture or carious lesion opened the pulp at an earlier stage, the tooth died, and wear exposed the already open chamber later (Pedersen, 1938; Hilming & Pedersen, 1940). There is no evidence that the high-crowned ungulates, with their much faster wear, ever open their pulp to infection by wear alone.

More widespread inflammatory lesions
A periapical lesion, or an injury, may occasionally lead to a more widespread infection of the jaw. A mixed bacterial community is usually involved, and proliferates into the trabecular spaces deep in the bone (Soames & Southam, 1993; Wood & Goaz, 1997). An inflammatory reaction (*osteomyelitis*) occurs, with accumulation of exudate. Pressure forces it through the channels and spaces within the bone, which is unable to expand and accommodate it. Blood supply is compromised by compression of veins and, together with the toxins produced by the bacteria, this causes areas of bone to die (*necrosis*). The necrotic bone is resorbed and pus fills the spaces inside the bone. Often, a mass of necrotic bone (a *sequestrum*) is left isolated. Channels (*cloacae*) carrying pus to the surface of the bone are formed and the periosteum balloons out. This may be lined by new bone, forming a shell (*involucrum*). At the surface, the cloacae may be seen and the bone may appear swollen. In X-rays, the widespread bone loss shows as broad radiolucencies, with an irregular 'moth-eaten' appearance. Structures of this kind occasionally occur in archaeological human skeletons (Alexandersen, 1967b).

In humans, the soft tissues of the jaw may become infected by *Actinomyces israelii*, a component of normal plaque. This causes a chronic, pus-forming inflammation, which does not normally affect the bone. The mode of infection is not known. A similar disease exists in cattle, due to infection by *Actinomyces bovis*. This is called 'lumpy jaw' and involves a honeycomb remodelling of the bone over a wide area, with spongy cavities, rather than bone loss at a single focal point (Baker & Brothwell, 1980). A similar disease is found in sheep, deer, horse, camelids, wallabies and kangaroos (Miles & Grigson, 1990). In Australia, *Fusobacterium* seems to be largely implicated, but in American zoos *Actinomyces* is again the culprit (Robinson, 1979).

Patterns of bone loss in the alveolar process
Periodontitis near the cervix of the tooth, and chronic periapical periodontitis, cause localised loss of bone which can be seen in human archaeological specimens. The situation is, however, complicated by slow continuous eruption of teeth throughout life. Three structural elements of the alveolar process are involved:

- the thick cortical bone plate on the buccal and lingual sides
- the alveolar bone which, strictly speaking, is just the thin lining of the tooth sockets
- the trabecular bone which forms the core of the process.

The buccal and lingual plates come together with the alveolar bone at the tops of the sockets to form the crest of the alveolar process (*alveolar crest* or AC). In between neighbouring teeth, the two sockets sandwich a narrow zone of trabecular bone between them and form an *interdental wall*. To buccal and lingual of a row of sockets, a similar sandwich is formed between the alveolar bone and outer cortical bone plate to make a wall on the buccal and lingual sides. Initially, eruption brings the newly formed teeth into the occlusal plane and they are described as 'erupted'. This is not the end of the matter, however, because as teeth wear down they continue to erupt, to maintain occlusion. In high-crowned mammals, this process may be rapid. Amongst wild apes, continuous eruption adjusts for wear without a reduction in the height of the alveolar process (Dean *et al.*, 1992b). Even in humans, archaeological assemblages show evidence of rapid wear and matching rapid continuous eruption (Whittaker *et al.*, 1982, 1985; Levers & Darling, 1983; Danenberg *et al.*, 1991). Living people wear their teeth much more slowly and, although the continuous eruption is also slower, it outstrips the wear so that older people have slightly taller faces than they did when they were younger (Whittaker *et al.*, 1990). Continuous eruption is partly achieved by deposition of cement on the roots and, in tall-crowned animals, by persistent growth of the crown and root, but after this is completed most of it occurs through remodelling of the bone in the jaws. The sockets gradually migrate upwards through the alveolar process, carrying the teeth with them. Teeth also wear approximately (p. 214), so they move forward in the jaw at the same time, again by migration of the sockets – so-called *mesial drift* (Begg, 1954; Corruccini, 1990, 1991; Kaul & Corruccini, 1992). As a result of continuous eruption, the roots of the teeth are progressively exposed, and in humans the buccal bone wall of the alveolar process tends to be thinned in older individuals.

Periodontal disease initially causes loss only of the alveolar bone lining the socket, to make a crater-like depression which extends, next to the root, down into the alveolar process from its crest. In effect, it is a widening of the tooth socket. This can be on any side of the tooth and, where there has been a larger periodontal pocket, may involve two or more sides. Often, it involves particularly the interdental wall between teeth. On an X-ray, this is seen as a radiolucency next to the tooth, or overlying the root, and is described as an 'intra-bony pocket' or 'vertical bone loss', because the line of the alveolar crest can still be seen continuing above it (Goaz & White, 1994; Wood & Goaz, 1997). Loss of alveolar bone is, however, only the first stage. In many cases, the buccal or lingual bone plate is remodelled later, down to the floor of the pocket, so there is a general reduction in the height of the alveolar process at that point. That may initially happen more on one side of a tooth than the other, forming a slope, but later all sides are usually affected and the whole alveolar process becomes lower. On X-rays, this is seen as exposure of the roots which can be measured in humans by the distance between the alveolar crest (AC) to the cervical boundary of the crown (CEJ). A CEJ-AC measurement of 2 mm is considered normal in living people. More than that is described as 'horizontal bone loss'. This approach is acceptable for living people because periodontal disease is very common, and continuous eruption is very slow. In archaeological populations with heavy wear rate, continuous eruption seems to have been much more rapid; it is difficult to disentangle the two processes because there is no independent way of measuring the extent to which the alveolar process has been reduced in height. As a result, the only way to diagnose the modern human pattern of periodontal disease for certain is to find the crater-like loss of alveolar bone before further remodelling has taken place.

Karn *et al.* (1984) developed a terminology for describing the pattern of bone loss. Loss of the alveolar bone lining the tooth socket is described as a crater, trench or moat. A crater is the defect produced by loss from just one side of the root, a trench from two or three sides, and a moat from all sides. Where defects involve loss of the buccal or lingual walls, as well as the interdental wall, these are described as ramps or planes. A ramp is the sloping surface caused by uneven bone loss, and a plane is where it is even. Sometimes a crater, trench or moat is combined with a ramp or a plane. In an attempt to isolate bone loss specifically due to periodontal disease, Kerr (1991) developed a scoring system based only on the interdental walls of the alveolar process. The top of this wall in the alveolar crest of healthy jaws is smooth, and little marked by foramina or grooves. It bulges up slightly between the anterior teeth, and runs flat between cheek teeth. Under gingivitis, the top of the interdental wall becomes porous, with an accumulation of foramina and uneven openings, although it maintains its normal outline between teeth. Under periodontitis, this outline breaks down as the alveolar bone is lost and the interdental wall is lowered in a crater- or trench-like defect. More aggressive periodontitis results in a deeper defect, with the wall lowered by more than 3 mm. At the time of writing, Kerr's scheme is the most practical method for recording periodontal disease.

A lateral periodontal abscess (above) is also associated with disruption to the periodontal ligament, and associated loss of bone. It tends to involve deep intra-bony defects although, in X-rays, these are not really distinguishable from other types of periodontal disease (Whaites, 1992). Periodontal disease in non-human Primates and ungulates (Miles & Grigson, 1990; Lovell, 1991) commonly seems to involve abscesses (above), and the pattern of bone loss particularly affects interdental walls. Several of Sir Frank Colyer's specimens in the Royal College of Surgeons of England show such severe interdental bone loss that a notch has opened, right through the side of the alveolar process (Miles & Grigson, 1990). In other specimens, there may be more general trench- and moat-like widenings of the socket (Levitan, 1985; Grant, 1975a, 1975b, 1977), or sometimes wider chambers inside the jaw, and deposition of new bone at the surface (Figure 5.4).

Periapical granulomata (Figure 5.5) are represented by a small, smooth-walled chamber around the apex of the root. These are usually just 2–3 mm in diameter, although they may be as large as 6 mm across (Dias & Tayles, 1997). Radicular cysts (above) produce cavities of a similar appearance and, although they tend to be larger, many are less than 5 mm in diameter. Differential diagnosis in dry bone can therefore be difficult. Really large smooth-walled cavities 16 mm or more in diameter are most likely to be cysts, particularly if centred on the upper anterior teeth in humans. Occasionally, the cavities of granulomata and cysts are centred, not on the root apex, but around a lateral root canal at the side (Clarke & Hirsch, 1991).

Sometimes, granuloma/cyst cavities are detected only as radiolucencies on X-rays of the bone below a tooth with a gross carious lesion, a fracture, or an exposed pulp chamber. The radiolucency is usually (but not always) sharply delineated because of the smooth wall of compact bone lining the cavity. More commonly, the cavity is exposed through a window in the buccal or lingual wall of the alveolar process. This happens when the cavity has grown large enough, or continuous eruption has moved the root apex close enough to either the buccal or lingual wall. The bony opening often has wafer-thin edges although, with a large cavity close to the alveolar crest, there may be widespread remodelling of bone which rounds the edges. Where there is an isolated opening in the alveolar process side, it is known as a *fenestration* (Muller & Perizonius, 1980; Clarke & Hirsch, 1991). Where the opening extends down from the alveolar crest, it is called a *dehiscence*. Care is needed in archaeological material, because both dehiscence and fenestration may result from post-mortem damage (this may show as bone of a different colour exposed in the edge). In addition, the thinning of the buccal wall with continuous eruption may expose the root in an opening anyway, even without the presence of a granuloma or cyst.

Bony cavities exposed by a fenestration, associated with a tooth in which the pulp chamber is open, are common in archaeological human remains. The great majority probably represent granulomata or cysts, but they are commonly described in the literature as 'abscesses'. The distinguishing feature of an abscess is the presence of pus. Just occasionally, it is possible to see a clear sinus opening (Dias & Tayles, 1997), where there is a tube-like channel in the bone and evidence of new bone formation under the periosteum at the surface. In living patients, if an abscess is associated with a cavity in the bone, then it has most likely developed from a pre-existing granuloma. In archaeological material, this is difficult to test, although where the wall of the cavity is rough rather than smooth, this may indicate such an abscess. In addition, where there is localised deposition of new bone at the surface, this might indicate the soft tissue inflammation that often accompanies the sinus opening of an abscess.

Trauma

Damage to teeth and jaws may be seen in all mammals and occurs quite widely in archaeological material.

Fractured teeth: Either the crown alone may be fractured, or the root only, or both. Crown fractures may involve just chipping of the enamel (Turner & Cadien, 1969), which may cause little further trouble. If dentine is involved, the ends of the dentinal tubules are exposed, opening a pathway for infection, leading to pulpitis (above). A more major fracture may expose the pulp directly but, in cases where the root alone has been broken, the pulp usually survives. Healing occurs, either by deposition of secondary dentine joining the fragments, or by formation of a wall of bone or soft tissue between them (Pindborg, 1970). In humans, the most common site of fracture is the upper incisor. Human dental fractures have been noted in archaeological material (Alexandersen, 1967b). In animals with large canines, such as many Primates and carnivores, fractures of these teeth are common and have also been noted in archaeological material (Siegel, 1976). Carnassial teeth of carnivores may be similarly affected (Miles & Grigson, 1990). Where teeth have been subjected to severe twisting force, tiny fractures may occur in the cement. These are usually along its incremental layers and can be detected only by histology (Pindborg, 1970).

Fractured jaws: Damage to the bone of the jaws is really beyond the scope of this book, except that it may involve the occlusion of teeth. After a fracture has occurred, healing follows a well-established path. First a blood clot (haematoma) forms around the fracture, bulging out the fibrous periosteum which covers the bone. Next, connective tissue cells proliferate within the clot. They deposit a 'callus' of special 'woven bone' to bind the different parts of the fracture together. The next stage is consolidation, in which woven bone is replaced by normal, lamellar bone, and finally, the fracture site remodels. Fractures of the jaws are frequently compound – that is, they communicate with the surface of the oral mucosa, because the line of fracture tends to track along the tooth sockets. This may result in further complications if the lesion is infected. The effect of a jaw fracture on the occlusion depends on the extent of damage and the amount of displacement which occurs before consolidation. The impact or other force that causes the fracture may damage the teeth directly. Similarly, the bone resorption and deposition of the healing process may cause intact teeth to be evulsed. Displacement in mandible fractures occurs as a result of the pull of jaw muscles. In many animals, this is checked by the opposition of upper and lower sets of cheek teeth, but in animals such as the Felidae, the blade-like cheek teeth do not meet in this way, so that large displacements may occur (Miles & Grigson, 1990). Jaw fractures have been noted in archaeological material, for example in ancient Egyptians (Alexandersen, 1967a) and roe deer (Grant, 1975a) from Portchester Castle, England.

Damage to developing teeth: Tooth germs (Chapter 3) may be injured in jaw fractures, or by displacement of already erupted deciduous teeth. This may result in the death of tooth-forming cells and a complete halt in tooth development. In other cases, the tooth germ may be distorted, so that a deformed tooth is eventually erupted. Trauma may also cause a localised hypoplasia (Chapter 2), visible in the enamel. Injury may affect the growth zone of persistently erupting teeth. Animals such as pigs, hippopotamus and elephant may have deformed tusks as a result (Miles & Grigson, 1990).

Problems of teeth of persistent growth: The erupted portion of continuously growing teeth remains an almost constant length, because of similarly continuous attrition from opposing teeth. If these opposite numbers are lost by trauma, developmental anomalies or other pathology, there is no check on crown height. Persistently growing teeth usually grow in an arc or spiral. When growth is unchecked the tooth may protrude clear of the mouth, but it may also re-enter skull and jaw tissues, inflicting damage. Such overgrowth has been noted in rodents, lagomorphs, pigs and hippopotamus (Miles & Grigson, 1990), but has not yet been described in archaeological material.

Anomalies of eruption, resorptions and abrasions

Normal variation in eruption timing may be exceeded in some cases – premature or retarded. Deciduous teeth may fail to be shed, often where the succeeding permanent tooth is not formed. Such retained teeth remain in the dentition and are made more noticeable by their much greater degree of attrition than surrounding teeth. The other common anomaly of eruption is for teeth to be retained within the jaw. This is called *impaction* and, in humans, is most common in the canines and third molars (Pindborg, 1970). Various kinds of impaction occur. Teeth may remain completely enclosed in bone, or may be pressed up against another tooth, with the side of the crown or a number of cusps showing. In some forms of impaction, the tooth is almost fully erupted. Both impaction and retention of deciduous teeth can be seen quite commonly in archaeological human material. Teeth may also be reimpacted, submerged in the periodontium after a previously normal eruption. Where teeth have become impacted by either process, it is not uncommon for the tooth to be ankylosed – fused with the bone lining the socket (Pindborg, 1970).

Resorption of the roots of deciduous teeth is normal, as the permanent teeth are erupted underneath them. The crown may, however, also be resorbed. This often occurs in impacted teeth. It may also be related to trauma or infection, but a proportion of cases have no apparent cause. This type of lesion is observed in humans, and has also been noted in the sperm whale *Physeter* and in elephants (Miles & Grigson, 1990).

Anomalies of abrasion and attrition occur in a variety of mammals, due to malocclusions, or other factors. In humans, various types of defect may be caused by deliberate filing of the tooth (Milner & Larsen, 1991), or by use of the teeth as tools for holding objects. Cleaning, especially vigorous brushing of roots exposed by periodontitis, may cause erosions.

Cysts, odontomes and tumours

Cysts: A cyst is a fluid- or semi-fluid-filled cavity, lined with epithelium, and is more common in the jaws than elsewhere in the skeleton. In *odontogenic cysts*, the epithelial tissue has its origins in the development of the teeth. Almost all cysts in the jaws are odontogenic, and by far the most common form is the radicular cyst described above. After them, the most common form is the dentigerous cyst, where all or part of the crown of an unerupted tooth is included in the cavity. These have been described for sheep (Miles & Grigson, 1990) as well as humans (Soames & Southam, 1993). *Non-odontogenic cysts* are far less common but, of them, the most frequent is the *nasopalatine duct cyst*, which develops at the midline of the anterior palate.

Odontome is a term of long usage (Broca, 1867), and has taken on several meanings. In this book it is used to mean non-neoplastic (below) developmental anomalies which contain enamel and dentine (Soames & Southam, 1993). The dividing line between normal variation and anomaly is not at all clear. By far the most common odontome in human teeth is the *enameloma,* or *enamel pearl*, which is a small nodule of dentine with an enamel cap (Risnes, 1989) on the root surface. This may be an entirely separate nodule, or be joined to the crown at the cervix by a narrow strip of enamel (Lasker, 1951). Sometimes this strip, without a nodule, may extend right between the furcation of the roots. Such *enamel extensions* are especially common in molars and are recorded in the ASUDAS system (p. 274). Another common form is the *invaginated odontome*, where there is a deep infolding from the crown into the root (in effect an abnormal infundibulum in human teeth). Upper second incisors are most affected, especially barrel-shaped incisors. Similar features have been described for ungulates (Miles & Grigson, 1990). An *evaginated odontome* is an isolated, teat-like extra cusp in the middle of a premolar crown, also scored in the ASUDAS. A *geminated odontome* (Cawson, 1970) is two teeth which are partially joined together. Typically, the occlusal parts of the crowns and apical parts of the roots are separated, but all kinds of variation can be seen. *Compound odontomes* are complexes of fully divided small 'toothlets'. All odontomes may be confined within the jaw, or erupt partially or fully into the mouth. A *cementoma* is an irregular mass of cement deposited over an otherwise normal enamel and dentine tooth form. These are seen, for example, in horses (Miles & Grigson, 1990). *Complex odontomes* are irregular masses of enamel, dentine and cement which do not have any recognisable elements of normal tooth structure. Once again, they are found rarely in a wide range of mammals. A variety of odontomes has also been noted in archaeological remains (Baker & Brothwell, 1980).

Neoplasms are localised tissue growths that are no longer subject to normal developmental controls. By far the most common odontogenic neoplasm is the *ameloblastoma*. This is a benign tumour of epithelial cells, which appear to have their origin in ameloblasts (Chapter 2). No enamel is formed but, as the tumour expands, bone is resorbed, often into multiple chambers. Several examples exist in archaeological human material (Brothwell, 1981a), but tumours of dental origin are rare in wild non-human mammals (Robinson, 1979).

Conclusion – palaeoepidemiology and recording

Palaeoepidemiology is a study of the ancient distribution of disease. Archaeological investigations are usually confined to the most common diseases. Even in living populations, many diseases are relatively rare. In archaeological material, where samples are small and of uncertain origin, such diseases turn up only occasionally and, when they do, little can be said about their overall distribution. In human remains, the most common dental diseases are those related to plaque deposits – caries, calculus, periodontal disease, and periapical inflammation. These diseases are common enough, at least in Neolithic and later humans, to produce frequencies large enough to be compared statistically between sites and periods. In this way it is possible to look at their distribution in space, time and between different groups of a population. The distribution of plaque-related diseases can be used to reconstruct aspects of the plaque biology during life. Calculus and periodontal disease confirm the presence of long-standing plaque deposits, and dental caries yields information about plaque pH. From this, the amount and type of carbohydrate in the diet can be inferred so, together with stable isotopes, caries makes a useful tool for reconstructing diet. In all this, however, it is important to take account of the role of attrition, continuous eruption and the strongly age-related nature of all these phenomena. Dental pathology is much less frequently reported in non-human mammals, but it does occur and, in view of its high prevalence in some modern domestic animals, could be of great interest in relation to changes in the type of fodder used, and other aspects of husbandry.

The basis for any palaeoepidemiological study of dental disease is a reliable database of appropriate observations. From the start, it is essential to keep separate observations for each different tooth type, and for each lesion site on the tooth. Caries rates are best expressed as percentages of each site, for each tooth, with a carious lesion. Presence and absence of all the different lesion sites (or other ways in which a site may be obscured from view) need to be recorded meticulously, even if they do not have a carious lesion developing in them, because otherwise there will be no total count of sites present from which the caries rates can be calculated. Failure to distinguish between different types of caries, for example occlusal fissure versus root surface caries, invalidates a study because of the widely differing aetiologies. It is also essential to record dental eruption state, and both occlusal and approximal attrition, not only as a means of separating material into different age groups, but because they have a profound effect on the pattern of dental caries. As far as possible, it is also important to record any markers of age at death (Buikstra & Ubelaker, 1994) which may be independent of the teeth, for example pubic symphysis (Suchey, 1979; Katz & Suchey, 1986, 1989; Brooks & Suchey, 1990), auricular surface (Lovejoy *et al.*, 1985) and sternal end of ribs (Iscan & Loth, 1984; Iscan & Loth, 1986a, 1986b; Iscan *et al.*, 1987; Russell *et al.*, 1993). Also, where possible, separate records need to be kept for males and females. Recording for dental pathology in archaeological material is discussed in detail in Hillson (2000, 2001), where recording schemes and forms are also given. There are advantages to using paper forms, rather than direct recording into a computer database. Bitter experience has shown that data can be lost, and computers suffer in dusty laboratories, museum basements, or in the field. Direct data entry into a computer has an advantage too: it will in any case be necessary to type all the data into a computer eventually, the forms are heavy and bulky, and databases can be set to check records for anomalies as they are entered. Different researchers may, however, require different types of data and, if material has been reburied, there is no chance to re-examine it. Once again, personal experience has shown that recording forms may allow appropriate counts to be made, even if this is not possible from the database. One last, increasingly serious, problem is time-proofing computer databases. Not only does this involve storing them in a form which will not deteriorate over time, but also in a way that will be readable by computers and software of the future. Even after only 30 years or so of computer records in archaeology, stories abound of databases which are unusable because no machine nowadays is capable of running the version of software required, or nobody remembers what the different fields in the database represent.

APPENDIX A

THE GRANT DENTAL ATTRITION AGE ESTIMATION METHOD

This appendix includes charts (Figures A.1–A.3) of tooth wear stages (TWS), reproduced with kind permission from Annie Grant (1982).

The first, second and third molars of one half of the lower dentition are assigned either one of the tooth wear stages (a to p below) or one of the following stages of eruption:

        C   perforation in crypt visible
        V   tooth visible in crypt but below head of bone
        E   tooth erupting through bone
        1/2 tooth half erupted
        U   tooth almost at full height but unworn.

Each tooth is then assigned a score, using the system in Table A.1. The mandible wear stage (MWS) is calculated by adding the scores for all three molars together. Depending on the species involved, MWS varies from 1 to 50 or more.

Table A.1

Eruption/TWS	Score
C	1
V	2
E	3
1/2	4
U	5
a	6
b	7
c	8
d	9
e	10
f	11
g	12
h	13
j	14
k	15
l	16
M	17
n	18
o	19
p	20

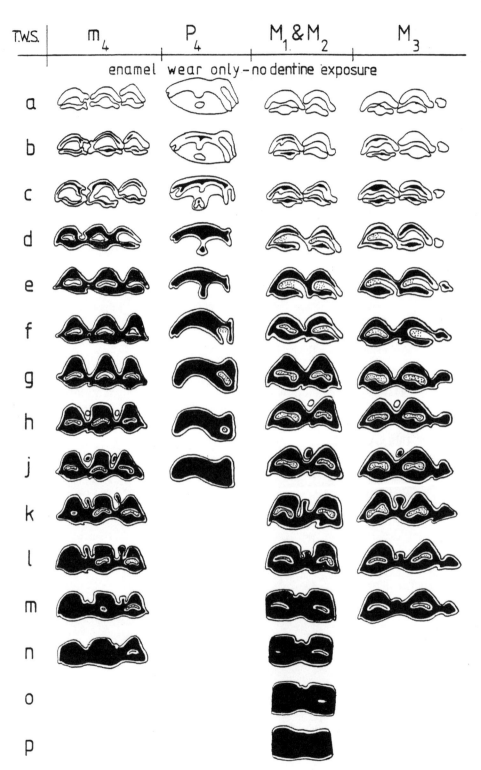

Figure A.1 Tooth wear stages of cattle teeth, showing deciduous lower fourth premolar (m_4 in the figure), permanent lower fourth premolar, first, second and third molar.

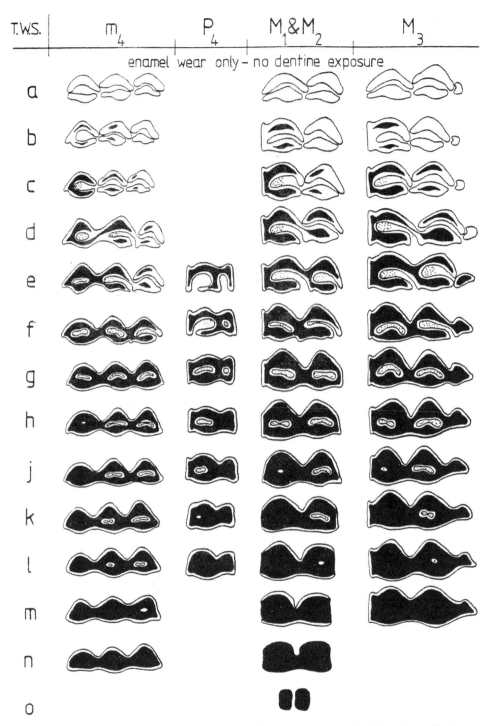

Figure A.2 Tooth wear stages of sheep/goat teeth, showing deciduous lower fourth premolar (m_4 in the figure), permanent lower fourth premolar, first, second and third molar.

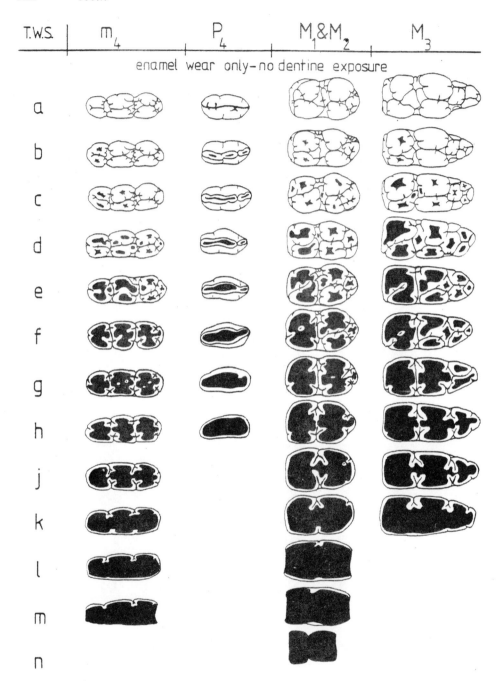

Figure A.3 Tooth wear stages of pig teeth, showing deciduous lower fourth premolar (m_4 in the figure), permanent lower fourth premolar, first, second and third molar.

REFERENCES

Aas, I. H. (1983). Variability of a dental morphological trait. *Acta Odontologica Scandinavica*, **41**, 257–63.
Aitken, M. J. (1990). *Science-based Dating in Archaeology*. Longman Archaeology Series. London: Longman.
Aitken, R. J. (1975). Cementum layers and tooth wear as criteria for ageing roe deer (*Capreolus capreolus*). *Journal of Zoology (London)*, **175**, 15–28.
Alexandersen, V. (1967a). The evidence for injuries to the jaws. In Brothwell, D. R. & Sandison, A. T. (eds.), *Diseases in Antiquity*. Springfield: Thomas, pp. 623–9.
 (1967b). The pathology of the jaws and temporomandibular joint. In Brothwell, D. R. & Sandison, A. T. (eds.), *Diseases in Antiquity*. Springfield: Thomas, pp. 551–95.
Allen, T. (2003). *Manual of Equine Dentistry*. St Louis: Mosby.
Alvesalo, L. (1971). The influence of sex chromosome genes on tooth size in man. *Suomen Hammaslääkäriseunan Toimituksia (Proceedings of the Finnish Dental Society)*, **67**, 3–54.
Alvesalo, L. & Tigerstedt, P. M. A. (1974). Heritabilities of human tooth dimensions. *Hereditas*, **77**, 311–18.
Ambrose, S. H. (1993). Isotopic analysis of paleodiets: methodological and interpretive considerations. In Sandford, M. K. (ed.), *Investigations of Ancient Human Tissue: Chemical Analyses in Anthropology*. Food and Nutrition in History and Anthropology Volume 10. Langhorne, Pennsylvania: Gordon & Breach, pp. 59–130.
Andresen, V. (1898). Die Querstreifung des Dentins. *Deutsche Monatsschrift für Zahnheilkunde*, **16**, 386–9.
Andrews, A. H. (1982). The use of dentition to age young cattle. In Wilson, B., Grigson, C., & Payne, S. (eds.), *Ageing and Sexing Animal Bones from Archaeological Sites*. British Archaeology Reports British Series 109. Oxford: British Archaeological Reports, pp. 141–53.
Andrews, A. H. & Noddle, B. (1975). Absence of premolar teeth from ruminant mandibles found at archaeological sites. *Journal of Archaeological Science*, **2**, 137–44.
Anemone, R. L. (1995). Dental development in chimpanzees of known chronological age: implications for understanding the age at death of Plio-Pleistocene hominids. In Moggi-Cecchi, J. (ed.), *Aspects of Dental Biology: Palaeontology, Anthropology and Evolution*. Florence: International Institute for the Study of Man, pp. 201–15.
Anemone, R. L., Mooney, M. P., & Siegel, M. I. (1996). Longitudinal study of dental development in chimpanzees of known chronological age: implications for understanding the age at death of Plio-Pleistocene hominids. *American Journal of Physical Anthropology*, **99**, 119–34.
Angel, J. L. (1969). The bases of paleodemography. *American Journal of Physical Anthropology*, **30**, 427–37.
Antoine, D. M., Dean, M. C., & Hillson, S. W. (1999). The periodicity of incremental structures in dental enamel based on the developing dentition of post-Medieval known-age children. In Mayhall, J. T. & Heikinnen, T. (eds.), *Dental Morphology '98*. Oulu: Oulu University Press, pp. 48–55.
Armitage, P. (1975). The extraction and identification of opal phytoliths from the teeth of ungulates. *Journal of Archaeological Science*, **2**, 187–97.

Arya, B. S., Thomas, D. R., Savara, B. S., & Clarkson, Q. D. (1974). Correlations among tooth sizes in a sample of Oregon caucasoid children. *Human Biology*, **46**, 693–8.

Asper, H. (1916). Ueber die 'Braune Retzius'sche Parallelstreifung' im Schmelz der Menschlichen Zähne. *Schweizerische Vrtljschrift für Zahnheilkunde*, **26**, 275.

Aufderheide, A. C. (1989). Chemical analysis of skeletal remains. In Iscan, M. Y. & Kennedy, K. A. R. (eds.), *Reconstruction of Life from the Skeleton*. New York: Alan R Liss, pp. 237–60.

Bäckman, B. (1989). *Amelogenesis imperfecta. An epidemiologic, genetic, morphologic and clinical study*. MD Dissertation, University of Umeå, Sweden.

(1997). Inherited enamel defects. In Chadwick, D. J. & Cardew, G. (eds.), *Dental Enamel*. Ciba Foundation Symposium 205. Chichester: Wiley, pp. 175–86.

Bader, R. S. (1965). Heritability of dental characters in the house mouse. *Evolution*, **19**, 378–84.

Bader, R. S. & Lehmann, W. H. (1965). Phenotypic and genotypic variation in odontometric traits of the house mouse. *American Midland Naturalist*, **74**, 28–38.

Baker, J. & Brothwell, D. R. (1980). *Animal Diseases in Archaeology*. London: Academic Press.

Balasse, M., Bocherens, H., Mariotti, A., & Ambrose, S. H. (2001). Detection of dietary changes by intra-tooth carbon and nitrogen isotopic analysis: an experimental study of dentine collagen of cattle (*Bos taurus*). *Journal of Archaeological Science*, **28**, 235–45.

Bang, G. & Ramm, E. (1970). Determination of age in humans from root dentine transparency. *Acta Odontologica Scandinavica*, **28**, 3–35.

Bang, P. & Dahlstrom, P. (1972). *Collins Guide to Animal Tracks and Signs*. London: Collins.

Barnosky, A. D. (1990). Evolution of dental traits since latest Pleistocene in Meadow Voles (*Microtus pennsylvanicus*) from Virginia. *Paleobiology*, **16**, 370–83.

Barrett, M. J., Brown, T., & Macdonald, M. R. (1963). Dental observations on Australian aborigines: mesiodistal crown diameters of permanent teeth. *Australian Dental Journal*, **8**, 150–6.

Baume, L. J., Horowitz, H. S., Summers, C. J., Dirks, O. B., Brown, W. A. B., Carlos, J. P., Cohen, L. K., Freer, T. J., Harvold, E. P., Moorrees, C. F. A., Satzmann, J. A., Schmuth, G., Solow, B., & Taatz, H. (1970). A method for measuring occlusal traits. Developed by the F. D. I. Commission on Classification and Statistics for Oral Conditions (COCSTOC) Working Group 2 on Dentofacial Anomalies, 1969–72. *International Dental Journal*, **23**, 530–7.

Beasley, M. J., Brown, W. A. B., & Legge, A. J. (1992). Incremental banding in dental cementum: methods of preparation for teeth from archaeological sites and for modern comparative specimens. *International Journal of Osteoarchaeology*, **2**, 37–50.

(1993). Metrical discrimination between mandibular first and second molars in domestic cattle. *International Journal of Osteoarchaeology*, **3**, 303–14.

Beeley, J. G. & Lunt, D. A. (1980). The nature of the biochemical changes in softened dentine from archaeological sites. *Journal of Archaeological Science*, **7**, 371–7.

Begg, P. R. (1954). Stone Age man's dentition. *American Journal of Orthodontics*, **40**, 298–383, 462.

Bell, L. S. (1990). Palaeopathology and diagenesis: an SEM evaluation of structural changes using backscattered electron imaging. *Journal of Archaeological Science*, **17**, 85–102.

Bell, L. S., Boyde, A., & Jones, S. J. (1991). Diagenetic alteration to teeth in situ illustrated by backscattered electron imaging. *Scanning*, **13**, 173–83.

Bell, L. S., Skinner, M. F., & Jones, S. J. (1996). The speed of post mortem change to the human skeleton and its taphonomic significance. *Forensic Science International*, **82**, 129–40.

Benecke, N. (1987). Studies on early dog remains from northern Europe. *Journal of Archaeological Science*, **14**, 31–49.

Bermann, M., Edwards, L. F., & Kitchin, P. C. (1939). Effect of artificially induced hyperpyrexia on tooth structure in the rabbit. *Proceedings of the Society for Experimental Biology & Medicine*, **41**, 113–15.

Bermudez de Castro, J. M., Durand, A. I., & Ipiña, S. L. (1993). Sexual dimorphism in the human dental sample from the SH site (Sierra de Atapuerca, Spain): a statistical approach. *Journal of Human Evolution*, **24**, 43–56.

Berry, A. C. (1978). Anthropological and family studies on minor variants of the dental crown. In Butler, P. M. & Joysey, K. A. (eds.), *Development, Function and Evolution of Teeth*. London: Academic Press, pp. 81–98.

Berry, R. J. (1968). The biology of non-metrical variation in mice and men. In Brothwell, D. R. (ed.), *The Skeletal Biology of Earlier Human Populations*. Symposia of the Society for the Study of Human Biology. Oxford: Pergamon Press, pp. 104–133.

  (1969). History in the evolution of *Apodemus sylvaticus* (Mammalia) at one edge of its range. *Journal of Zoology*, **159**, 311–28.

  (1975). On the nature of genetical distance and island races of the *Apodemus sylvaticus*. *Journal of Zoology*, **176**, 293–6.

Berry, R. J. & Jakobson, M. E. (1975). Ecological genetics of an island population of the house mouse. *Journal of Zoology*, **175**, 523–40.

Berry, R. J., Jakobson, M. E., & Peters, J. (1977). The house mice of the Faroe Islands: a study in microdifferentiation. *Journal of Zoology*, **185**, 73–92.

Berry, R. J. & Rose, F. E. N. (1975). Islands and the evolution of *Microtus arvalis* (Microtinae). *Journal of Zoology*, **177**, 395–409.

Berten, J. (1895). Hypoplasie des Schmelzes (Congenitale Schmelzdefecte; Erosionen). *Deutsche Monatsschrift für Zahnheilkunde*, **13**, 425–39.

Beynon, A. D., Dean, M. C., & Reid, D. J. (1991). Histological study on the chronology of the developing dentition in gorilla and orangutan. *American Journal of Physical Anthropology*, **86**, 189–203.

Biggerstaff, R. H. (1973). Heritability of Carabelli's cusp in twins. *Journal of Dental Research*, **52**, 40–4.

  (1975). Cusp size, sexual dimorphism, and heritability of cusp size in twins. *American Journal of Physical Anthropology*, **42**, 127–40.

  (1976). Cusp size, sexual dimorphism, and heritability of maximum molar cusp size in twins. *Journal of Dental Research*, **55**, 189–95.

Bishop, M. P. (1982). *The Early Middle Pleistocene Mammal Fauna of Westbury-sub-Mendip*. Special Papers in Palaeontology 28. London: Palaeontological Association.

Black, T. K. (1978). Sexual dimorphism in the tooth-crown diameters of deciduous teeth. *American Journal of Physical Anthropology*, **48**, 77–82.

Bocquet-Appel, J. & Masset, C. (1982). Farewell to paleodemography. *Journal of Human Evolution*, **11**, 321–33.

Boessneck, J. & von den Driesch, A. (1978). The significance of measuring animals bones from archaeological sites. In Meadow, R. H. & Zeder, M. A. (eds.), *Approaches to Faunal Analysis in the Middle East*. Peabody Museum Bulletin 2. Cambridge, Mass.: Harvard University, pp. 25–39.

Bogin, B. (1999). *Patterns of Human Growth*. Cambridge Studies in Biological Anthropology 23. Cambridge: Cambridge University Press.

Bourque, B. J., Morris, K., & Speiss, A. (1978). Determining the season of death of mammal teeth from archaeological sites: a new sectioning technique. *Science*, **199**, 530–31.

Bover, P. & Alcover, J. A. (1999). The evolution and ontogeny of the dentition of *Myotragus balearicus* Bate, 1909 (Artiodactyla, Caprinae): evidence from new fossil data. *Biological Journal of the Linnean Society*, **68**, 401–28.

Bowen, W. H. & Pearson, S. K. (1993). Effect of milk on cariogenesis. *Caries Research*, **27**, 461–6.

Boyde, A. (1963). Estimation of age at death of young human skeletal remains from incremental lines in dental enamel. London: Third International Meeting in Forensic Immunology, Medicine, Pathology and Toxicology, Plenary Session 11A.

(1969). Electron microscopic observations relating to the nature and development of prism decussation in mammalian dental enamel. *Bulletin du Groupement International pour les Recherches Scientifique en Stomatologie*, **12**, 151–207.

(1971). Comparative histology of mammalian teeth. In Dahlberg, A. A. (ed.), *Dental Morphology and Evolution*. Chicago: University of Chicago Press, pp. 81–94.

(1978). Development of the structure of the enamel of the incisor teeth in the three classical subordinal groups of the Rodentia. In Butler, P. M. & Joysey, K. A. (eds.), *Development, Function and Evolution of Teeth*. London: Academic Press, pp. 43–58.

(1979). Carbonate concentration, crystal centres, core dissolution, caries, cross striations, circadian rhythms and compositional contrast in the SEM. *Journal of Dental Research*, **58**, 981–3.

(1980). Histological studies of dental tissues in Odontocetes. In Perrin, W. F. & Myrick, A. C. (eds.), *Growth of Odontocetes and Sirenians: Problems in Age Determination. Proceedings of the International Conference on Determining Age of Odontocete Ceteans (and Sirenians), La Jolla, California, September 5–19, 1978*. Report of the International Whaling Commission, Special Issue 3. Cambridge: International Whaling Commission, pp. 65–87.

(1984). Methodology of calcified tissue specimen preparation for scanning electron microscopy. In Dickson, G. R. (ed.), *Methods of Calcified Tissue Preparation*. Amsterdam: Elsevier, pp. 251–307.

(1989). Enamel. In Berkovitz, B. K. B., Boyde, A., Frank, R. M., Höhling, H. J., Moxham, B. J., Nalbandian, J., & Tonge, C. H. (eds.), *Teeth*. Handbook of Microscopic Anatomy. New York: Springer, pp. 309–473.

(1990). Developmental interpretations of dental microstructure. In DeRousseau, C. J. (ed.), *Primate Life History and Evolution*. Monographs in Primatology Volume 14. New York: Wiley-Liss, pp. 229–67.

Boyde, A. & Fortelius, M. (1986). Development, structure and function of rhinoceros enamel. *Zoological Journal of the Linnean Society*, **87**, 181–214.

Boyde, A., Fortelius, M., Lester, K. S., & Martin, L. B. (1988). Basis of the structure and development of mammalian enamel as seen by scanning electron microscopy. *Scanning Microscopy*, **2**, 1479–90.

Boyde, A. & Jones, S. J. (1983). Backscattered electron imaging of dental tissues. *Anatomy and Embryology*, **5**, 145–50.

(1995). Skeletal connective tissues. In Williams, P. L. & Bannister, L. H. (eds.), *Gray's Anatomy: the Anatomical Basis of Medicine and Surgery*. London: Churchill Livingstone, pp. 443–84.

Boyde, A. & Lester, K. S. (1967). The structure and development of Marsupial enamel tubules. *Zeitschrift für Morphologie und Anthropologie*, **82**, 558–76.

Boyde, A. & Martin, L. (1982). Enamel microstructure determination in hominoid and cercopithecoid Primates. *Anatomy and Embryology*, **165**, 193–212.

(1987). Tandem scanning reflected light microscopy of primate enamel. *Scanning Microscopy*, **1**, 1935–48.

Brace, C. L. (1964). The probable mutation effect. *American Naturalist*, **97**, 39–49.

Brace, C. L. & Mahler, P. E. (1971). Post-Pleistocene changes in the human dentition. *American Journal of Physical Anthropology*, **34**, 191–204.

Brace, C. L., Rosenberg, K. R., & Hunt, K. D. (1987). Gradual change in human tooth size in the late Pleistocene and post-Pleistocene. *Evolution*, **41**, 705–20.

Brace, C. L. & Ryan, A. S. (1980). Sexual dimorphism and human tooth size differences. *Journal of Human Evolution*, **9**, 417–35.

Brace, C. L., Shao, Z-Q., & Zhang, Z-B. (1984). Biological and cultural change in the European Late Pleistocene and Early Holocene. In Smith, F. H. & Spencer, F. (eds.), *The Origin of Modern Humans*. New York: Alan R. Liss, pp. 485–516.

Brace, C. L., Smith, S. L., & Hunt, K. D. (1991). What big teeth you had Grandma! Human tooth size, past and present. In Kelley, M. A. & Larsen, C. S. (eds.), *Advances in Dental Anthropology*. New York: Wiley-Liss, pp. 33–57.

Bradford, E. W. (1967). Microanatomy and histochemistry of dentine. In Miles, A. E. W. (ed.), *Structural and Chemical Organization of Teeth*. London: Academic Press, pp. 3–34.

Brescia, N. J. (1961). *Applied Dental Anatomy*. St Louis: Mosby.

Broca, P. (1867). Recherches sur un nouveau group de tumeurs désigné sous le nom d'ontomes. *Comptes Rendus Hebdomadaires des Séances de l'Académie de Sciences*, **65**, 1117–21.

Bromage, T. G. (1991). Enamel incremental periodicity in the pig-tailed macaque: a polychrome fluorescent labeling study of dental hard tissues. *American Journal of Physical Anthropology*, **86**, 205–14.

Bromage, T. G. & Dean, M. C. (1985). Re-evaluation of the age at death of immature fossil hominids. *Nature*, **317**, 525–7.

Brooks, S. & Suchey, J. M. (1990). Skeletal age determination based on the os pubis: a comparison of the Acsádi-Nemeskéri and Suchey-Brooks methods. *Human Evolution*, **5**, 227–38.

Brothwell, D. R. (1963a). *Dental Anthropology*. London: Pergamon Press.

  (1963b). *Digging up Bones*. London & Oxford: British Museum & Oxford University Press.

  (1963c). The macroscopic dental pathology of some earlier human populations. In Brothwell, D. R. (ed.), *Dental Anthropology*. London: Pergamon Press, pp. 272–87.

  (1981a). *Digging up Bones*. Third edition. London & Oxford: British Museum & Oxford University Press.

  (1981b). The Pleistocene and Holocene archaeology of the house mouse and related species. *Symposia of the Zoological Society of Great Britian*, **47**, 1–13.

  (1989). The relationship of tooth wear to aging. In Iscan, M. Y. (ed.), *Age Markers in the Human Skeleton*. Springfield: Charles C. Thomas, pp. 303–16.

Brothwell, D. R., Carbonell, V. M., & Goose, D. H. (1963). Congenital absence of teeth in human populations. In Brothwell, D. R. (ed.), *Dental Anthropology*. London: Pergamon Press, pp. 179–89.

Brown, D. & Anthony, D. (1998). Bit wear, horseback riding and the Botai Site in Kazakhstan. *Journal of Archaeological Science*, **25**, 331–47.

Brown, W. A. B. & Chapman, N. G. (1990). The dentition of fallow deer (*Dama dama*): a scoring scheme to assess age from wear of the permanent molariform teeth. *Journal of Zoology*, **221**, 659–82.

  (1991a). The dentition of red deer (*Cervus elaphus*): a scoring scheme to assess age from wear of the permanent molariform teeth. *Journal of Zoology*, **224**, 519–36.

  (1991b). The dentition of red deer (*Cervus elaphus*): from a scoring scheme based on radiographs of developing permanent molariform teeth. *Journal of Zoology*, **224**, 85–97.

Brown, W. A. B., Christofferson, D. V. M., Massler, M., & Weiss, M. B. (1960). Postnatal tooth development in cattle. *American Journal of Veterinary Research XXI*, **80**, 7–34.

Brudevold, F. & Söremark, R. (1967). Chemistry of the mineral phase of enamel. In Miles, A. E. W. (ed.), *Structural and Chemical Organization of Teeth*. London: Academic Press, pp. 247–78.

Buikstra, J. E. & Konigsberg, L. W. (1985). Paleodemography: critiques and controversies. *American Anthropologist*, **87**, 316–33.

Buikstra, J. E. & Mielke, J. H. (1985). Demography, diet, and health. In Gilbert, R. I. & Mielke, J. H. (eds.), *Analysis of Prehistoric Diets*. New York: Academic Press, pp. 359–422.

Buikstra, J. E. & Ubelaker, D. H. (1994). *Standards for Data Collection from Human Skeletal Remains*. Fayetteville: Arkansas Archeological Survey.

Bull, G. & Payne, S. (1982). Tooth eruption and epiphyseal fusion in pigs and wild boar. In Wilson, B., Grigson, C., & Payne, S. (eds.), *Ageing and Sexing Animal Bones from Archaeological Sites*. British Archaeology Reports British Series 109. Oxford: British Archaeology Reports.

Bullock, D. & Rackham, J. (1982). Epiphyseal fusion and tooth eruption of feral goats from Moffatdale, Dumfries and Galloway, Scotland. In Wilson, B., Grigson, C., & Payne, S. (eds.), *Ageing and Sexing Animal Bones from Archaeological Sites*. British Archaeology Reports British Series 109. Oxford: British Archaeology Reports, pp. 73–80.

Burke, A. & Castanet, J. (1995). Histological observations of cementum growth in horse teeth and their application to archaeology. *Journal of Archaeological Science*, **22**, 479–93.

Burns, K. R. & Maples, W. R. (1976). Estimation of age from individual adult teeth. *Journal of Forensic Sciences*, **21**, 343–56.

Burton, J. H. & Price, T. (1999). Evaluation of bone strontium as a measure of seafood consumption. *International Journal of Osteoarchaeology*, **9**, 233–6.

Burton, J. H. & Wright, L. E. (1995). Nonlinearity in the relationship between bone Sr/Ca and diet: paleodietary implications. *American Journal of Physical Anthropology*, **96**, 273–82.

Butler, P. M. (1939). Studies in the mammalian dentition – and of differentiation of the postcanine dentition. *Proceedings of the Zoological Society, London, B*, **107**, 103–32.

  (1978). Molar cusp nomenclature and homology. In Butler, P. M. & Joysey, K. A. (eds.), *Development, Function and Evolution of Teeth*. London: Academic Press, pp. 439–54.

  (1982). Some problems of the ontogeny of tooth patterns. In Kurtén, B. (ed.), *Teeth, Form Function and Evolution*. Columbia: University Press, pp. 44–51.

Calcagno, J. M. (1989). *Mechanisms of Human Dental Reduction. A Case Study from Post-Pleistocene Nubia*. University of Kansas Publications in Anthropology 18. Lawrence: University of Kansas.

Cameron, N. (1998). Radiographic assessment. In Ulijaszek, S. J., Johnston, F. E., & Preece, M. A. (eds.), *The Cambridge Encyclopedia of Human Growth and Development*. Cambridge: Cambridge University Press, pp. 42–4.

Campbell, T. D. (1937). Dental observations on the teeth of Australian aborigines, Hermannsberg, Central Australia. *Australian Journal of Dentistry*, **41**, 1–6.

Campbell, T. D. & Gray, J. H. (1936). Observations on the teeth of Australian aborigines. *Australian Journal of Dentistry*, **40**, 290–5.

Carlson, D. S. & van Gerven, D. P. (1977). Masticatory function and post-Pleistocene evolution in Nubia. *American Journal of Physical Anthropology*, **46**, 495–506.

Carrick, R. & Ingham, S. E. (1962). Studies on the southern elephant seal, *Mirounga leonina* (L.). II Canine tooth structure in relation to function and age determination. *Commonwealth Scientific and Industrial Research Organization Wildlife Research*, **7**, 102–18.

Carter, J. T. (1922). On the structure of enamel in Primates and some other mammals. *Proceedings of the Zoological Society of London*, **1922**, 599–608.

Carter, R. J. (1997). Age estimation of the roe deer (*Capreolus capreolus*) mandibles from the Mesolithic site of Star Carr, Yorkshire, based on radiographs of mandibular tooth development. *Journal of Zoology*, **241**, 495–502.

  (1998). Reassessment of seasonality at the Early Mesolithic site of Star Carr, Yorkshire, based on radiographs of mandibular tooth development in red deer (*Cervus elaphus*). *Journal of Archaeological Science*, **25**, 851–6.

  (2000). New evidence for seasonal human presence at the Early Mesolithic site of Thatcham, Berkshire, England. *Journal of Archaeological Science*, **28**, 1055–60.

Castanet, J. (1981). Quelques remarques sur la méthode squelettochronologique chez les vertébrés supérieurs (oiseaux et mammifères). *Bulletin Société Zoologique de France*, **105**, 371–6.

Casteel, R. W. (1976). Incremental growth zones in mammals and their archaeological value. *Papers of the Kroeber Anthropological Society (Berkeley)*, **47**, 1–27.

Caughley, G. (1965). Horn rings and tooth eruption as criteria of age in the Himalayan Thar, *Hemitragus jemlahicus*. *New Zealand Journal of Science*, **8**, 333–51.

Cawson, R. A. (1970). *Essentials of Dental Surgery and Pathology*. Edinburgh: Churchill Livingstone.

Charles, D. K., Condon, K., Cheverud, J. M., & Buikstra, J. E. (1986). Cementum annulation and age determination in *Homo sapiens*. I. Tooth variability and observer error. *American Journal of Physical Anthropology*, **71**, 311–20.
  (1989). Estimating age at death from growth layer groups in cementum. In Iscan, M. Y. (ed.), *Age Markers in the Human Skeleton*. Springfield: Charles C. Thomas, pp. 277–316.
Clarke, N. G. (1990). Periodontal defects of pulpal origin: evidence in early man. *American Journal of Physical Anthropology*, **82**, 371–6.
Clarke, N. G. & Hirsch, R. S. (1991). Physiological, pulpal, and periodontal factors influencing alveolar bone. In Kelley, M. A. & Larsen, C. S. (eds.), *Advances in Dental Anthropology*. New York: Wiley-Liss, pp. 241–66.
Clement, A. J. (1963). Variations in the microstructure and biochemistry of human teeth. In Brothwell, D. R. (ed.), *Dental Anthropology*. London: Pergamon Press, pp. 245–69.
Clutton-Brock, J. (1987). *A Natural History of Domesticated Mammals*. London & Cambridge: British Museum (Natural History) & Cambridge University Press.
Clutton-Brock, J., Dennis-Bryan, K., Armitage, P., & Jewell, P. A. (1990). Osteology of Soay sheep. *Bulletin of the British Museum of Natural History*, **56**, 1–56.
Colby, G. R. (1996). Analysis of dental sexual dimorphism in two Western Gulf of Mexico pre-contact populations utilizing cervical measurements (abstract). *American Journal of Physical Anthropology*, Supplement **22**, 87.
Colyer, J. F. (1936). *Variations and Diseases of the Teeth of Animals*. London: John Bale, Sons & Danielsson.
Commission on Oral Health (1982). An epidemiological index of developmental defects of dental enamel (DDE Index). *International Dental Journal*, **32**, 159–67.
Condon, K. W. (1981). Correspondence of developmental enamel defects between the mandibular canine and first premolar. *American Journal of Physical Anthropology*, **54**, 211.
Condon, K., Becker, J., Condon, C., & Hoffman, J. R. (1994). Dental and skeletal indicators of a congenital treponematosis. *American Journal of Physical Anthropology*, Supplement **18**, 70.
Condon, K., Charles, D. K., Cheverud, J. M., & Buikstra, J. E. (1986). Cementum annulation and age determination in *Homo sapiens*. II. Estimates and accuracy. *American Journal of Physical Anthropology*, **71**, 321–30.
Conroy, G. C. & Vannier, M. W. (1987). Dental development of the Taung skull from computerized tomography. *Nature*, **329**, 625–7.
Cook, D. C. (1981). Mortality, age-structure, and status in interpretation of stress indicators in prehistoric skeletons: a dental example from the lower Illinois Valley. In Chapman, R., Kinnes, I., & Randsborg, K. (eds.), *The Archaeology of Death*. Cambridge: Cambridge University Press, pp. 133–44.
  (1984). Subsistence and health in the Lower Illinois Valley: osteological evidence. In Cohen, M. N. & Armelagos, G. J. (eds.), *Palaeopathology at the Origins of Agriculture*. New York: Academic Press, pp. 235–69.
Cooper, A. & Poinar, H. (2000). Ancient DNA: do it right or not at all. *Science*, **289**, 1139.
Coote, G. E. (1988). Fluorine diffusion profiles in archaeological human teeth: a method for relative dating of burials? In Prescott, J. R. (ed.), *Archaeometry: Australian Studies 1988*. Adelaide: Department of Physics, University of Adelaide, pp. 99–104.
Coote, G. E. & Nelson, P. (1987). Diffusion profiles of fluorine in archaeological bones and teeth: their measurement and application. In Ward, G. K. (ed.), *Archaeology at ANZAAS*. Canberra: Canberra Archaeological Society, Australian National University, pp. 22–7.
Coote, G. E. & Sparks, R. J. (1981). Fluorine concentration profiles in archaeological bones. *New Zealand Journal of Archaeology*, **3**, 21–32.
Corbet, G. B. (1975). Examples of short-and long-term changes of dental pattern in Scottish voles (Rodentia; Microtinae). *Mammal Review*, **5**, 17–21.

(1978). *The Mammals of the Palaearctic Region: a Taxonomic Review*. London and Ithaca: British Museum of Natural History and Cornell University Press.
Corbet, G. B. & Hill, J. E. (1990). *World List of Mammalian Species*. London: Natural History Museum.
Corbett, M. E. & Moore, W. J. (1976). Distribution of dental caries in ancient British populations: IV The 19th century. *Caries Research*, **10**, 401–14.
Corruccini, R. S. (1977a). Crown component variation in hominoid lower third molars. *Zeitschrift für Morphologie und Anthropologie*, **68**, 14–25.
 (1977b). Crown component variation in the hominoid lower second premolar. *Journal of Dental Research*, **56**, 1093–6.
 (1978). Crown component analysis of the hominoid upper first premolar. *Archives of Oral Biology*, **23**, 491–4.
 (1990). Australian aboriginal tooth succession, interproximal attrition, and Begg's theory. *American Journal of Orthodontics & Dentofacial Orthopaedics*, **97**, 349–57.
 (1991). Anthropological aspects of orofacial and occlusal variations and anomalies. In Kelley, M. A. & Larsen, C. S. (eds.), *Advances in Dental Anthropology*. New York: Wiley-Liss, pp. 295–323.
Corruccini, R. S. & Potter, R. H. Y. (1980). Genetic analysis of occlusal variation in twins. *American Journal of Orthodontics*, **78**, 140–54.
Corruccini, R. S., Sharma, K., & Potter, R. H. Y. (1986). Comparative genetic variance and heritability of dental occlusal variables in US and Northwest Indian twins. *American Journal of Physical Anthropology*, **70**, 293–9.
Corruccini, R. S. & Shimada, I. (2002). Dental relatedness corresponding to mortuary patterning at Huaca Loro, Peru. *American Journal of Physical Anthropology*, **117**, 113–21.
Corruccini, R. S., Shimada, I., & Shinoda, K. (2002). Dental and mtDNA relatedness among thousand-year-old remains from Huaca Loro, Peru. *Dental Anthropology*, **16**, 9–14.
Corruccini, R. S., Townsend, G. C., & Brown, T. (1990). Occlusal variation in Australian aboriginals. *American Journal of Physical Anthropology*, **82**, 257–65.
Coy, J. P., Jones, R. T., & Turner, K. A. (1982). Absolute ageing in cattle from tooth sections and its relevance to archaeology. In Wilson, B., Grigson, C., & Payne, S. (eds.), *Ageing and Sexing Animal Bones from Archaeological Sites*. British Archaeological Reports, British Series 109. Oxford: British Archaeological Reports, pp. 127–40.
Crossner, C. G. & Mansfield, L. (1983). Determination of dental age in adopted non-European children. *Swedish Dental Journal*, **7**, 1–10.
Crowcroft, P. (1957). *The Life of the Shrew*. London: Reinhardt.
Cutress, T. W. & Schroeder, H. E. (1982). Histopathology of periodontitis (broken-mouth) in sheep: a further consideration. *Research in Veterinary Science*, **33**, 64–9.
d'Errico, F. & Vanhaeren, M. (2002). Criteria for identifying red deer (*Cervus elaphus*) age and sex from their canines. Application to the study of Upper Palaeolithic and Mesolithic ornaments. *Journal of Archaeological Science*, **29**, 211–32.
Dahlberg, A. A. (1945). The changing dentition of man. *Journal of the American Dental Association*, **6**, 676–90.
 (1947). The evolutionary significance of the protostylid. *American Journal of Physical Anthropology*, **8**, 15–25.
 (1961). Relationship of tooth size to cusp number and groove conformation of occlusal surface patterns of lower molar teeth. *Journal of Dental Research*, **44**, 476–9.
 (1963). Analysis of the American Indian dentition. In Brothwell, D. R.(ed.), *Dental Anthropology*. London: Pergamon Press, pp. 149–78.
Danenberg, P. J., Hirsch, R. S., Clarke, N. G., Leppard, P. I., & Richards, L. C. (1991). Continuous tooth eruption in Australian aboriginal skulls. *American Journal of Physical Anthropology*, **85**, 305–12.

Daniel, G. (1978). *150 Years of Archaeology*. London: Duckworth.
Darling, A. I. (1959). The pathology and prevention of caries. *British Dental Journal*, **107**, 287–96.
  (1963). Microstructural changes in early dental caries. In Soggnaes, R. F. (ed.), *Mechanisms of Hard Tissue Destruction*. Washington: American Association for the Advancement of Science, pp. 171–85.
Dart, R. A. (1925). *Australopithecus africanus*: the man-ape of South Africa. *Nature*, **115**, 195–9.
Davies, T. G. H. & Pedersen, P. O. (1955). The degree of attrition of the deciduous teeth and the first permanent molars of primitive and urbanised Greenland natives. *British Dental Journal*, **99**, 35–43.
Davis, S. J. M. (1980). A note on the dental and skeletal ontogeny of *Gazella*. *Israel Journal of Zoology*, **29**, 129–34.
  (1981). The effects of temperature change and domestication on the body size of late Pleistocene to Holocene mammals in Israel. *Paleobiology*, **7**, 101–14.
  (1983). The age profiles of gazelles predated by ancient man in Israel: possible evidence for a shift from seasonality to sedentism in the Natufian. *Paléorient*, **9**, 55–62.
  (1984). Tiny elephants and giant mice. *New Scientist*, **105**, 25–7.
  (1987). *The Archaeology of Animals*. London: Batsford.
Day, M. H. & Molleson, T. I. (1973). The Trinil femora. In Day, M. H. (ed.), *Human Evolution*. London: Taylor and Francis, pp. 127–54.
Dayan, T. (1994). Early domesticated dogs of the Near East. *Journal of Archaeological Science*, **21**, 633–40.
De Vito, C. & Saunders, S. R. (1990). A discriminant function analysis of deciduous teeth to determine sex. *Journal of Forensic Sciences*, **35**, 845–58.
Dean, M. C. (1987). Growth layers and incremental markings in hard tissues; a review of the literature and some preliminary observations about enamel structure in *Paranthropus boisei*. *Journal of Human Evolution*, **16**, 157–72.
  (1995). The nature and periodicity of incremental lines in primate dentine and their relationship to periradicular bands in OH 16 (*Homo habilis*). In Moggi-Cecchi, J. (ed.), *Structure, Function and Evolution of Teeth. Dental Morphology Meeting, Florence, September 1992*. Florence: International Institute for the Study of Man, pp. 239–65.
  (1998a). A comparative study of cross striation spacings in cuspal enamel and of four methods of estimating the time taken to grow molar cuspal enamel in *Pan*, *Pongo* and *Homo*. *Journal of Human Evolution*, **35**, 449–62.
  (1998b). Comparative observations on the spacing of short-period (von Ebner's) lines in dentine. *Archives of Oral Biology*, **43**, 1009–21.
Dean, M. C. & Beynon, A. D. (1991). Histological reconstruction of crown formation times and initial root formation times in a modern human child. *American Journal of Physical Anthropology*, **86**, 215–28.
Dean, M. C., Beynon, A. D., & Reid, D. J. (1992a). Microanatomical estimates of rates of root extension in a modern human child from Spitalfields, London. In Smith, P. & Tchernov, E. (eds.), *Structure, Function and Evolution of Teeth*. London & Tel Aviv: Freund Publishing House, pp. 311–34.
Dean, M. C., Beynon, A. D., Thackeray, J. F., & Macho, G. A. (1993). Histological reconstruction of dental development and age at death of a juvenile *Paranthropus robustus* specimen, SK 63, from Swartkrans, South Africa. *American Journal of Physical Anthropology*, **91**, 401–20.
Dean, M. C., Jones, M. E., & Pilley, J. R. (1992b). The natural history of tooth wear, continuous eruption and periodontal disease in wild shot great apes. *Journal of Human Evolution*, **22**, 23–39.
Dean, M. C., Leakey, M. G., Reid, D. J., Schrenk, F., Schwartz, G., Stringer, C. B., & Walker, A. (2001). Growth processes in teeth distinguish modern humans from *Homo erectus* and earlier hominins. *Nature*, **414**, 628–31.

Dean, M. C. & Scandrett, A. E. (1996). The relation between long-period incremental markings in dentine and daily cross-striations in enamel in human teeth. *Archives of Oral Biology*, **41**, 233–41.

Degerbøl, M. (1961). On a find of Preboreal domestic dog (*Canis familiaris*, L.) from Star Carr, Yorkshire, with remarks on other Mesolithic dogs. *Proceedings of the Prehistoric Society*, **27**, 35–55.

Demetsopoullos, I. C., Burleigh, R., & Oakley, K. P. (1983). Relative and absolute dating of the human skull and skeleton from Galley Hill, Kent. *Journal of Archaeological Science*, **10**, 129–34.

Demirjian, A., Goldstein, H., & Tanner, J. M. (1973). A new system of dental age assessment. *Human Biology*, **45**, 211–27.

Dempsey, P., Townsend, G. C., Martin, N., & Neale, M. (1995). Genetic covariance structure of incisor crown size in twins. *Journal of Dental Research*, **74**, 1389–98.

Deniz, E. & Payne, S. (1979). Eruption and wear in the mandibular dentition of Turkish Angora goats in relation to ageing sheep/goat mandibles from archaeological sites. In Kubasiewicz, M. (ed.), *Archaeozoology Volume I*. Szezecin: Agricultural Academy, pp. 153–63.

  (1982). Eruption and wear in the mandibular dentition as a guide to ageing Turkish angora goats. In Wilson, B., Grigson, C., & Payne, S. (eds.), *Ageing and Sexing Animal Bone from Archaeological Sites*. British Archaeological Reports British Series 109. Oxford: British Archaeological Reports.

Dent, V. E. & Marsh, P. D. (1981). Evidence for a basic plaque microbial community on the tooth surface in animals. *Archives of Oral Biology*, **26**, 171–9.

Deutsch, D., Dafni, A., Palmon, A., Held, A. J., Young, M. F., & Fisher, L. W. (1997). Tuftelin: enamel mineralization and amelogenesis imperfecta. In Chadwick, D. J. & Cardew, G. (eds.), *Dental Enamel*. Ciba Foundation Symposium 205. Chichester: Wiley, pp. 135–55.

Dias, G. & Tayles, N. (1997). 'Abscess cavity' – a misnomer. *International Journal of Osteoarchaeology*, **7**, 548–54.

DiBennardo, R. & Bailit, H. L. (1978). Stress and dental asymmetry in a population of Japanese children. *American Journal of Physical Anthropology*, **48**, 89–94.

Dirks, W. (1998). Histological reconstruction of dental development and age at death in a juvenile gibbon (*Hylobates lar*). *Journal of Human Evolution*, **35**, 411–25.

Dirks, W., Reid, D. J., Jolly, C. J., Phillips-Conroy, J. E., & Brett, F. L. (2002). Out of the mouths of baboons: stress, life history, and dental development in the Awash National Park Hybrid Zone, Ethiopia. *American Journal of Physical Anthropology*, **118**, 239–52.

Ditch, L. E. & Rose, J. C. (1972). A multivariate sexing technique. *American Journal of Physical Anthropology*, **37**, 61–4.

Doberenz, A. R., Miller, M. F., & Wyckoff, R. W. G. (1969). Fossil enamel protein. *Calcified Tissue Research*, **3**, 93–5.

Doberenz, A. R. & Wyckoff, R. W. G. (1967). The microstructure of fossil teeth. *Journal of Ultrastructure Research*, **18**, 166–75.

Dobney, K. & Brothwell, D. (1986). Dental calculus: its relevance to ancient diet and oral ecology. In Cruwys, E. & Foley, R. A. (eds.), *Teeth and Anthropology*. B. A. R. International Series 291. Oxford: British Archaeological Reports, pp. 55–82.

Dobney, K. & Ervynck, A. (2000). Interpreting developmental stress in archaeological pigs: the chronology of linear enamel hypoplasia. *Journal of Archaeological Science*, **27**, 597–607.

Ducos, P. (1968). *L'origine des Animaux Domestiques en Palestine*. Bordeaux: Travaux de l'Université de Bordeaux 6.

  (1969). Methodology and results of the study of the earliest domesticated animals in the Near East (Palestine). In Ucko, P. J. & Dimbleby, G. W. (eds.), *The Domestication and Exploitation of Plants and Animals*. London: Duckworth, pp. 265–75.

Duncan, W. J., Persson, G. R., Sims, T. J., Braham, P., Pack, A. R. C., & Page, R. C. (2003). Ovine periodontitis as a potential model for periodontal studies. Cross-sectional anlayais of clinical, microbiological, and serum immunological parameters. *Journal of Clinical Periodontology*, **30**, 63–72.

Ekstrand, K. R., Kuzmina, I., Bjørndal, L., & Thylstrup, A. (1995). Relationship between external and histologic features of progressive stages of caries in the occlusal fossa. *Caries Research*, **29**, 243–50.

Elliott, J. C. (1997). Structure, crystal chemistry and density of enamel apatites. In Chadwick, D. J. & Cardew, G. (eds.), *Dental Enamel*. Ciba Foundation Symposium 205. Chichester: Wiley, pp. 54–84.

Elvery, M. W., Savage, N. W., & Wood, W. B. (1998). Radiographic study of the Broadbeach aboriginal dentition. *American Journal of Physical Anthropology*, **107**, 211–19.

Enlow, D. H. & Hansen, B. F. (1996). *Essentials of Facial Growth*. London: Saunders.

Espelid, I., Tveit, A. B., & Fjelltveit, A. (1994). Variations among dentists in radiographic detection of occlusal caries. *Caries Research*, **28**, 169–75.

Evans, K., Hindell, M. A., Robertson, K., Lockyer, C., & Rice, D. (2002). Factors affecting the precision of age determination of sperm whales (*Physeter macrocephalus*). *Journal of Cetacean Research and Management*, **4**, 193–201.

Ewbank, J. M., Phillipson, D. W., & Whitehouse, R. D. (1964). Sheep in the Iron Age: a method of study. *Proceedings of the Prehistoric Society*, **30**, 423–6.

Ezzo, J. A., Johnson, C. M., & Price, T. D. (1997). Analytical perspectives on prehistoric migration: a case study from East-Central Arizona. *Journal of Archaeological Science*, **24**, 447–66.

Falguères, C. (2003). ESR dating and the human evolution: contribution to the chronology of the earliest humans in Europe. *Quaternary Science Reviews*, **22**, 1345–51.

Falin, L. I. (1961). Histological and histochemical studies of human teeth of the Bronze and Stone Ages. *Archives of Oral Biology*, **5**, 5–13.

Ferembach, D., Schwidetzky, I., & Stloukal, M. (1980). Recommendations for age and sex diagnoses of skeletons. *Journal of Human Evolution*, **9**, 517–49.

Fernandez, P. & Legendre, S. (2003). Mortality curves for horses from the Middle Palaeolithic site of Bau de l'Aubesier (Vaucluse, France): methodological, palaeo-ethnological, and palaeo-ecological approaches. *Journal of Archaeological Science*, **30**, 1577–98.

Fincham, A. G. & Simmer, J. P. (1997). Amelogenin proteins of developing dental enamel. In Chadwick, D. J. & Cardew, G. (eds.), *Dental Enamel*. Ciba Foundation Symposium 205. Chichester: Wiley, pp. 118–46.

Fitzgerald, C. & Rose, J. (2000). Reading between the lines: dental development and subadult age assessment using the microstructural growth markers of teeth. In Katzenberg, M. A. & Saunders, S. R. (eds.), *Biological Anthropology of the Human Skeleton*. New York: John Wiley, pp. 163–86.

Fitzgerald, C. M. (1998). Do enamel microstructures have regular time dependency? Conclusions from the literature and a large-scale study. *Journal of Human Evolution*, **35**, 371–86.

Foti, B., Adalian, P., Signoli, M., Ardagna, Y., Dutour, O., & Leonetti, G. (2001). Limits of the Lamendin method in age determination. *Forensic Science International*, **122**, 101–6.

Frank, R. M. (1978). Les stries brunes de Retzius en microscopie électronique à balayage. *Journal de Biologie Buccale*, **6**, 139–51.

Frank, R. M. & Brendel, A. (1966). Ultrastructure of the approximal dental plaque. *Archives of Oral Biology*, **11**, 883–912.

Frank, R. M. & Nalbandian, J. (1989). Structure and ultrastructure of dentine. In Berkovitz, B. K. B., Boyde, A., Frank, R. M., Höhling, H. J., Moxham, B. J., Nalbandian, J., & Tonge, C. H. (eds.), *Teeth*. Handbook of Microscopic Anatomy. New York: Springer, pp. 173–247.

Frayer, D. W. (1978). *Evolution of the Dentition in Upper Palaeolithic and Mesolithic Europe*. University of Kansas Publications in Anthropology 10. Lawrence: University of Kansas.

(1980). Sexual dimorphism and cultural evolution in the Late Pleistocene and Holocene of Europe. *Journal of Human Evolution*, **9**, 399–415.

(1984). Biological and cultural change in the European Late Pleistocene and Early Holocene. In Smith, F. H. & Spencer, F. (eds.), *The Origin of Modern Humans*. New York: Alan R. Liss, pp. 211–50.

Frison, G. C. & Reher, C. A. (1970). Appendix I: age determination of buffalo by teeth eruption and wear. In Frison, G. C. (ed.), *The Glenrock Buffalo Jump, 48CO304: Late Prehistoric Period Buffalo Procurement and Butchering on the Northwestern Plains*. Plains Anthropologist Memoir 7, pp. 46–47.

Fujita, T. (1939). Neue Feststellungen uber Retzius'schen Parallelstreifung des Zahnschmelzes. *Anatomischer Anzeiger*, **87**, 350–5.

Fukuhara, T. (1959). Comparative anatomical studies of the tooth growth lines in the enamel of mammalian teeth (in Japanese). *Acta Anatomica Nipponica*, **34**, 322–32.

Funmilayo, O. (1976). Age determination, age distribution and sex ratio in mole populations. *Acta Theriologica*, **21**, 207–15.

Garn, S. M., Cole, P. E., & van Astine, W. L. (1979a). Sex discriminatory effectiveness using combinations of root lengths and crown diameters. *American Journal of Physical Anthropology*, **50**, 115–18.

Garn, S. M., Cole, P. E., Wainwright, R. L., & Guire, K. E. (1977). Sex discriminatory effectiveness using combinations of permanent teeth. *Journal of Dental Research*, **56**, 697.

Garn, S. M., Lewis, A. B., Dahlberg, A. A., & Kerewsky, R. S. (1966a). Interaction between relative molar size and relative number of cusps. *Journal of Dental Research*, **45**, 1240.

Garn, S. M., Lewis, A. B., & Kerewsky, R. S. (1966b). The meaning of bilateral asymmetry in the permanent dentition. *Angle Orthodontist*, **36**, 55–62.

(1967a). Sex difference in tooth shape. *Journal of Dental Research*, **46**, 1470.

(1968). The magnitude and implications of the relationship between tooth size and body size. *Archives of Oral Biology*, **13**, 129–31.

Garn, S. M., Lewis, A. B., Kerewsky, R. S., & Dahlberg, A. A. (1966c). Genetic independence of Carabelli's trait from tooth size or crown morphology. *Archives of Oral Biology*, **11**, 745–7.

Garn, S. M., Lewis, A. B., Swindler, D. R., & Kerewsky, R. S. (1967b). Genetic control of dimorphism in tooth size. *Journal of Dental Research*, **46**, 963–72.

Garn, S. M., Osborne, R. H., & McCabe, K. D. (1979b). The effect of prenatal factors on crown dimensions. *American Journal of Physical Anthropology*, **51**, 665–78.

Garniewicz, R. C. (2000). Age and sex determination from the mandibular dentition of raccoons: techniques and applications. *Archaeozoologia*, **XI**, 223–38.

Gasaway, W. C., Harkness, D. B., & Rausch, R. A. (1978). Accuracy of moose age determinations from incisor cementum layers. *Journal of Wildlife Management*, **42**, 558–63.

Getty, R. (1975). *Sisson and Grossman's The Anatomy of the Domestic Animals*. Fifth edition. Philadelphia: Saunders.

Gifford-Gonzalez, D. (1992). Examining and refining the quadratic crown height method of age estimation. In Stiner, M. C. (ed.), *Human Predators and Prey Mortality*. Boulder: Westview Press, pp. 41–78.

Gilbert, F. F. & Stolt, S. L. (1970). Variation and aging white-tailed deer by tooth wear characteristics. *Journal of Wildlife Management*, **34**, 532–5.

Gilkeston, C. F. (1997). Tubules in Australian marsupials. In von Koenigswald, W. & Sander, P. M. (eds.), *Tooth Enamel Microstructure*. Rotterdam: A. A. Balkema, pp. 113–21.

Gingerich, P. D. (1977). Correlation of tooth size and body size in living hominoid Primates, with a note on the relative brain size in Aegyptopithecus and Proconsul. *American Journal of Physical Anthropology*, **47**, 395–8.

Gipson, P. S., Ballard, W. B., Nowak, R. M., & Mech, L. D. (2000). Accuracy and precision of estimating age of gray wolves by tooth wear. *Journal of Wildlife Management*, **64**, 752–8.

Gittleman, J. L. & van Valkenburgh, B. (1997). Sexual dimorphism in the canines and skulls of carnivores: effects of size, phylogeny, and behavioural change. *Journal of Zoology*, **242**, 97–117.

Glimcher, M. J., Cohensolal, L., Kossiva, D., & Dericqles, A. (1990). Biochemical analyses of fossil enamel and dentin. *Paleobiology*, **16**, 219–32.

Goaz, P. W. & White, S. C. (1994). *Oral Radiology: Principles and Interpretation*. St Louis: Mosby.

Gobetz, K. E. & Bozarth, S. R. (2001). Implications for Late Pleistocene Mastodon diet from opal phytoliths in tooth calculus. *Quaternary Research*, **55**, 115–22.

Goldstein, J. I., Newbury, D. E., Echlin, P., Joy, D. C., Romig, A. D., Lyman, C. E., Fiori, C., & Lifshin, E. (1992). *Scanning Electron Microscopy and X-ray Microanalysis*. New York: Plenum Press.

Goodman, A. H., Armelagos, G. J., & Rose, J. C. (1980). Enamel hypoplasias as indicators of stress in three prehistoric populations from Illinois. *Human Biology*, **52**, 515–28.

(1984a). The chronological distribution of enamel hypoplasias from prehistoric Dickson Mounds populations. *American Journal of Physical Anthropology*, **65**, 259–66.

Goodman, A. H., Lallo, J., Armelagos, G. J., & Rose, J. C. (1984b). Health changes at Dickson Mounds, Illinois (A.D. 950–1300). In Cohen, M. N. & Armelagos, G. J. (eds.), *Palaeopathology at the Origins of Agriculture*. New York: Academic Press, pp. 271–306.

Goodman, A. H., Martinez, C., & Chavez, A. (1991). Nutritional supplementation and the development of linear enamel hypoplasias in children from Tezonteopan, Mexico. *American Journal of Clinical Nutrition*, **53**, 773–81.

Goodman, A. H. & Rose, J. C. (1990). Assessment of systemic physiological perturbations from dental enamel hypoplasias and associated histological structures. *Yearbook of Physical Anthropology*, **33**, 59–110.

Goodman, A. H., Thomas, R. B., Swedlund, A. C., & Armelagos, G. J. (1988). Biocultural perspectives of stress in prehistoric, historical and contemporary population research. *Yearbook of Physical Anthropology*, **31**, 169–202.

Goose, D. H. (1963). Dental measurement: an assessment of its value in Anthropological studies. In Brothwell, D. R. (ed.), *Dental Anthropology*. London: Pergamon Press, pp. 125–48.

(1971). The inheritance of tooth size in British families. In Dahlberg, A. A. (ed.), *Dental Morphology and Evolution*. Chicago: University of Chicago Press, pp. 263–70.

Goose, D. H. & Appleton, J. (1982). *Human Dentofacial Growth*. Oxford: Pergamon.

Goose, D. H. & Lee, G. R. T. (1971). The mode of inheritance of Carabelli's trait. *Human Biology*, **43**, 64–9.

Goose, D. H. & Roberts, E. E. (1982). Size and morphology of children's teeth in North Wales. In Kurtén, B. (ed.), *Teeth: Form, Function, and Evolution*. New York: Columbia University Press, pp. 228–36.

Gordon, K. D. (1988). A review of methodology and quantification in dental microwear analysis. *Scanning Microscopy*, **2**, 1139–47.

Gould, S. J. (1975). On the scaling of tooth size in mammals. *American Zoologist*, **15**, 351–62.

Grant, A. (1975a). Fauna remains. In Cunliffe, B. (ed.), *Excavations at Porchester Castle: II Saxon*. Reports of the Research Committee of the Society of Antiquaries of London, 33. London: Thames & Hudson, pp. 262–96.

(1975b). The animal bones. In Cunliffe, B. (ed.), *Excavations at Portchester Castle: I Roman*. Reports of the Research Committee of the Society of Antiquaries of London, 32. London: Thames & Hudson, pp. 378–408.

(1975c). The use of tooth wear as a guide to the age of domestic animals. In Cunliffe, B. (ed.), *Excavations at Portchester Castle: I Roman*. Reports of the Research Committee of the Society of Antiquaries of London, 32. London: Thames & Hudson, pp. 437–50.

(1977). Mammals. In Cunliffe, B. (ed.), *Excavations at Portchester Castle: III Medieval*. Reports of the Research Committee of the Society of Antiquaries of London, 34. London: Thames & Hudson, pp. 213–332.

(1978). Variation in dental attrition in animals and its relevance to age estimation. In Brothwell, D., Thomas, K. D., & Clutton-Brock, J. (eds.), *Research Problems in Zooarchaeology,* Occasional Publications 3. London: University of London, Institute of Archaeology, pp. 103–6.

(1982). The use of tooth wear as a guide to the age of domestic animals. In Wilson, B., Grigson, C., & Payne, S. (eds.), *Ageing and Sexing Animal Bones from Archaeological Sites*. British Archaeological Reports, British Series 109. Oxford: British Archaeological Reports, pp. 91–108.

Gray, S. W. & Lamons, F. P. (1959). Skeletal development and tooth eruption in Atlanta children. *American Journal of Orthodontics*, **45**, 272–7.

Grayson, D. K. (1984). *Quantitative Zooarchaeology*. New York: Academic Press.

Greene, D. L. (1984). Fluctuating dental asymmetry and measurement error. *American Journal of Physical Anthropology*, **65**, 283–9.

Grigson, C. (1982a). Sex and age determination of bones and teeth of domestic cattle: a review of the literature. In Wilson, B., Grigson, C., & Payne, S. (eds.), *Ageing and Sexing Animal Bones from Archaeological Sites*. British Archaeological Reports, British Series 109. Oxford: British Archaeological Reports, pp. 7–73.

(1982b). Sexing Neolithic cattle skulls and horncores. In Wilson, B., Grigson, C., & Payne, S. (eds.), *Ageing and Sexing Animal Bones from Archaeological Sites*. British Archaeological Reports, British Series 109. Oxford: British Archaeological Reports, pp. 24–35.

(1989). Size and sex: morphometric evidence for the domestication of cattle in the Near East. In Milles, A., Williams, D., & Gardner, N. (eds.), *The Beginnings of Agriculture*. BAR International Series 496. Oxford: British Archaeological Reports, pp. 77–109.

Grine, F. E. (1986). Dental evidence for dietary differences in *Australopithecus* and *Paranthropus*: a quantitative analysis of permanent molar microwear. *Journal of Human Evolution*, **15**, 783–822.

(1987). Quantitative analysis of occlusal microwear in *Australopithecus* and *Paranthropus*. *Scanning Microscopy*, **1**, 647–56.

Grine, F. E., Fosse, G., Krause, D. W., & Jungers, W. L. (1986). Analysis of enamel ultrastructure in archaeology: the identification of *Ovis aries* and *Capra hircus* dental remains. *Journal of Archaeological Science*, **13**, 579–95.

Grine, F. E., Krause, D. W., Fosse, G., & Jungers, W. L. (1987). Analysis of individual, intraspecific and interspecific variability of caprine tooth enamel structure. *Acta Odontologica Scandinavica*, **45**, 1–23.

Grün, R. & Stringer, C. B. (2000). Tabun revisited: revised ESR chronology and new ESR and U-series analyses of dental material from Tabun C1. *Journal of Human Evolution*, **39**, 601–12.

Grüneberg, H. (1963). *The Pathology of Development*. Oxford: Blackwell Scientific Publications.

(1965). Genes and genotypes affecting the teeth of the mouse. *Journal of Embryology and Experimental Morphology*, **14**, 137–59.

Guatelli-Steinberg, D. (2000). Linear enamel hypoplasia in gibbons (*Hylobates lar carpenteri*). *American Journal of Physical Anthropology*, **112**, 395–410.

(2001). What can developmental defects of enamel reveal about physiological stress in non-human primates? *Evolutionary Anthropology*, **10**, 138–51.

(2003). Macroscopic and microscopic analyses of linear enamel hypoplasia in Plio-Pleistocene South African hominins with respect to aspects of enamel development and morphology. *American Journal of Physical Anthropology*, **120**, 309–22.

Guatelli-Steinberg, D. & Lukacs, J. R. (1998). Preferential expression of linear enamel hypoplasia on the sectorial premolars of Rhesus monkeys (*Macaca mulatta*). *American Journal of Physical Anthropology*, **107**, 179–86.

(1999). Interpreting sex differences in enamel hypoplasia in human and non-human primates: developmental, environmental and cultural considerations. *Yearbook of Physical Anthropology*, **42**, 73–126.

Guatelli-Steinberg, D. & Skinner, M. (2000). Prevalence and etiology of linear enamel hypoplasia in monkeys and apes from Asia and Africa. *Folia Primatologia*, **71**, 115–32.

Gustafson, A. G. (1955). The similarity between contralateral pairs of teeth. *Odontologisk Tidskrift*, **63**, 245–8.

Gustafson, A. G. & Persson, P. (1957). The relationship between Sharpey's fibres and the deposition of cementum. *Odontologisk Tidskrift*, **65**, 457–63.

Gustafson, G. (1950). Age determinations on teeth. *Journal of the American Dental Association*, **41**, 45–54.

(1966). *Forensic Odontology*. London: Staples Press.

Gustafson, G. & Gustafson, A. G. (1967). Microanatomy and histochemistry of enamel. In Miles, A. E. W. (ed.), *Structural and Chemical Organization of Teeth*. London: Academic Press, pp. 135–62.

Gustafson, G. & Koch, G. (1974). Age estimation up to 16 years of age based on dental development. *Odontologisk Revy*, **25**, 297–306.

Gysi, A. (1931). Metabolism in adult enamel. *Dental Digest*, **37**, 661–8.

Habermehl, K. H. (1961). *Die Alterbestimmung bei Haustieren, Pelztieren und beim jagdbaren Wildtieren*. Berlin: Paul Parey.

(1975). *Die Alterbestimmung bei Haus- und Labortieren*. Berlin: Paul Parey.

Haeussler, A. M. & Turner, C. G. (1992). The dentition of Soviet Central Asians and the quest for New World ancestors. In Lukacs, J. R. (ed.), *Culture, Ecology and Dental Anthropology*. Journal of Human Ecology Special Issue 2. Delhi: Kamla-Raj Enterprises, pp. 273–97.

Hägg, U. & Matsson, L. (1985). Dental maturity as an indicator of chronological age: the accuracy and precision of three methods. *European Journal of Orthodontics*, **7**, 24–34.

Haikel, Y., Frank, R. M., & Voegel, J. C. (1983). Scanning electron microscopy of the human enamel surface layer of incipient carious lesions. *Caries Research*, **17**, 1–13.

Hall, E. R. (1981). *The Mammals of North America*. New York: Wiley.

Halse, A. & Selvig, K. A. (1974). Incorporation of iron in rat incisor enamel. *Scandinavian Journal of Dental Research*, **82**, 47.

Halstead, P. & Collins, P. (2002). Sorting the sheep from the goats: morphological distinctions between the mandibles and mandibular teeth of adult *Ovis* and *Capra*. *Journal of Archaeological Science*, **29**, 545–53.

Hamilton, J. (1978). A comparison of the age structure at mortality of some Iron Age and Romano-British cattle and sheep populations. In Parrington, M. (ed.), *The Excavation of Iron Age Settlement, Bronze Age Ring Ditches and Roman Features at Ashville Trading Estate, Abingdon (Oxfordshire)*. CBA Research Reports 28. London: Council for British Archaeology, pp. 126–33.

(1982). Re-examination of a sample of Iron Age sheep mandibles from Ashville Trading Estate, Abingdon Oxfordshire. In Wilson, B., Grigson, C., & Payne, S. (eds.), *Ageing and Sexing Animal Bones from Archaeology Sites*. British Archaeology Reports, British Series 109. Oxford: British Archaeology Reports, pp. 215–22.

Hamlin, K. L., Pac, D. F., Sime, C. A., DeSimone, R. M., & Dusek, G. L. (2000). Evaluating the accuracy of ages obtained by two methods for Montana ungulates. *Journal of Wildlife Management*, **64**, 441–9.

Hancox, N. M. (1972). *Biology of Bone*. Cambridge: Cambridge University Press.

Hardie, J. M. & Bowden, G. H. (1974). The normal microbial flora of the mouth. In Skinner, F. A. & Carr, J. G. (eds.), *The Normal Microbial Flora of Man*. London: Academic Press, pp. 47–83.

Harris, E. F. (1992). Laterality in human odontometrics: analysis of a contemporary American White series. In Lukacs, J. R. (ed.), *Culture, Ecology & Dental Anthropology*. Journal of Human Ecology Special Issue 2. Delhi: Kamla-Raj Enterprises, pp. 157–70.
  (1998a). Dental maturation. In Ulijaszek, S. J., Johnston, F. E., & Preece, M. A. (eds.), *The Cambridge Encyclopedia of Human Growth and Development*. Cambridge: Cambridge University Press, pp. 45–8.
  (1998b). Ontogenetic and intraspecific patterns of odontometric associations in humans. In Lukacs, J. R. (ed.), *Human Dental Development, Morphology and Pathology: and Tribute to Albert, A. Dahlberg*. University of Oregon Anthropological Papers 54. Eugene: University of Oregon, pp. 299–346.
  (2001). Deciduous tooth size distributions in recent humans: a world-wide survey. In Brook, A. H. (ed.), *Dental Morphology 2001 Sheffield*. Sheffield: Sheffield Academic Press, pp. 13–30.
  (2003). Where's the variation? Variance components in tooth sizes of the permanent dentition. *Dental Anthropology*, **16**, 84–94.
Harris, E. F. & Bailit, H. L. (1980). The metaconule: a morphologic and familial analysis of a molar cusp in humans. *American Journal of Physical Anthropology*, **53**, 349–58.
  (1988). A principal components analysis of human odontometrics. *American Journal of Physical Anthropology*, **75**, 87–99.
Harris, E. F. & Buck, A. L. (2002). Tooth mineralization: a technical note on the Moorrees-Fanning-Hunt standards. *Dental Anthropology*, **16**, 15–21.
Harris, E. F. & Rathbun, T. A. (1991). Ethnic differences in the apportionment of tooth sizes. In Kelley, M. A. & Larsen, C. S. (eds.), *Advances in Dental Anthropology*. New York: Wiley-Liss, pp. 121–42.
Harrison, R. J. & King, J. E. (1980). *Marine Mammals*. London: Hutchinson.
Harshyne, W. A., Diefenbach, D. R., Alt, G. R., & Matson, G. M. (1998). Analysis of error from cementum-annuli age estimates of known-age Pennsylvania black bears. *Journal of Wildlife Management*, **62**, 1281–91.
Hartley, W. G. (1979). *Hartley's Microscopy*. Charlbury, UK: Senecio Publishing.
Hartman, G. D. (1995). Age determination, age structure, and longevity in the mole, *Scalopus aquaticus* (Mammalia: Insectivora). *Journal of Zoology*, **237**, 107–22.
Hartman, S. E. (1989). Stereophotogrammetric analysis of occlusal morphology of extant hominoid molars: phenetics and function. *American Journal of Physical Anthropology*, **80**, 145–66.
Haynes, G. (1991). *Mammoths, Mastodonts, and Elephants*. Cambridge: Cambridge University Press.
Healy, W. B., Cutress, T. W., & Michie, C. (1967). Wear of sheep's teeth IV. Reduction of soil ingestion and tooth wear by supplementary feeding. *New Zealand Journal of Agricultural Research*, **10**, 201–9.
Healy, W. B. & Ludwig, T. G. (1965). Wear of sheep's teeth I. The role of ingested soil. *New Zealand Journal of Agricultural Research*, **8**, 737–52.
Hedges, R. E. M. & Wallace, C. J. A. (1978). The survival of biochemical information from archaeological bone. *Journal of Archaeological Science*, **5**, 377–86.
Henderson, A. M. & Corruccini, R. S. (1976). Relationship between tooth size and body size in American Blacks. *Journal of Dental Research*, **54**, 94–6.
Henderson, P., Marlow, C. A., Molleson, T. I., & Williams, C. T. (1983). Patterns of chemical change during fossilization. *Nature*, **306**, 358–60.
Hershkovitz, P. (1971). Basic crown patterns and cusp homologies of mammalian teeth. In Dahlberg, A. A. (ed.), *Dental Morphology and Evolution*. Chicago: Chicago University Press, pp. 95–105.
Hewer, H. E. (1964a). *British Seals*. London: Collins.
  (1964b). The determination of age, sexual maturity, longevity and a life table in the grey seal (*Halichoerus grypus*). *Proceedings of the Zoological Society of London*, **142**, 593–624.

Higham, C. W. F. (1967). Stock rearing as a cultural factor in prehistoric Europe. *Proceeding of the Prehistoric Society*, **33**, 84–106.
  (1968). Size trends in prehistoric European domestic fauna, and the problem of local domestication. *Acta Zoologica Fennica*, **120**, 3–21.
Hillson, S. W. (1979). Diet and dental disease. *World Archaeology*, **11**, 147–62.
  (1986). *Teeth*. Cambridge Manuals in Archaeology. Cambridge: Cambridge University Press.
  (1992a). Impression and replica methods for studying hypoplasia and perikymata on human tooth crown surfaces from archaeological sites. *International Journal of Osteoarchaeology*, **2**, 65–78.
  (1992b). Studies of growth in dental tissues. In Lukacs, J. R. (ed.), *Culture, Ecology and Dental Anthropology*. Journal of Human Ecology Special Issue 2. Delhi: Kamla-Raj Enterprises, pp. 7–23.
  (1996). *Dental Anthropology*. Cambridge: Cambridge University Press.
  (1998). Crown diameters, tooth crown development, and environmental factors in growth. In Lukacs, J. R. (ed.), *Human Dental Development, Morphology and Pathology: and Tribute to Albert, A. Dahlberg*. University of Oregon Anthropological Papers 54. Eugene: University of Oregon, pp. 17–28.
  (2000). Dental pathology. In Katzenberg, M. A. & Saunders, S. R. (eds.), *Biological Anthropology of the Human Skeleton*. New York: John Wiley, pp. 249–86.
  (2001). Recording dental caries in archaeological human remains. *International Journal of Osteoarchaeology*, **11**, 249–89.
Hillson, S. W., Antoine, D. M., & Dean, M. C. (1999). A detailed developmental study of the defects of dental enamel in a group of post-Medieval children from London. In Mayhall, J. T. & Heikinnen, T. (eds.), *Dental Morphology '98*. Oulu: Oulu University Press, pp. 102–11.
Hillson, S. W. & Bond, S. (1996). A scanning electron microscope study of bone, cement, dentine and enamel. In Bell, M., Fowler, P., & Hillson, S. (eds.), *The Experimental Earthwork Project, 1960–1992*. CBA Research Report. York: Council for British Archaeology, pp. 185–94.
  (1997). Relationship of enamel hypoplasia to the pattern of tooth crown growth: a discussion. *American Journal of Physical Anthropology*, **104**, 89–104.
Hillson, S. & Fitzgerald, C. (2003). Tooth size variation and dental reduction in Europe, the Middle East and North Africa between 120,000 and 5000 BP. *American Journal of Physical Anthropology*, Supplement **36**, 114.
Hillson, S. W., Fitzgerald, C. M., & Flinn, H. M. (in press). Alternative dental measurements – proposals and relationships with other measurements. *American Journal of Physical Anthropology*.
Hillson, S. W., Grigson, C., & Bond, S. (1998). The dental defects of congenital syphilis. *American Journal of Physical Anthropology*, **107**, 25–40.
Hillson, S. W. & Jones, B. K. (1989). Instruments for measuring surface profiles: an application in the study of ancient human tooth crown surfaces. *Journal of Archaeological Science*, **16**, 95–105.
Hilming, F. & Pedersen, P. O. (1940). Über die Paradentalverhältnisse und die Abrasion bei rezenten ostgrönländischen Eskimos. *Paradentium*, **12**, 69–78.
Hinton, R. J. (1981). Form and patterning of anterior tooth wear among aboriginal human groups. *American Journal of Physical Anthropology*, **54**, 555–64.
  (1982). Differences in interproximal and occlusal tooth wear among prehistoric Tennessee Indians: implications for masticatory function. *American Journal of Physical Anthropology*, **57**, 103–15.
Hlusko, L. & Mahaney, M. C. (2003). Genetic contributions to expression of the baboon cingular remnant. *Archives of Oral Biology*, **48**, 663–72.
Horowitz, S. L., Osborne, R. H., & de George, F. V. (1958). Hereditary factors in tooth dimensions, a study of the anterior teeth in twins. *Angle Orthodontist*, **28**, 87–93.
Howells, W. W. (1973). *Cranial Variation in Man. A Study by Multivariate Analysis of Patterns of Difference Among Recent Human Populations*. Papers of the Peabody Museum of Archaeology and Ethnology 67. Cambridge, Massachusetts: Harvard University.

(1989). *Skull Shapes and the Map. Craniometric Analyses in the Dispersion of Modern Homo*. Papers of the Peabody Museum of Archaeology and Ethnology 79. Cambridge, Massachusetts: Harvard University.

(1995). *Ethnic Identification of Crania from Measurements*. Papers of the Peabody Museum of Archaeology and Ethnology 82. Cambridge, Massachusetts: Harvard University.

Hughes, T., Dempsey, P., Richards, L., & Townsend, G. C. (2000). Genetic analysis of deciduous tooth size in Australian twins. *Archives of Oral Biology*, **45**, 997–1004.

Hunt, E. E. & Gleiser, I. (1955). The estimation of age and sex of preadolescent children from bones and teeth. *American Journal of Physical Anthropology*, **13**, 479–87.

Hunter, J. (1771). *The Natural History of the Teeth*. London.

Hunter, W. S. & Priest, W. R. (1960). Errors and discrepancies in measurement of tooth size. *Journal of Dental Research*, **39**, 405–8.

Hutchinson, D. L., Larsen, C. S., & Choi, I. (1997). Stressed to the max? Physiological perturbation in the Krapina Neandertals. *Current Anthropology*, **38**, 904–14.

Infante, P. F. & Gillespie, G. M. (1974). An epidemiologic study of linear enamel hypoplasia of deciduous anterior teeth in Guatemalan children. *Archives of Oral Biology*, **19**, 1055–61.

International Whaling Commission (1969). Report of the meeting on age determination in whales. *Report of the International Whaling Commission*, **19**, 131–7.

Irish, J. D. (1997). Characteristic high- and low-frequency dental traits in Sub-Saharan African populations. *American Journal of Physical Anthropology*, **102**, 455–68.

Iscan, M. Y. & Loth, S. R. (1984). Metamorphosis at the sternal rib end: a new method to estimate age at death in white males. *American Journal of Physical Anthropology*, **65**, 147–56.

(1986a). Determination of age from the sternal rib in White females: a test of the phase method. *Journal of Forensic Sciences*, **31**, 990–9.

(1986b). Determination of age from the sternal rib in White males: a test of the phase method. *Journal of Forensic Sciences*, **31**, 122–32.

Iscan, M. Y., Loth, S. R., & Wright, R. K. (1987). Racial variation in the sternal extremity of the rib and its effect on age determination. *Journal of Forensic Sciences*, **32**, 452–66.

Jacobi, K. P., Collins Cook, D., Corruccini, R. S., & Handler, J. S. (1992). Congenital syphilis in the past: slaves at Newton Plantation, Barbados, West Indies. *American Journal of Physical Anthropology*, **89**, 145–58.

James, W. W. (1960). *The Jaws and Teeth of Primates*. London: Pitman Medical Publishing.

Jernvall, J. & Thesleff, I. (2000a). Reiterative signaling and patterning during mammalian tooth morphogenesis. *Mechanisms of Development*, **92**, 19–29.

(2000b). Return of lost structure in the developmental control of tooth shape. In Teaford, M. F., Meredith Smith, M., & Ferguson, M. W. J. (eds.), *Development, Function and Evolution of Teeth*. Cambridge: Cambridge University Press, pp. 13–21.

Johanson, G. (1971). Age determination from human teeth. *Odontologisk Revy*, **22**, 1–126.

Jones, S. J. (1974). Coronal cementogenesis in the horse. *Archives of Oral Biology*, **19**, 605–14.

(1981). Cement. In Osborn, J. W. & Johns, R. B. (eds.), *Dental Anatomy and Embryology*. Oxford: Blackwell Scientific Publications, pp. 193–205.

(1987). The root surface: an illustrated review of some scanning electron microscope studies. *Scanning Microscopy*, **1**, 2003–18.

Jones, S. J. & Boyde, A. (1972). A study of human root cementum surfaces as prepared for and examined in the scanning electron microscope. *Zeitschrift für Zellforschung und Microskopische Anatomie*, **130**, 318–37.

(1984). Ultrastructure of dentin and dentinogenesis. In Linde, A. (ed.), *Dentin and Dentinogenesis*. Boca Raton: CRC Press, pp. 81–134.

(1987). Scanning microscopic observations on dental caries. *Scanning Microscopy*, **1**, 1991–2002.

(1988). The resorption of dentine and cementum *in vivo* and *in vitro*. In Davidovitch, Z. (ed.), *The Biological Mechanisms of Tooth Eruption and Root Resorption*. Birmingham, AL: EBSCO Media, pp. 335–54.
Jonsgard, A. (1969). Age determination of marine animals. In Andersen, H. T. (ed.), *The Biology of Marine Animals*. London: Academic Press, pp. 1–30.
Jordan, R. E., Abrams, L., & Kraus, B. S. (1992). *Kraus' Dental Anatomy and Occlusion*. St Louis: Mosby.
Jubb, K. V. F. & Kennedy, P. C. (1963). *Pathology of Domestic Animals*. New York and London: Academic Press.
Kaestle, F. A. & Horsburgh, K. A. (2002). Ancient DNA in anthropology: methods, applications, and ethics. *Yearbook of Physical Anthropology*, **45**, 92–130.
Kaifu, Y., Kasai, K., Townsend, G. C., & Richards, L. C. (2003). Tooth wear and the 'design' of the human dentition: a perspective from evolutionary medicine. *Yearbook of Physical Anthropology*, **46**, 47–61.
Källestål, C. & Wall, S. (2002). Socio-economic effect on caries. Incidence data among Swedish 12–14-year-olds. *Community Dentistry and Oral Epidemiology*, **30**, 108–14.
Karn, K. W., Shockett, H. P., Moffitt, W. C., & Gray, J. L. (1984). Topographic classification of deformities of the alveolar process. *Journal of Periodontology*, **55**, 336–40.
Katz, D. & Suchey, J. M. (1986). Age determination of the male Os pubis. *American Journal of Physical Anthropology*, **69**, 427–36.
  (1989). Race differences in pubic symphyseal aging patterns in the male. *American Journal of Physical Anthropology*, **80**, 167–72.
Kaul, S. S. & Corruccini, R. S. (1992). Dental arch length reduction through interproximal attrition in modern Australian aborigines. In Lukacs, J. R. (ed.), *Culture, Ecology and Dental Anthropology*. Journal of Human Ecology Special Issue 2. Delhi: Kamla-Raj Enterprises, pp. 195–200.
Kawai, N. (1955). Comparative anatomy of the bands of Schreger. *Okajimas Folia Anatomica Japonica*, **27**, 115–31.
Kawano, S., Tsukamoto, T., Ohtaguro, H., Tustsumi, H., Takahashi, T., Miura, I., Mukoyama, R., Aboshi, H., & Komuro, T. (1995). Sex determination from dental calculus by polymerase chain reaction (PCR) (In Japanese). *Nippon Hoigaku Zasshi*, **49**, 193–8.
Kawasaki, K., Tanaka, S., & Isikawa, T. (1980). On the daily incremental lines in human dentine. *Archives of Oral Biology*, **24**, 939–43.
Kay, M. (1974). Dental annuli age determination on white-tailed deer from archaeological sites. *Plains Anthropologist*, **19**, 224–7.
Kerr, P. F. (1959). *Optical Mineralogy*. New York: McGraw-Hill.
Kerr, N. W. (1991). Prevalence and natural history of periodontal disease in Scotland – the mediaeval period (900–1600 AD). *Journal of Periodontal Research*, **26**, 346–54.
Kierdorf, U. & Becher, J. (1997). Mineralization and wear of mandibular first molars in red deer (*Cervus elaphus*) of known age. *Journal of Zoology*, **241**, 135–43.
Kieser, J. A. (1990). *Human Adult Odontometrics*. Cambridge Studies in Biological Anthropology 4. Cambridge: Cambridge University Press.
Kieser, J. A. & Groeneveld, H. T. (1991). The reliability of human odontometric data. *Journal of the Dental Association of South Africa*, **46**, 267–70.
  (1998). Fluctuating dental asymmetry and prenatal exposure to tobacco smoke. In Lukacs, J. R. (ed.), *Human Dental Development, Morphology and Pathology: and Tribute to Albert, A. Dahlberg*. University of Oregon Anthropological Papers 54. Eugene: University of Oregon, pp. 287–97.
Kieser, J. A., Preston, C. B., & Evans, W. G. (1983). Skeletal age at death: an evaluation of the Miles method of ageing. *Journal of Archaeological Science*, **10**, 9–12.
Kilgore, L. (1995). Patterns of dental decay in African great apes. In Cockburn, E. (ed.), *Papers on Paleopathology presented at the 22nd Annual Meeting*, p. 6. Paleopathology Association.

King, C. M. (1991). A review of age determination methods for the stoat *Mustela erminea*. *Mammal Review*, **21**, 31–49.

King, J. E. (1964). *Seals of the World*. London: British Museum Natural History.

(1983). *Seals of the World*. London and Oxford: British Museum Natural History and Oxford University Press.

King, T., Hillson, S., & Humphrey, L. T. (2002). A detailed study of enamel hypoplasia in a post-medieval adolescent of known age and sex. *Archives of Oral Biology*, **47**, 29–39.

Kingdon, J. (1979). *East African Mammals: An Atlas of Evolution in Africa*. London: Academic Press.

Klein, R. G. (1978). Stone age predation on large African bovids. *Journal of Archaeological Science*, **5**, 195–217.

(1981). Stone age predation on small African bovids. *South African Archaeological Bulletin*, **36**, 55–65.

Klein, R. G., Allwarden, K., & Wolf, C. (1983). The calculation and interpretation of ungulate age profiles from dental crown heights. In Bailey, G. (ed.), *Hunter Gatherer Economy in Prehistory*. Cambridge: Cambridge University Press.

Klein, R. G. & Cruz-Uribe, K. (1984). *The Analysis of Animal Bones from Archaeological Sites*. Chicago: University of Chicago Press.

Klein, R. G., Wolf, C., Freeman, L. G., & Allwarden, K. (1981). The use of dental crown heights for construction age profiles of red deer and similar species in archaeological samples. *Journal of Archaeological Science*, **8**, 1–32.

Klevezal, G. A. (1996). *Recording Structures of Mammals: Determination of Age and Reconstruction of Life History*. Rotterdam: A. A. Balkema.

Klevezal, G. A. and Kleinenberg, S. E. (1967). *Age Determination of Mammals from Annual Layers in Teeth and Bones*. Springfield, Academy of Sciences USSR translated 1969 Department of the Interior and National Sciences Foundation, US Department of Commerce, Clearinghouse for Federal Scientific and Technical Information.

Klevezal, G. A. & Stewart, B. S. (1994). Patterns and calibration of layering in tooth cementum of female Northern elephant seals, *Mirounga angustirostris*. *Journal of Mammalogy*, **75**, 483–7.

Konigsberg, L. W. & Frankenberg, S. R. (1994). Paleodemography: 'not quite dead'. *Evolutionary Anthropology*, **3**, pp. 92–105.

(2002). Deconstructing death in paleodemography. *American Journal of Physical Anthropology*, **117**, 297–309.

Korvenkontio, V. A. (1934). Mikroskopische Untersuchungen an Nagerincisiven unter Hinweis auf die Schmelztruktur der Backenzähne. *Annales Societatis Zoologici Botanicae Fennici Vanamo*, **2**, 1–274.

Kratochvil, Z. (1981). Tierknochenfunde aus der grossmahrischen siedlung Mikulcice: I Das Hausschwein. *Studie Archeologickeho u stavu Ceskoslovenske Akademi ved v Brno Pocnik*, **IX** 3, Academia Praha.

Kreshover, S. J. (1944). The pathogenesis of enamel hypoplasia: an experimental study. *Journal of Dental Research*, **23**, 231–8.

(1960). Metabolic disturbances in tooth formation. *Annals of the New York Academy of Sciences*, **85**, 161–7.

Kreshover, S. J. & Clough, O. W. (1953). Prenatal influences on tooth development II. Artificially induced fever in rats. *Journal of Dental Research*, **32**, 565–72.

Kreshover, S. J., Clough, O. W., & Hancock, J. A. (1954). Vaccinia infection in pregnant rabbits and its effect on maternal and fetal dental tissues. *Journal of the American Dental Association*, **49**, 549–62.

Krings, M., Stone, A., Schmitz, R. W., Krainitz, H., Stoneking, M., & Pääbo, S. (1997). Neandertal DNA sequences and the origin of modern humans. *Cell*, **90**, 19–30.

Kronfeld, R. & Schour, I. (1939). Neonatal dental hypoplasia. *Journal of the American Dental Association*, **26**, 18–32.

Kubota, K. (1963). Morphological observations of the deciduous dentition of the fur seal (*Callorhinus ursinus*). *Bulletin of the Tokyo Medical & Dental University*, **63**, 75–81.

Kubota, K. & Matsumoto, K. (1963). On the deciduous teeth shed into the amnion of the fur seal embryo. *Bulletin of the Tokyo Medical & Dental University*, **10**, 90–3.

Kurtén, B. (1955). Sex dimorphism and size trends in the cave bear, *Ursus spelaeus* Rosenmüller and Heinroth. *Acta Zoologica Fennica*, **90**, 1–48.

  (1968). *Pleistocene Mammals of Europe*. London: Weidenfield & Nicholson.

  (1976). *The Cave Bear Story*. New York: Columbia University Press.

  (1979). The stilt-legged deer *Sangamona* of the North-American Pleistocene. *Boreas*, **8**, 313–21.

Kurtén, B. & Anderson, E. (1980). *Pleistocene Mammals of North America*. New York: Columbia University Press.

Kuykendall, K. L. (1996). Dental development in chimpanzees (*Pan troglodytes*): the timing of tooth calcification stages. *American Journal of Physical Anthropology*, **99**, 135–58.

Kvaal, S. I. & During, E. M. (1999). A dental study comparing age estimations of the human remains from the Swedish warship *Vasa*. *International Journal of Osteoarchaeology*, **9**, 170–81.

Kvaal, S. I., Kolltveit, K. M., Thomsen, I. O., & Solheim, T. (1995). Age estimation of adults from dental radiographs. *Forensic Science International*, **74**, 175–85.

Lahr, M. M. (1996). *The Evolution of Modern Human Diversity: a Study of Cranial Variation*. Cambridge: Cambridge University Press.

Lamendin, H., Baccino, E., Humbert, J. F., Tavernier, J. C., Nossintchouk, R. M., & Zerilli, A. (1992). A simple technique for age estimation in adult corpses: the two criteria dental method. *Journal of Forensic Sciences*, **37**, 1373–9.

Landon, D. B. (1993). Testing a seasonal slaughter model for colonial New England using tooth cementum increment analysis. *Journal of Archaeological Science*, **20**, 439–55.

Landon, D. B., Waite, C. A., Peterson, R. O., & Mech, L. D. (1998). Evaluation of age determination techniques for gray wolves. *Journal of Wildlife Management*, **62**, 674–82.

Lane, G. (1981). Small animal dentistry. *In Practice*, **3**, 23–30.

Larsen, C. S. (1995). Biological changes in human populations with agriculture. *Annual Review of Anthropology*, **24**, 185–213.

  (1997). *Bioarchaeology*. Cambridge Studies in Biological Anthropology. Cambridge: Cambridge University Press.

Larsen, C. S., Hutchinson, D. L., Schoeninger, M. J., & Norr, L. (2001). Food and stable isotopes in La Florida: diet and nutrition before and after contact. In Larsen, C. S. (ed.), *Bioarchaeology of Spanish Florida*. Gainesville: University Press of Florida, pp. 52–81.

Larsen, C. S., Shavit, R., & Griffin, M. C. (1991). Dental caries evidence for dietary change: an archaeological context. In Kelley, M. A. & Larsen, C. S. (eds.), *Advances in Dental Anthropology*. New York: Wiley-Liss, pp. 179–202.

Lasker, G. W. (1951). Genetic analysis of racial traits of the teeth. *Cold Spring Harbor Symposia on Quantitative Biology*, **XV**, 191–203.

Lavelle, C. L. B. & Moore, W. J. (1973). The incidence of agenesis and polygenesis in the primate dentition. *American Journal of Physical Anthropology*, **38**, 671–80.

Laws, R. M. (1952). A new method of age determination for mammals. *Nature*, **169**, 972–3.

  (1953a). A new method of age determination in mammals with special reference to the elephant seal (*Mirounga lenonina*, L.). *Falkland Islands Dependencies Survey Scientific Reports*, **2**, 1–11.

  (1953b). The elephant seal (*Mirounga leonina*, L.). Growth and age. *Falkland Islands Dependencies Survey Scientific Reports*, **8**, 1–62.

(1966). Age criteria for the African elephant *Loxodonta africana*. *East African Wildlife Journal*, **4**, 1–37.

(1968). Dentition and ageing of the hippopotamus. *East African Wildlife Journal*, **6**, 19–52.

Laws, R. M., Baird, A., & Bryden, M. M. (2002). Age estimation in crabeater seals (*Lobodon carcinophagus*). *Journal of Zoology*, **258**, 197–203.

Leek, F. F. (1972). Teeth and bread in ancient Egypt. *Journal of Egyptian Archaeology*, **58**, 126–32.

Legge, A. J. & Rowley-Conwy, P. (1988). *Star Carr revisited – a re-analysis of the large mammals*. London: Centre for Extra-Mural Studies, Birkbeck College.

Lehner, T. (1992). *Immunology of Oral Diseases*. Oxford: Blackwell Scientific Publications.

Lester, K. S. & Hand, S. J. (1987). Chiropteran enamel structure. *Scanning Microscopy*, **1**, 421–36.

Leutenegger, W. & Kelly, J. T. (1977). Relationship of sexual dimorphism in canine size and body size to social, behavioural and ecological correlates in anthropoid primates. *Primates*, **18**, 117–36.

Levers, B. G. H. & Darling, A. I. (1983). Continuous eruption of some adult human teeth of ancient populations. *Archives of Oral Biology*, **28**, 401–8.

Levine, M. A. (1982). The use of crown height measurements and eruption-wear sequences to age horse teeth. In Wilson, B., Grigson, C., & Payne, S. (eds.), *Ageing and Sexing Animal Bones from Archaeological Sites*. British Archaeological Reports, British Series 109. Oxford: British Archaeological Reports, pp. 223–50.

(1983). Mortality models and the interpretation of horse population structure. In Bailey, G. (ed.), *Hunter Gatherer Economy in Prehistory*. Cambridge: Cambridge University Press.

Levitan, B. M. (1985). A methodology for recording the pathology and other anomalies of ungulate mandibles from archaeological sites. In Fieller, N. R. J., Gilbertson, D. D., & Ralph, N. G. A. (eds.), *Palaeobiological Investigations: Research Design, Methods and Data Analysis*. British Archaeology Reports International Series 266. Oxford: British Archaeology Reports, pp. 41–54.

Lieberman, D. E. (1993a). Life history variables preserved in dental cementum microstructure. *Science*, **261**, 1162–4.

(1993b). The rise and fall of seasonal mobility among hunter-gatherers. *Current Anthropology*, **34**, 599–631.

(1994). The biological basis for seasonal increments in dental cementum and their application to archaeological research. *Journal of Archaeological Science*, **21**, 525–39.

Lieberman, D. E., Deacon, T. W., & Meadow, R. H. (1990). Computer image enhancement and analysis of cementum increments as applied to teeth of *Gazella gazella*. *Journal of Archaeological Science*, **17**, 519–33.

Lieberman, D. E. & Meadow, R. H. (1992). The biology of cementum increments (with an archaeological application). *Mammal Review*, **22**, 57–78.

Limbrey, S. (1975). *Soil Science and Archaeology*. Studies in Archaeological Science. London: Academic Press.

Lindemann, G. (1958). Forekomsten af emaljehypoplasi hos børn, som har lidt af mave – tramsygdomme. *Odontologisk Tidskrift*, **66**, 101–26.

Lingström, P., Birkhed, D., Granfeldt, Y., & Björck, I. (1993). pH measurements of human dental plaque after consumption of starchy foods using the microtouch and the sampling method. *Caries Research*, **27**, 394–401.

Lipsinic, F. E., Paunovitch, E., Houston, G. D., & Robinson, S. F. (1986). Correlation of age and incremental lines in the cementum of human teeth. *Journal of Forensic Sciences*, **31**, 982–9.

Lister, A. M. (1996). The morphological distinction between bones and teeth of Fallow Deer (*Dama dama*) and Red Deer (*Cervus elaphus*). *International Journal of Osteoarchaeology*, **6**, 119–43.

Lister, A. M. & Bahn, P. (2000). *Mammoths. Giants of the Ice Age*. London: Marshall Publishing.

Listgarten, M. A. (1968). A light and electron microscopic study of coronal cementogenesis. *Archives of Oral Biology*, **13**, 93–114.

Liversidge, H. M. (1994). Accuracy of age estimation from developing teeth of a population of known age (0 to 5.4 years). *International Journal of Osteoarchaeology*, **4**, 37–46.

Liversidge, H. M. & Molleson, T. I. (1999). Developing permanent tooth length as an estimate of age. *Journal of Forensic Sciences*, **44**, 917–20.

Locker, D. (2000). Deprivation and oral health: a review. *Community Dentistry & Oral Epidemiology*, **28**, 161–9.

Logan, W. H. G. & Kronfeld, R. (1933). Development of the human jaws and surrounding structures from birth to the age of fifteen years. *Journal of the American Dental Association*, **20**, 379–427.

Lovejoy, C. O., Meindl, R. S., Pryzbeck, T. R., & Mensforth, R. P. (1985). Chronological metamorphosis of the auricular surface of the ilium: a new method for the determination of adult skeletal age at death. *American Journal of Physical Anthropology*, **68**, 15–28.

Lovell, N. C. (1990). Skeletal and dental pathology of free-ranging mountain gorillas. *American Journal of Physical Anthropology*, **81**, 399–412.

 (1991). An evolutionary framework for assessing illness and injury in nonhuman primates. *Yearbook of Physical Anthropology*, **34**, 117–55.

Low, W. A. & Cowan, I. M. (1963). Age determination of deer by annular structure of dental cementum. *Journal of Wildlife Management*, **27**, 466–71.

Lowe, V. P. W. (1967). Teeth as indicators of age with special reference to Red Deer (*Cervus elaphus*) of known age from Rhum. *Journal of Zoology (London)*, **152**, 136–53.

 (1971). Root development of molar in the bank vole (*Clethrionomys glareolus*). *Journal of Animal Ecology*, **40**, 49–61.

Luan, W. M., Baelum, V., Chen, X., & Fejerskov, O. (1989). Dental caries in adult and elderly Chinese. *Journal of Dental Research*, **68**, 1771–6.

Lubell, D., Jackes, M., Schwarcz, H., Knyf, M., & Meiklejohn, C. (1994). The Mesolithic-Neolithic transition in Portugal: isotopic and dental evidence of diet. *Journal of Archaeological Science*, **21**, 201–16.

Lubinski, P. M. (2001). Estimating age and season of death of pronghorn antelope (*Antilocapra americana* Ord) by means of tooth eruption and wear. *International Journal of Osteoarchaeology*, **11**, 218–30.

Lubinski, P. M. & O'Brien, C. J. (2001). Observations on seasonality and mortality from a recent catastrophic death assemblage. *Journal of Archaeological Science*, **28**, 833–42.

Lucy, D., Aykroyd, R. G., & Pollard, A. M. (2002). Nonparametric calibration for age estimation. *Applied Statistics*, **51**, 183–96.

Lucy, D., Aykroyd, R. G., Pollard, A. M., & Solheim, T. (1996). A Bayesian approach to adult human age estimation from dental observations by Johanson's age changes. *Journal of Forensic Sciences*, **41**, 5–10.

Lucy, D. & Pollard, A. M. (1995). Further comments on the estimation of error associated with the Gustafson dental age estimation method. *Journal of Forensic Sciences*, **40**, 222–7.

Lucy, D., Pollard, A. M., & Roberts, C. A. (1995). A comparison of three dental techniques for estimating age at death in humans. *Journal of Archaeological Science*, **22**, 417–28.

Lukacs, J. R., Retief, D. H., & Jarrige, J. F. (1985). Dental disease in prehistoric Baluchistan. *National Geographic Research*, Spring 1985, 184–97.

Lundström, A. (1948). *Tooth Size and Occlusion in Twins*. Basel: Karger.

Lunt, D. A. (1969). An odontometric study of Medieval Danes. *Acta Odontologica Scandinavica*, **27**, 55–113.

 (1978). Molar attrition in Medieval Danes. In Butler, P. M. & Joysey, K. A. (eds.), *Development, Function and Evolution of Teeth*. London: Academic Press, pp. 465–82.

Lussi, A. (1996). Impact of including or excluding cavitated lesions when evaluating methods for the diagnosis of occlusal caries. *Caries Research*, **30**, 389–93.

Macdonald, D. (1984). *The Encyclopedia of Mammals*. London: George Allen & Unwin.

MacGregor, A. (1985). *Bone, Antler, Ivory and Horn*. London: Croom Helm and Totowa, NJ: Barnes and Noble.
Magner, L. N. (1979). *A History of the Life Sciences*. New York: Marcel Dekker.
Mainland, I. L. (2003). Dental microwear in grazing and browsing Gotland sheep (*Ovis aries*) and its implications for dietary reconstruction. *Journal of Archaeological Science*, **30**, 1513–27.
Manji, F., Fejerskov, O., & Baelum, V. (1989). Pattern of dental caries in an adult rural population. *Caries Research*, **23**, 55–62.
Manji, F., Fejerskov, O., Baelum, V., Luan, W. M., & Chen, X. (1991). The epidemiological features of dental caries in African and Chinese populations: implications for risk assessment. In Johnson, N. W. (ed.), *Volume 1. Dental Caries. Markers of High and Low Risk Groups and Individuals*. Cambridge: Cambridge University Press, pp. 62–99.
Mann, A. (1988). The nature of Taung dental maturation. *Nature*, **333**, 123.
Mann, A., Lampl, M., & Monge, J. (1990). Patterns of ontogeny in human evolution: evidence from dental development. *Yearbook of Physical Anthropology*, **33**, pp. 111–50.
Mann, A. E. (1975). *Some Paleodemographic Aspects of the South African Australopithecines*. University of Pennsylvania Publications in Anthropology 1. Philadelphia: University of Pennsylvania.
Mann, A. E., Lampl, M., & Monge, J. (1987). Maturational patterns in early hominids. *Nature*, **328**, 673–4.
Mann, A. E., Monge, J. M., & Lampl, M. (1991). Investigation into the relationship between perikymata counts and crown formation times. *American Journal of Physical Anthropology*, **86**, 175–88.
Mansfield, A. W. & Fisher, H. D. (1960). Age determination in the harbour seal *Phoca vitulina* L. *Nature*, **186**, 92–3.
Maples, W. R. (1978). An improved technique using dental histology for the estimation of adult age. *Journal of Forensic Sciences*, **23**, 764–70.
Maples, W. R. & Rice, P. M. (1979). Some difficulties in the Gustafson dental age estimations. *Journal of Forensic Sciences*, **24**, 118–72.
Marsh, H. (1980). Age determination of the Dugong (*Dugong dugon* (Müller)) in northern Australia and its biological implications. In Perrin, W. F. & Myrick, A. C. (eds.), *Growth of Odontocetes and Sirenians: Problems in Age Determination. Proceedings of the International Conference on Determining Age of Odontocete Ceteans (and Sirenians), La Jolla, California, September 5–19, 1978*. Report of the International Whaling Commission, Special Issue 3. Cambridge: International Whaling Commission, pp. 181–201.
Marsh, P. & Martin, M. (1992). *Oral Microbiology*. London: Chapman & Hall.
Martin, T. (1997). Incisor enamel microstructure and systematics in rodents. In von Koenigswald, W. & Sander, P. M. (eds.), *Tooth Enamel Microstructure*. Rotterdam: A. A. Balkema, pp. 163–75.
Massler, M., Schour, I., & Poncher, H. (1941). Developmental pattern of the child as reflected in the calcification pattern of the teeth. *American Journal of Diseases of Children*, **62**, 33–67.
Masters, P. M. (1987). Preferential preservation of noncollagenous protein during bone diagenesis: implications for chronometric and stable isotopic measurements. *Geochimica et Cosmochimica Acta*, **51**, 3209–14.
Matschke, G. H. (1967). Ageing European wild hogs by dentition. *Journal of Wildlife Management*, **31**, 103–13.
Mayhall, J. T. (1992). Techniques for the study of dental morphology. In Saunders, S. R. & Katzenberg, M. A. (eds.), *Skeletal Biology of Past Peoples: Research Methods*. New York: Wiley-Liss, pp. 59–78.
Mayhall, J. T. & Alvesalo, L. (1992). Sexual dimorphism in the three-dimensional determinations of the maxillary first molar: cusp height, area, volume and position. In Smith, P. & Tchernov, E. (eds.), *Structure, Function and Evolution of Teeth*. London: Freund Publishing House, pp. 425–36.

Mayhall, J. T. & Kanazawa, E. (1989). Three-dimensional analysis of the maxillary first molar crowns of Canadian Inuit. *American Journal of Physical Anthropology*, **78**, 73–8.

Mayhall, J. T. & Saunders, S. R. (1986). Dimensional and discrete dental trait asymmetry relationships. *American Journal of Physical Anthropology*, **69**, 403–11.

Mayhew, D. F. (1978). Age structure of a sample of subfossil beavers (*Castor fiber*, L.). In Butler, P. M. & Joysey, K. A. (eds.), *Development, Function and Evolution of Teeth*. London: Academic Press, pp. 495–506.

Mays, S. A. (2002). The relationship between molar wear and age in an early 19th century AD archaeological human skeletal series of documented age at death. *Journal of Archaeological Science*, **29**, 861–71.

Mays, S. A., de la Rua, C., & Molleson, T. I. (1995). Molar crown height as a means of evaluating existing dental wear scales for estimating age at death in human skeletal remains. *Journal of Archaeological Science*, **22**, 659–70.

Mazak, V. (1963). Eruption of permanent dentition in the genera *Mustela* Linnaeus, 1758 and *Putorius* Cuvier, 1817, with a note on the genus *Martes* Pinel 1972. *Vestnik Ceskoslovenske Spolecnosit Zoologicke*, **27**, 328–34.

McCance, R. A., Ford, E. H. R., & Brown, W. A. B. (1961). Severe undernutrition in growing and adult animals 7. Development of the skull, jaws and teeth in pigs. *British Journal of Nutrition*, **15**, 213–24.

McKinley, J. I. (1994). *The Anglo-Saxon Cemetery at Spong Hill, North Elmham. Part VIII: the Cremations*. East Anglian Archaeology Report 69. Dereham: Field Archaeology Division, Norfolk Museums Service.

Meade, G. E. (1961). The saber-toothed cat, *Dinobastis serus*. *Bulletin of the Texas Memorial Museum*, **2**, 24–60.

Meikle, M. C. (2002). *Craniofacial Development, Growth and Evolution*. Bressingham: Bateson Publishing.

Mellanby, M. (1929). *Diet and Teeth: an Experimental Study. Part I. Dental Structure in Dogs*. Medical Research Council, Special Report Series, 140. London: His Majesty's Stationery Office.

(1930). *Diet and Teeth: an Experimental Study. Part II. A. Diet and Dental Disease. B. Diet and Dental Structure in Mammals other than the Dog*. Medical Research Council, Special Report Series, 153. London: His Majesty's Stationery Office.

(1934). *Diet and Teeth: an Experimental Study. Part III. The Effect of Diet on the Dental Structure and Disease in Man*. Medical Research Council, Special Report Series, 191. London: His Majesty's Stationery Office.

Merriam, J. C. & Stock, C. (1932). The Felidae of Rancho la Brea. *Carnegie Institution of Washington Publications*, **422**, 1–232.

Miles, A. E. W. (1958). The assessment of age from the dentition. *Proceedings of the Royal Society of Medicine*, **51**, 1057–60.

(1962). Assessment of the ages of a population of Anglo-Saxons from their dentitions. *Proceedings of the Royal Society of Medicine*, **55**, 881–6.

(1963a). Dentition and the estimation of age. *Journal of Dental Research*, **42**, 255–63.

(1963b). The dentition in the assessment of individual age in skeletal material. In Brothwell, D. R. (ed.), *Dental Anthropology*. London: Pergamon Press, pp. 191–209.

(1978). Teeth as an indicator of age in man. In Butler, P. M. & Joysey, K. A. (eds.), *Development, Function and Evolution of Teeth*. London: Academic Press, pp. 455–62.

(2001). The Miles method of assessing age from tooth wear revisited. *Journal of Archaeological Science*, **28**, 973–82.

Miles, A. E. W. & Grigson, C. (1990). *Colyer's Variations and Diseases of the Teeth of Animals*. Revised Edition. Cambridge: Cambridge University Press.

Miller, C. S., Dove, S. B., & Cottone, J. A. (1988). Failure of use of cemental annulations in teeth to determine the age of humans. *Journal of Forensic Sciences*, **33**, 137–43.

Miller, F. L. (1972). Eruption and attrition of mandibular teeth in barren-ground caribou. *Journal of Wildlife Management*, **36**, 606–12.

(1974). Biology of the Kaminuriak population of barren ground caribou Part 2. *Canadian Wildlife Service Report Series*, **31**.

Miller, G. S. (1912). *A Catalogue of the Mammals of Western Europe (Europe exclusive of Russia) in the Collection of the British Museum*. London: British Museum (Natural History).

Milner, G. R. & Larsen, C. S. (1991). Teeth as artifacts of human behavior: intentional mutilation and accidental modification. In Kelley, M. A. & Larsen, C. S. (eds.), *Advances in Dental Anthropology*. New York: Wiley-Liss, pp. 357–78.

Mimura, T. (1939). Horoshitsu ni mirareru Seicho-sen no shuki (The periodicity of growth lines seen in the enamel). *Kobyo-shi*, **13**, 454–5.

Mincer, H. H., Harris, E. F., & Berryman, H. E. (1993). The A.B.F.O. study of third molar development and its use as an estimator of chronological age. *Journal of Forensic Sciences*, **38**, 379–90.

Mitchell, B. (1963). Determination of age in Scottish red deer from growth layers in dental cement. *Nature*, **198**, 350–1.

(1967). Growth layers in dental cement for determining the age of Red Deer (*Cervus elaphus* L.). *Journal of Animal Ecology*, **36**, 279–93.

Moffitt, S. A. (1998). Aging bison by the incremental cementum growth layers in teeth. *Journal of Wildlife Management*, **62**, 1276–80.

Møller, I. J. (1982). Fluorides and dental fluorosis. *International Dental Journal*, **32**, 135–47.

Molleson, T. I. (1993). The human remains. In Farwell, D. E. & Molleson, T. I. (eds.), *Excavations at Poundbury 1966–80*. Dorset Natural History and Archaeological Society Monograph Series 11. Dorchester: Dorset Natural History and Archaeological Society, pp. 142–214.

Molleson, T. I., Jones, K., & Jones, S. (1993). Dietary change and the effects of food preparation on microwear patterns in the Late Neolithic of abu Hureyra, northern Syria. *Journal of Human Evolution*, **24**, 455–68.

Molnar, S. (1971). Human tooth wear, tooth function and cultural variability. *American Journal of Physical Anthropology*, **34**, 175–90.

Moody, J. E. H. (1960). The dental and periodontal conditions of aborigines at settlements in Arnhem Land and adjacent areas. In Mountford, C. R. (ed.), *Records of the American–Australian Scientific Expedition to Arnhem Land: Anthropology and Nutrition*. Melbourne: Melbourne University Press, pp. 60–71.

Moore, W. J. (1974). *Growth of the Facial Skeleton in Hominoidea*. London: Academic Press.

Moore, W. J. & Corbett, M. E. (1971). Distribution of dental caries in ancient British populations: I Anglo-Saxon period. *Caries Research*, **5**, 151–68.

(1973). Distribution of dental caries in ancient British populations: II Iron Age, Romano-British and Medieval periods. *Caries Research*, **7**, 139–53.

(1975). Distribution of dental caries in ancient British populations: III The 17th century. *Caries Research*, **9**, 163–75.

Moorrees, C. F. A., Fanning, E. A., & Hunt, E. E. (1963). Age variation of formation stages for ten permanent teeth. *Journal of Dental Research*, **42**, 1490–502.

Moorrees, C. F. A. & Reed, R. B. (1964). Correlations among crown diameters of human teeth. *Archives of Oral Biology*, **9**, 685–97.

Morales, A. & Rodríguez, J. (1997). Black rats (*Rattus rattus*) from medieval Mertola (Baixo Alentejo, Portugal). *Journal of Zoology*, **241**, 623–42.

Moran, N. C. & O'Connor, T. P. (1994). Age attribution in domestic sheep by skeletal and dental maturation: a pilot study of available sources. *International Journal of Osteoarchaeology*, **4**, 267–86.

Morris, P. (1972). A review of mammalian age determination methods. *Mammal Review*, **2**, 69–104.
  (1978). The use of teeth for estimating the age of wild mammals. In Butler, P. M. & Joysey, K. A. (eds.), *Development, Function and Evolution of Teeth*. London: Academic Press, pp. 483–94.
Mountain, J. L. (1998). Molecular evolution and modern human origins. *Evolutionary Anthropology*, **4**, 53–63.
Muller, D. & Perizonius, W. R. K. (1980). The scoring of defects of the alveolar process in human crania. *Journal of Human Evolution*, **9**, 113–16.
Mundorff, S. A., Featherstone, J. D. B., Bibby, B. G., Curzon, M. E. J., Eisenberg, A. D., & Espeland, M. A. (1990). Cariogenic potential of foods. I. Caries in the rat model. *Caries Research*, **24**, 344–55.
Mundorff-Shrestha, S. A., Featherstone, J. D. B., Eisenberg, A. D., Cowles, E., Curzon, M. E. J., Espeland, M. A., & Shields, C. P. (1994). Cariogenic potential of foods. II. Relationship of food composition, plaque microbial counts, and salivary parameters to caries in the rat model. *Caries Research*, **28**, 106–15.
Munson, P. J. (1984). Teeth of juvenile woodchucks as seasonal indicators on archaeological sites. *Journal of Archaeological Science*, **11**, 395–404.
Murphy, T. (1959a). Gradients of dentine exposure in human molar tooth attrition. *American Journal of Physical Anthropology*, **17**, 179–86.
  (1959b). The changing pattern of dentine exposure in human tooth attrition. *American Journal of Physical Anthropology*, **17**, 167–78.
Nadachowski, A. (1991). Systematics, geographic variation, and evolution of snow voles (*Chionomys*) based on dental characters. *Acta Theriologica*, **36**, 1–45.
Nalbandian, J. & Soggnaes, R. F. (1960). Structural age changes in human teeth. In Shock, N. W. (ed.), *Ageing – Some Social and Biological Aspects. Symposia presented at the Chicago meeting, December 29–30, 1959*. American Association for the Advancement of Science Publication 65. Washington DC: American Association for the Advancement of Science, pp. 367–82.
Navia, J. M. (1977). *Animal Models in Dental Research*. Alabama: University of Alabama Press.
  (1994). Carbohydrates and dental health. *American Journal of Clinical Nutrition*, **59**, 719S–27S.
Naylor, J. W., Miller, W. G., Stokes, G. N., & Stott, G. G. (1985). Cemental annulation enhancement: a technique for age determination in man. *American Journal of Physical Anthropology*, **68**, 197–200.
Nichol, C. R. (1989). Complex segregation analysis of dental morphological variants. *American Journal of Physical Anthropology*, **78**, 37–59.
Noddle, B. (1974). Ages of epiphyseal closure in feral and domestic goats and ages of dental eruption. *Journal of Archaeological Science*, **1**, 195–204.
Nowak, R. M. & Paradiso, J. L. (1983). *Walker's Mammals of the World*. Baltimore: Johns Hopkins University Press.
Nowell, G. W. (1978). An evaluation of the Miles method of ageing using the Tepe Hissar dental sample. *American Journal of Physical Anthropology*, **49**, 271–6.
Nyvad, B. & Fejerkov, O. (1982). Root surface caries: clinical, histopathological and microbiological features and clinical implications. *International Dental Journal*, **32**, 311–26.
O'Brien, C. J. (2000). A re-evaluation of dental increment formation in East African mammals: implications for wildlife biology and zooarchaeology. *Archaeozoologia*, **XI**, 43–6.
O'Higgins, P. & Johnson, D. R. (1988). The quantitative description and comparison of biological forms. *CRC Critical Reviews in Anatomical Sciences*, **1**, 149–70.
Oakley, K. P. (1969). Analytical methods of dating bones. In Brothwell, D. R. & Higgs, E. S. (eds.), *Science in Archaeology*. London: Thames & Hudson, pp. 35–45.

Ogilvie, M. D., Curran, B. K., & Trinkaus, E. (1989). Incidence and patterning of dental enamel hypoplasia among the Neandertals. *American Journal of Physical Anthropology*, **79**, 25–41.

Ognev, S. I. (1948). *Mammals of USSR and Adjacent Countries: Volume 6 Rodents*. Moscow: Translated 1962 Israel Programme for Scientific Translation.

Olsen, S. J. (1985). *Origins of the Domestic Dog*. Tucson: University of Arizona Press.

Osborn, D. J. and Helmy, I. (1980). *Contemporary Land Mammals of Egypt (Including Sinai)*. Fieldiana: Zoology New Series 5. Chicago, Field Museum of Natural History.

Osborn, H. F. (1907). *Evolution of Mammalian Molar Teeth. To and from the triangular type, including collected and revised researches on trituberculy and new sections on the forms and homologies of the molar teeth in the different orders of mammals*. Biological Studies and Addresses 1. New York: Macmillan.

Osborn, J. W. (1973). Variations in structure and development of enamel. *Oral Science Reviews*, **3**, 3–83.

(1978). Morphogenetic gradients: fields versus clones. In Butler, P. M. & Joysey, K. A. (eds.), *Development, Function and Evolution of Teeth*. London: Academic Press, pp. 171–99.

(ed.) (1981). *Dental Anatomy and Eembryology*. Oxford: Blackwell Scientific Publications.

Osborn, J. W. & Ten Cate, A. R. (1983). *Advanced Dental Histology*. Dental Practitioner Handbook, 6. Bristol: John Wright.

Osborne, R. H. (1963). Respective role of twin, sibling, family, and population methods in dentistry and medicine. *Journal of Dental Research*, **42**, 1276–87.

Osborne, R. H., Horowitz, S. L., & de George, F. V. (1958). Genetic variation of tooth dimensions: a twin study of the permanent anterior teeth. *American Journal of Human Genetics*, **10**, 350–6.

Ovchinnikov, I. V., Götherström, A., Romanova, G. P., Kharitonov, V. M., Lidén, K., & Goodwin, W. (2000). Molecular analysis of Neanderthal DNA from the northern Caucasus. *Nature*, **404**, 490–3.

Owen, R. (1845). *Odontography or a treatise on the comparative anatomy of the teeth: their physiological relations, mode of development and microscopic structure in the vertebrate animals*. London: Hyppolyte Baillière.

Page, R. C. & Schroeder, H. E. (1982). *Periodontitis in Man and Other Animals: a Comparative Review*. Basel and New York: Karger.

Passmore, R., Peterson, R. L., & Cringan, A. T. (1955). A study of mandibular tooth wear as an index to age of moose. In Peterson, R. L. (ed.), *North American Moose*. Toronto: University of Toronto Press, pp. 223–98.

Pastor, R. F. (1994). A multivariate dental microwear analysis of prehistoric groups from the Indian subcontinent (abstract). *American Journal of Physical Anthropology*, Supplement **18**, 158–9.

Payne, S. (1973). Kill-off patterns in sheep and goats: the mandibles from Asvan Kale. *Anatolian Studies*, **23**, 281–303.

(1985). Morphological distinctions between the mandibular teeth of young sheep, Ovis, and Goats, Capra. *Journal of Archaeological Science*, **12**, 139–47.

(1987). Reference codes for wear states in the mandibular cheek teeth of sheep and goats. *Journal of Archaeological Science*, **14**, 609–14.

(1991). Early Holocene equids from Tall-i-Mushki (Iran) and Can Hasan III (Turkey). In Meadow, R. H. & Uerpmann, H.-P. (eds.), *Equids in the Ancient World*. Beihefte zum Tübinger Atlas des Vorderen Orients, Reihe A (Naturwissenschaften) Nr 19/2. Wiesbaden: Dr Ludwig Reichert Verlag, pp. 132–77.

Payne, S. & Bull, G. (1988). Components of variation in measurements of pig bones and teeth, and the use of measurements to distinguish wild from domestic pig remains. *Archaeozoologia*, **II**, 27–65.

Pedersen, P. O. (1938). Investigations into the dental conditions of about 3000 ancient and modern Greenlanders. *Dental Record*, **58**, 191–8.
  (1947). Dental investigations of Greenland Eskimos. *Proceedings of the Royal Society of Medicine*, **40**, 726–32.
  (1949). The East Greenland Eskimo dentition. *Meddelelser om Grønland*, **142**, 1–244.
Penniman, T. K. (1952). *Pictures of Ivory and Other Animal Teeth, Bone and Antler, with a Brief Commentary on their Use in Identification*. Occasional Papers on Technology, 5. Oxford: Pitt Rivers Museum, University of Oxford.
Penning, C., van Amerongen, J. P., Seef, R. E., & Ten Cate, J. M. (1992). Validity of probing for fissure caries diagnosis. *Caries Research*, **26**, 445–9.
Pérez-Barbería, F. J. (1994). Determination of age in Cantabrian chamois (*Rupicapra rupicapra parva*) from jaw tooth-row eruption and wear. *Journal of Zoology*, **233**, 649–56.
Perrin, W. F. & Myrick, A. C. (eds.) (1980). *Growth of Odontocetes and Sirenians: Problems in Age Determination. Proceedings of the International Conference on Determining Age of Odontocete Ceteans (and Sirenians), La Jolla, California, September 5–19, 1978*. Report of the International Whaling Commission, Special Issue 3. Cambridge: International Whaling Commission.
Perzigian, A. J. (1981). Allometric analysis of dental variation in a human population. *American Journal of Physical Anthropology*, **54**, 341–5.
Peters, H. & Balling, R. (1999). Teeth: where and how to make them. *Trends in Genetics*, **15**, 59–65.
Philippas, G. G. & Applebaum, E. (1966). Age factors in secondary dentin formation. *Journal of Dental Research*, **45**, 778–89.
Pike-Tay, A., Morcomb, C. A., & O'Farrell, M. (2000). Reconsidering the Quadratic Crown Height Method of age estimation for *Rangifer* from archaeological sites. *Archaeozoologia*, **XI**, 145–74.
Pindborg, J. J. (1970). *Pathology of the Dental Hard Tissues*. Philadelphia: Saunders.
  (1982). Aetiology of developmental enamel defects not related to fluorosis. *International Dental Journal*, **32**, 123–34.
Plavcan, J. M. (2001). Sexual dimorphism in Primate evolution. *Yearbook of Physical Anthropology*, **44**, 25–53.
Poole, D. F. G. & Tratman, E. K. (1978). Post-mortem changes on human teeth from late upper Palaeolithic/Mesolithic occupants of an English limestone cave. *Archives of Oral Biology*, **23**, 1115–20.
Portin, P. & Alvesalo, L. (1974). The inheritance of shovel shape in maxillary central incisors. *American Journal of Physical Anthropology*, **41**, 59–62.
Potter, R. H. & Nance, W. E. (1976). A twin study of dental dimension. I. Discordance, asymmetry, and mirror imagery. *American Journal of Physical Anthropology*, **44**, 391–6.
Potter, R. H., Nance, W. E., Yu, P. L., & Davis, W. B. (1976). A twin study of dental dimension. II. Independent genetic determinants. *American Journal of Physical Anthropology*, **44**, 397–412.
Potter, R. H. Y., Corruccini, R. S., & Green, L. J. (1981). Variance of occlusion traits in twins. *Journal of Craniofacial Genetics & Developmental Biology*, **1**, 217–27.
Potter, R. H. Y., Rice, J. P., Dahlberg, A. A., & Dahlberg, T. (1983). Dental size traits within families: path analysis for first molar and lateral incisors. *American Journal of Physical Anthropology*, **61**, 283–9.
Potter, R. H. Y., Yu, P. L., Dahlberg, A. A., Merritt, A. D., & Connelly, P. M. (1968). Genetic studies of tooth size factors in Pima Indian families. *American Journal of Human Genetics*, **20**, 89–100.
Poulakakis, N., Mylonas, M., Lymberakis, P., & Fassoulas, C. (2002). Origin and taxonomy of the fossil elephants of the island of Crete (Greece): problems and perspectives. *Palaeogeography, Palaeoclimatology, Palaeoecology*, **186**, 163–83.
Preiswerk, G. (1895). *Beiträge zur Kentniss der Schmelzstructur bei Säugetieren mit besonderer Berüksichtigung der Ungulaten*. Basel.

Price, T. D., Bentley, A., Lüning, J., Gronenborn, D., & Wahl, J. (2001). Prehistoric human migration in the *Linearbandkeramik* of Central Europe. *Antiquity*, **75**, 593–603.

Price, T. D., Grupe, G., & Schroter, P. (1998). Migration in the Bell Beaker period of central Europe. *Antiquity*, **72**, 405–11.

Quintero, L. A. & Köhler-Rollefson, I. (1997). The 'Ain Ghazal dog: a case for the Neolithic origin of *Canis familiaris* in the Near East. In Gebel, H. G. K., Kafafi, Z., & Rollefson, G. O. (eds.), *The Prehistory of Jordan. II. Perspectives from 1997*. Berlin: Ex Oriente, pp. 567–74.

Ramis, D. & Bover, P. (2001). A review of the evidence for domestication of *Myotragus balearicus* Bate 1909 (Artiodactyla, Caprinae) in the Balearic Islands. *Journal of Archaeological Science*, **28**, 265–82.

Reher, C. A. (1974). Population study of the Casper Site bison. In Frison, G. C. (ed.), *The Casper Site: a Hell Gap Bison Kill on the High Plains*. New York: Academic Press, pp. 113–24.

Reid, C., Reenen, J. R., & Groeneveld, H. T. (1991). Tooth size and the Carabelli trait. *American Journal of Physical Anthropology*, **84**, 427–32.

Reid, C., Van Reenen, J. F., & Groeneveld, H. T. (1992). The Carabelli trait and maxillary molar cusp and crown base areas. In Smith, P. & Tchernov, E. (eds.), *Structure, Function and Evolution of Teeth*. London: Freund Publishing House, pp. 451–66.

Reid, D. J., Beynon, A. D., & Ramirez-Rozzi, F. V. (1998). Histological reconstruction of dental development in four individuals from a medieval site in Picardie, France. *Journal of Human Evolution*, **35**, 463–77.

Reid, D. J. & Dean, M. C. (2000). The timing of linear hypoplasias on human anterior teeth. *American Journal of Physical Anthropology*, **113**, 135–40.

Rensberger, J. M. (1978). Scanning electron microscopy of wear and occlusal events in some small herbivores. In Butler, P. M. & Joysey, K. A. (eds.), *Development, Function and the Evolution of Teeth*. London: Academic Press, pp. 415–38.

(1997). Mechanical adaptation in enamel. In von Koenigswald, W. & Sander, P. M. (eds.), *Tooth Enamel Mmicrostructure*. Rotterdam: A. A. Balkema, pp. 259–66.

Report of the Workshop (1980). In Perrin, W. F. & Myrick, A. C. (eds.), *Growth of Odontocetes and Sirenians: Problems in Age Determination. Proceedings of the International Conference on Determining Age of Odontocete Ceteans (and Sirenians), La Jolla, California, September 5–19, 1978*. Report of the International Whaling Commission, Special Issue 3. Cambridge: International Whaling Commission, pp. 1–50.

Retzius, A. (1837). Bemerkungen über den inneren Bau der Zähne, mit besonderer Rücksicht auf den im Zahnknochen vorkommenden Röhrenbau. *(Müllers) Archiv Anat Phys*, 486–566.

Rice, P. M. & Maples, W. R. (1979). Some difficulties in the Gustafson dental age estimations. *Journal of Forensic Sciences*, **24**, 118–72.

Richards, M. P. & Hedges, R. E. M. (1999). Stable isotope evidence for similarities in the types of marine foods used by Late Mesolithic humans at sites along the Atlantic coast of Europe. *Journal of Archaeological Science*, **26**, 717–22.

Richards, M. P., Mays, S. A., & Fuller, B. T. (2002). Stable carbon and nitrogen isotope values of bone and teeth reflect weaning age at the Medieval Wharram Percy Site, Yorkshire, UK. *American Journal of Physical Anthropology*, **119**, 205–10.

Richards, M. P., Pettitt, P. B., Stiner, M. C., & Trinkaus, E. (2001). Stable isotope evidence for increasing dietary breadth in the European mid-Upper Paleolithic. *Proceedings of the National Academy of Sciences USA*, **98**, 6528–32.

Richards, M. P., Pettitt, P. B., Trinkaus, E., Smith, F. H., Paunovich, M., & Karavanic, I. (2000). Neanderthal diet at Vindija and Neanderthal predation: the evidence from stable isotopes. *Proceedings of the National Academy of Sciences USA*, **97**, 7663–6.

Risnes, S. (1979a). A scanning electron microscope study of aberrations in the prism pattern of rat incisor inner enamel. *American Journal of Anatomy*, **154**, 419–36.

(1979b). The prism pattern of rat molar enamel: a scanning electron microscope study. *American Journal of Anatomy*, **155**, 245–57.

(1984). Rationale for consistency in the use of enamel surface terms: perikymata and imbrications. *Scandinavian Journal of Dental Research*, **92**, 1–5.

(1985a). A scanning electron microscope study of the three-dimensional extent of Retzius lines in human dental enamel. *Scandinavian Journal of Dental Research*, **93**, 145–52.

(1985b). Circumferential continuity of perikymata in human dental enamel investigated by scanning electron microscopy. *Scandinavian Journal of Dental Research*, **93**, 185–91.

(1986). Enamel apposition rate and the prism periodicity in human teeth. *Scandinavian Journal of Dental Research*, **94**, 394–404.

(1989). Ectopic tooth enamel. An SEM study of the structure of enamel in enamel pearls. *Advances in Dental Research*, **3**, 258–64.

Robinette, W. L. & Archer, A. L. (1971). Notes on ageing criteria and reproduction of Thompson gazelle. *East African Wildlife Journal*, **9**, 83–98.

Robinson, C., Brookes, W. A., Bonass, W. A., Shore, R. C., & Kirkham, J. (1997). Enamel maturation. In Chadwick, D. J. & Cardew, G. (eds.), *Dental Enamel*. Ciba Foundation Symposium 205. Chichester: Wiley, pp. 156–74.

Robinson, C., Kirkham, J., Weatherell, J. A., & Strong, M. (1986). Dental enamel – a living fossil. In Cruwys, E. & Foley, R. A. (eds.), *Teeth and Anthropology*. B. A. R. International Series 291. Oxford: British Archaeological Reports, pp. 31–54.

Robinson, J. T. (1956). *The Dentition of the Australopithecinae*. Transvaal Museum Memoir 9. Pretoria: Transvaal Museum.

Robinson, P. T. (1979). A literature review of dental pathology and ageing by dental means in non domestic animals Part II. *Journal of Zoo Animal Medicine*, **10**, 81–91.

Rogers, A. H. (1981). The source of infection in the intrafamilial transfer of *Streptococcus mutans*. *Caries Research*, **15**, 26–37.

Romer, A. S. (1966). *Vertebrate Palaeontology*. Chicago: Chicago University Press.

Rose, J. C. (1977). Defective enamel histology of prehistoric teeth from Illinois. *American Journal of Physical Anthropology*, **46**, 439–46.

Rose, J. C. & Ungar, P. S. (1998). Gross dental wear and dental microwear in historical perspective. In Alt, K. W., Rösing, F. W., & Teschler-Nicola, M. (eds.), *Dental Anthropology: Fundamentals, Limits, and Prospects*. Wien; New York: Springer, pp. 349–86.

Rose, J. C., Armelagos, G. J., & Lallo, J. W. (1978). Histological enamel indicator of childhood stress in prehistoric skeletal samples. *American Journal of Physical Anthropology*, **49**, 511–16.

Rösing, F. W. (1983). Sexing immature human skeletons. *Journal of Human Evolution*, **12**, 149–55.

Roth, V. L. (1988). Dental identification and age determination in *Elephas maximus*. *Journal of Zoology*, **214**, 567–88.

(1989). Fabricational noise in elephant dentitions. *Paleobiology*, **15**, 165–79.

Rowles, S. L. (1967). Chemistry of the mineral phase of dentine. In Miles, A. E. W. (ed.), *Structural and Chemical Organization of Teeth*. London: Academic Press, pp. 201–46.

Rudney, J. D. (1983a). Dental indicators of growth disturbance in a series of ancient Lower Nubian populations: changes over time. *American Journal of Physical Anthropology*, **60**, 463–70.

(1983b). The age-related distribution of dental indicators of growth disturbance in ancient Lower Nubia: an etiological model from the ethnographic record. *Journal of Human Evolution*, **12**, 535–43.

Rudney, J. D. & Greene, D. L. (1982). Interpopulation differences in the severity of early childhood stress in ancient lower Nubia: implications for hypotheses of X-group origins. *Journal of Human Evolution*, **11**, 559–65.

Rugg-Gunn, A. J. (1993). *Nutrition and Dental Caries*. Oxford Medical Publications. Oxford: Oxford University Press.

Russell, K. F., Simpson, S. W., Genovese, J., Kinkel, K. D., Meindl, R. S., & Lovejoy, C. O. (1993). Independent test of the fourth rib aging technique. *American Journal of Physical Anthropology*, **92**, 53–62.

Sahni, A. & von Koenigswald, W. (1997). The enamel structure of some fossil and recent whales. In von Koenigswald, W. & Sander, P. M. (eds.), *Tooth Enamel Microstructure*. Rotterdam: A. A. Balkema, pp. 177–91.

Sakae, T., Suzuki, K., & Kozawa, Y. (1997). A short review of studies on chemical and physical properties of enamel crystallites. In von Koenigswald, W. & Sander, P. M. (eds.), *Tooth Enamel Microstructure*. Rotterdam: A. A. Balkema, pp. 31–9.

Sandford, M. K. (1992). A reconsideration of trace element analysis in prehistoric bone. In Saunders, S. R. & Katzenberg, M. A. (eds.), *Skeletal Biology of Past Peoples: Research Methods*. New York: Wiley-Liss, pp. 79–104.

Sandford, M. K. & Weaver, D. S. (2000). Trace element research in anthropology: new perspectives and challenges. In Katzenberg, M. A. & Saunders, S. R. (eds.), *Biological Anthropology of the Human Skeleton*. New York: John Wiley, pp. 329–50.

Sarnat, B. G. & Schour, I. (1941). Enamel hypoplasia (chronologic enamel aplasia) in relation to systemic disease: a chronologic, morphologic and etiologic classification. *Journal of the American Dental Association*, **28**, 1989–2000.

(1942). Enamel hypoplasia (chronologic enamel aplasia) in relation to systemic disease: a chronologic, morphologic and etiologic classification. *Journal of the American Dental Association*, **29**, 397–418.

Saxon, A. & Higham, C. W. F. (1968). Identification and interpretation of growth rings in the secondary dental cementum of *Ovis aries, L. Nature*, **219**, 634–5.

Scheffer, V. B. (1950). Growth layers on the teeth of Pinnipedia as an indication of age. *Science*, **112**, 309–11.

Scheffer, V. B. & Myrick, A. C. (1980). A review of studies to 1970 of growth layers in the teeth of marine mammals. In Perrin, W. F. & Myrick, A. C. (eds.), *Growth of Odontocetes and Sirenians: Problems in Age Determination. Proceedings of the International Conference on Determining Age of Odontocete Ceteans (and Sirenians), La Jolla, California, September 5–19, 1978*. Report of the International Whaling Commission, Special Issue 3. Cambridge: International Whaling Commission, pp. 51–63.

Schluger, S., Yuodelis, R., Page, R. C., & Johnson, R. H. (1990). *Periodontal Diseases: Basic Phenomena, Clinical Management, and Occlusal and Restorative Interrelationships*. Philadelphia: Lea & Febiger.

Schmidt, C. W. (2001). Dental microwear evidence for a dietary shift between two nonmaize-reliant prehistoric human populations from Indiana. *American Journal of Physical Anthropology*, **114**, 139–45.

Schmidt, W. J. & Keil, A. (1971). *Polarizing Microscopy of Dental Tissues. Theory, methods and results from the structural analysis of normal and diseased hard dental tissues and tissues associated with them in man and other vertebrates*. Oxford: Pergamon Press.

Schoeninger, M. J. (1995). Stable isotope studies in human evolution. *Evolutionary Anthropology*, **4**, 83–98.

Schour, I. (1936). Neonatal line in enamel and dentin of human deciduous teeth and first permanent molar. *Journal of the American Dental Association*, **23**, 1946–55.

Schour, I. & Kronfeld, R. (1938). Tooth ring analysis: IV. Neonatal dental hypoplasia analysis of the teeth of an infant with injury of the brain at birth. *Archives of Pathology*, **26**, 471–90.

Schour, I. & Massler, M. (1941). The development of the human dentition. *Journal of the American Dental Association*, **28**, 1153–60.

(1944). *Development of the Human Dentition*. Chicago: American Dental Association.

Schreger, D. (1800). Beitrag zur Geschichte der Zähne. *Beitr Zergliderungskunst*, **1**, 1–7.

Schüpbach, P., Guggenheim, B., & Lutz, F. (1989). Human root caries: histopathology of initial lesions in cementum and dentin. *Journal of Oral Pathology & Medicine*, **3**, 146–56.
  (1990). Human root caries: histopathology of advanced lesions. *Caries Research*, **24**, 145–58.
Schüpbach, P., Lutz, F., & Guggenheim, B. (1992). Human root caries: histopathology of arrested lesions. *Caries Research*, **26**, 153–64.
Schwarcz, H. P. & Schoeninger, M. J. (1991). Stable isotope analyses in human nutritional ecology. *Yearbook of Physical Anthropology*, **34**, 283–321.
Schwartz, G. T. & Dean, M. C. (2001). The ontogeny of canine dimorphism in extant hominoids. *American Journal of Physical Anthropology*, **115**, 269–83.
Sciulli, P. W. (1977). A descriptive and comparative study of the deciduous dentition of prehistoric Ohio Valley Amerindians. *American Journal of Physical Anthropology*, **47**, 71–80.
  (1978). Developmental abnormalities of the permanent dentition in prehistoric Ohio Valley Amerindians. *American Journal of Physical Anthropology*, **48**, 193–8.
  (2003). Dental asymmetry in a Late Archaic and Late Prehistoric skeletal sample of the Ohio Valley area. *Dental Anthropology*, **16**, 33–44.
Sciulli, P. W., Doyle, W. J., Kelley, C., Siegel, P., & Siegel, M. I. (1979). The interaction of stressors in the induction of increased levels of fluctuating asymmetry in the laboratory rat. *American Journal of Physical Anthropology*, **50**, 279–84.
Scott, D. B., Kaplan, H., & Wyckoff, R. W. G. (1949). Replica studies of changes in tooth surfaces with age. *Journal of Dental Research*, **28**, 31–47.
Scott, D. B. & Wyckoff, R. W. G. (1949a). Studies of tooth surface structure by optical and electron microscopy. *Journal of the American Dental Association*, **39**, 275–82.
  (1949b). Typical structures on replicas of apparently intact tooth surfaces. *Public Health Reports*, **61**, 1397–400.
Scott, G. R. & Turner, II, C. G. (1997). *The Anthropology of Modern Human Teeth. Dental Morphology and its Variation in Recent Human Populations*. Cambridge Studies in Biological Anthropology. Cambridge: Cambridge University Press.
Scott, J. H. & Symons, N. B. B. (1974). *Introduction to Dental Anatomy*. Edinburgh: Churchill Livingstone.
Sealy, J. C., Armstrong, R., & Schrire, C. (1995). Beyond lifetime averages: tracing life histories through isotopic analysis of different calcified tissues from archaeological human skeletons. *Antiquity*, **69**, 290–300.
Seeto, E. & Seow, W. K. (1991). Scanning electron microscopic analysis of dentin in vitamin D-resistant rickets – assessment of mineralization and correlation with clinical findings. *Pediatric Dentistry*, **13**, 43–8.
Sengupta, A., Whittaker, D. K., & Shellis, R. P. (1999). Difficulties in estimating age using root dentine translucency in human teeth of varying antiquity. *Archives of Oral Biology*, **44**, 889–99.
Sergeant, D. E. & Pimlott, D. H. (1959). Age determination in moose from sections of incisor teeth. *Journal of Wildlife Management*, **23**, 315–21.
Severinghaus, C. W. (1949). Tooth development and wear as criteria of age in white-tailed deer. *Journal of Wildlife Management*, **13**, 195–216.
Sharma, J. C. (1992). Dental morphology and odontometry of twins and the heritability of dental variation. In Lukacs, J. R. (ed.), *Culture, Ecology and Dental Anthropology*. Journal of Human Ecology Special Issue 2. Delhi: Kamla-Raj Enterprises, pp. 49–60.
Sharma, K. & Corruccini, R. S. (1986). Genetic basis of dental occlusal variation in northwest Indian twins. *European Journal of Orthodontics*, **8**, 91–7.
Sharpe, P. T. (2000). Homeobox genes in initiation and shape of teeth during development in mammalian embryos. In Teaford, M. F., Meredith Smith, M., & Ferguson, M. W. J. (eds.), *Development, Function and Evolution of Teeth*. Cambridge: Cambridge University Press, 3–12.

Sheiham, A. (1983). Sugars in dental decay. *Lancet*, **1**, 282–4.
  (1997). Impact of dental treatment on the incidence of dental caries in children and adults. *Community Dentistry & Oral Epidemiology*, **25**, 104–12.
Shellis, R. P. (1998). Utilization of periodic markings in enamel to obtain information on tooth growth. *Journal of Human Evolution*, **35**, 387–400.
Shipman, P., Foster, G., & Schoeninger, M. J. (1984). Burnt bones and teeth: an experimental study of color, morphology, crystal structure and shrinkage. *Journal of Archaeological Science*, **11**, 307–25.
Shorten, M. (1954). *Squirrels*. London: Collins.
Siegel, J. (1976). Animal palaeopathology: possibilities and problems. *Journal of Archaeological Science*, **3**, 349–84.
Siegel, M. I. & Doyle, W. (1975). The effects of cold stress on fluctuating asymmetry in the dentition of the mouse. *Journal of Experimental Zoology*, **193**, 385–9.
Siegel, M. I., Doyle, W., & Kelly, C. (1977). Heat stress, fluctuating asymmetry and prenatal selection in the laboratory rat. *American Journal of Physical Anthropology*, **46**, 121–6.
Sikes, S. K. (1966). The African elephant *Loxodonta africana*: a field method for estimation of age. *Journal of Zoology*, **150**, 279–95.
  (1968). The African elephant *Loxodonta africana:* a field method for estimation of age. *Journal of Zoology*, **154**, 235–318.
Sikorski, M. D. (1982). Non-metrical divergence of isolated populations of *Apodemus agrarius* in urban areas. *Acta Theriologica*, **27**, 169–80.
Sikorski, M. D. & Bernshtein, A. D. (1984). Geographical and intrapopulation divergence in *Clethrionomys glareolus*. *Acta Theriologica*, **29**, 219–30.
Sillen, A. & Kavanagh, M. (1982). Strontium and paleodietary research: a review. *Yearbook of Physical Anthropology*, **25**, 67–90.
Silver, I. A. (1969). The ageing of domestic animals. In Brothwell, D. & Higgs, E. S. (eds.), *Science in Archaeology*. London: Thames and Hudson, pp. 250–68.
Silverstone, L. M., Johnson, N. W., Hardie, J. M., & Williams, R. A. D. (1981). *Dental Caries: Aetiology, Pathology and Prevention*. London: Macmillan.
Sims, M. R. (1980). Angular changes in collagen cemental attachment during tooth movement. *Journal of Periodontal Research*, **15**, 638–45.
Slaughter, B. H., Pine, R. H., & Pine, N. E. (1974). Eruption of cheek teeth Insectivora and Carnivora. *Journal of Mammalogy*, **55**, 115–25.
Slijper, E. J. (1962). *Whales*. London: Hutchinson.
Smith, B. H. (1984). Patterns of molar wear in hunter-gatherers and agriculturalists. *American Journal of Physical Anthropology*, **63**, 39–56.
  (1986). Dental development in *Australopithecus* and early *Homo*. *Nature*, **323**, 327–30.
  (1987). Reply to 'Maturational patterns in early hominids' by A. E. Mann, M. Lampl, and J. Monge. *Nature*, **328**, 674–5.
  (1991a). Dental development and the evolution of life history in Hominidae. *American Journal of Physical Anthropology*, **86**, 157–74.
  (1991b). Standards of human tooth formation and dental age assessment. In Kelley, M. A. & Larsen, C. S. (eds.), *Advances in Dental Anthropology*. New York: Wiley-Liss, pp. 143–68.
Smith, B. H., Crummett, T. L., & Brandt, K. L. (1994). Ages of eruption of primate teeth: a compendium for ageing individuals and comparing life histories. *Yearbook of Physical Anthropology*, **37**, 177–232.
Smith, B. H., Garn, S. M., & Cole, P. E. (1982). Problems of the sampling and inference in the study of fluctuating dental asymmetry. *American Journal of Physical Anthropology*, **58**, 281–9.
Smith, G. F. H. (1972a). *Gem Stones*. London: Chapman and Hall.
Smith, P. (1972b). Diet and attrition in the Natufians. *American Journal of Physical Anthropology*, **37**, 233–8.

(1982). Dental reduction selection or drift? In Kurtén, B. (ed.), *Teeth: Form, Function and Evolution.* New York: Columbia University Press, 366–79.

Smith, P., Bar-Yosef, O., & Sillen, A. (1984). Archaeological and skeletal evidence for dietary change during the late Pleistocene/early Holocene in the Levant. In Cohen, M. N. & Armelagos, G. J. (eds.), *Palaeopathology at the Origins of Agriculture.* New York: Academic Press, pp. 101–36.

Smith, P., Wax, Y., Adler, F., Silberman, U., & Heinic, G. (1986). Post-Pleistocene changes in tooth root and jaw relationships. *American Journal of Physical Anthropology,* **70**, 339–48.

Smith, R. J. & Bailit, H. J. (1977). Variation in dental occlusion and arches among Melanesians of Bougainville Island, Papua New Guinea: I Methods, age changes, sex differences and population comparison. *American Journal of Physical Anthropology,* **47**, 195–208.

Smith, R. J., Kolakowski, D., & Bailit, H. J. (1978). Variation in dental occlusion and arches among Melanesians of Bougainville Island, Papua New Guinea: II Clinal variation, geographic microdifferentiation and synthesis. *American Journal of Physical Anthropology,* **48**, 331–42.

Smuts, G. L., Anderson, J. L., & Austin, J. C. (1978). Age determination of the African lion *Panthera leo. Journal of Zoology,* **185**, 115–46.

Soames, J. V. & Southam, J. C. (1993). *Oral Pathology.* Oxford: Oxford University Press.

Sofaer, J. A. (1969). The genetics and expression of a dental morphological variant in the mouse. *Archives of Oral Biology,* **14**, 1213–23.

Sofaer, J. A., Maclean, C. J., & Bailit, H. L. (1972). Heredity and morphological variation in early and late developing teeth of the same morphological class. *Archives of Oral Biology,* **17**, 811–16.

Soggnaes, R. F. (1950). Histological studies of ancient and recent teeth with special regard to differential diagnosis between intra-vitam and post-mortem characteristics. *American Journal of Physical Anthropology,* **8**, 269–70.

(1956). Histological evidence of developmental lesions in teeth originating from paleolithic, prehistoric, and ancient man. *American Journal of Pathology,* **32**, 547–77.

Solheim, T. (1992). Amount of secondary dentin as an indicator of age. *Scandinavian Journal of Dental Research,* **100**, 193–9.

Solounias, N. & Hayek, L.-A. (1993). New methods of tooth microwear analysis and application to dietary determination of two extinct antelopes. *Journal of Zoology,* **229**, 421–45.

Solounias, N., Teaford, M. F., & Walker, A. (1988). Interpreting the diet of extinct ruminants: the case of a non-browsing giraffid. *Paleobiology,* **14**, 287–300.

Spaan, A. (1996). *Hippopotamus creutzburgi*: the case of the Cretan *Hippopotamus*. In Reese, D. S. (ed.), *Pleistocene and Holocene Fauna of Crete and its First Settlers.* Monographs in World Archaeology 28. Madison: Prehistory Press, pp. 99–110.

Spence, C. A., Aitchison, G. U., Sykes, A. R., & Atkinson, P. J. (1980). Broken mouth (premature incisor loss) in sheep: the pathogenesis of periodontal disease. *Journal of Comparative Pathology,* **90**, 275–92.

Spencer, M. A. & Ungar, P. S. (2000). Craniofacial morphology, diet and incisor use in three Native American populations. *International Journal of Osteoarchaeology,* **10**, 229–41.

Spiess, A. E. (1976). Determining season of death of archaeological fauna by analysis of teeth. *Arctic,* **29**, 53–5.

(1979). *Reindeer and Caribou Hunters: An Archaeological Study.* New York: Academic Press.

Spinage, C. A. (1972). Age estimation of zebra. *East African Wildlife Journal,* **10**, 273–77.

(1973). A review of the age determination of mammals by means of teeth, with especial reference to Africa. *East African Wildlife Journal,* **11**, 165–87.

(1976). Incremental cementum lines in the teeth of tropical African mammals. *Journal of Zoology (London),* **178**, 117–31.

Stander, P. E. (1997). Field age determination of leopards by tooth wear. *African Journal of Ecology,* **35**, 156–61.

Stefen, C. (1997). Differentiation in Hunter-Schreger bands of carnivores. In von Koenigswald, W. & Sander, P. M. (eds.), *Tooth Enamel Microstructure*. Rotterdam: A. A. Balkema, pp. 123–37.

Stott, G. G., Sis, R. F., & Levy, B. M. (1980). Cemental annulation as an age criterion in the common marmoset (*Callithrix jaculus*). *Journal of Medical Primatology*, **9**, 274–85.

(1982). Cemental annulation as an age criterion in forensic dentistry. *Journal of Dental Research*, **61**, 814–17.

Stringer, C. B., Humphrey, L. T., & Compton, T. (1997). Cladistic analysis of dental traits in recent humans using a fossil outgroup. *Journal of Human Evolution*, **32**, 389–402.

Stringer, C. B. & McKie, R. (1996). *African Exodus: The Origins of Modern Humanity*. London: Pimlico.

Stuart, A. J. (1982). *Pleistocene Vertebrates in the British Isles*. London: Longman.

Suarez, B. K. (1974). Neandertal dental asymmetry and the probable mutation effect. *American Journal of Physical Anthropology*, **41**, 411–16.

Suchentrunk, F., Willing, R., & Hartl, G. B. (1994). Non-metrical polymorphism of the first lower premolar (P_3) in Austrian brown hares (*Lepus europaeus*): a study on regional differentiation. *Journal of Zoology*, **232**, 79–91.

Suchey, J. M. (1979). Problems in the aging of females using the Os pubis. *American Journal of Physical Anthropology*, **51**, 467–70.

Suckling, G. W., Nelson, D. G. A., & Patel, M. J. (1989). Macroscopic and scanning electron microscopic appearance and hardness values of developmental defects in human permanent tooth enamel. *Advances in Dental Research*, **3**, 219–33.

Suga, S. (1989). Enamel hypopmineralization viewed from the pattern of progressive mineralization of human and monkey developing enamel. *Advances in Dental Research*, **3**, 188–9.

Suwa, G., Wood, B. A., & White, T. D. (1994). Further analysis of mandibular molar crown and cusp areas in Pliocene and Early Pleistocene hominids. *American Journal of Physical Anthropology*, **93**, 407–26.

Swärdstedt, T. (1966) *Odontological Aspects of a Medieval Population in the Province of Jämtland/Mid Sweden*. Akademisk Avhandling som med vederbörligt tillstand av Odontologiska Fakulteten vid Lunds Universitet för vinnande av Odontologie Doktorgrad offentilgen försvaras i Tandläkarhöskolans Aula, Malmö, Fredagen den 9 December 1966 Kl 9 CT. Lund: Sweden.

Sweeney, E. A., Cabrera, J., Urritia, J., & Mata, L. (1969). Factors associated with linear hypoplasia of human deciduous incisors. *Journal of Dental Research*, **48**, 1275–9.

Sweeney, E. A., Saffir, A. J., & Leon, R. D. (1971). Linear hypoplasia of deciduous incisor teeth in malnourished children. *American Journal of Clinical Nutrition*, **24**, 29–31.

Swindler, D. R. (1976). *Dentition of Living Primates*. London: Academic Press.

Swindler, D. R. (2002). *Primate Dentition: An Introduction to the Teeth of Non-human Primates*. Cambridge: Cambridge University Press.

Tanner, J. M. (1973). Growing up. *Scientific American*, **229**, 34–43.

Tchernov, E. & Valla, F. F. (1997). Two new dogs, and other Natufian dogs, from the Southern Levant. *Journal of Archaeological Science*, **24**, 65–95.

Teaford, M. F. (1988). A review of dental microwear and diet in modern mammals. *Scanning Microscopy*, **2**, 1149–66.

(1991). Dental microwear: what can it tell us about diet and dental function. In Kelley, M. A. & Larsen, C. S. (eds.), *Advances in Dental Anthropology*. New York: Wiley-Liss, pp. 341–56.

Teaford, M. F., Larsen, C. S., Pastor, R. F., & Noble, V. E. (2001). Pits and scratches: microscopic evidence of tooth use and masticatory behavior in La Florida. In Larsen, C. S. (ed.), *Bioarchaeology of Spanish Florida*. Gainesville: University Press of Florida, pp. 82–112.

Teaford, M. F. & Lytle, J. D. (1996). Diet-induced changes in rates of human tooth microwear: a case study involving stone-ground maize. *American Journal of Physical Anthropology*, **100**, 143–8.

Teaford, M. F. & Oyen, O. J. (1989a). Differences in the rate of molar wear between monkeys raised on different diets. *Journal of Dental Research*, **68**, 1513–18.

(1989b). In vivo and in vitro turnover in dental microwear. *American Journal of Physical Anthropology*, **80**, 447–60.

Teaford, M. F. & Tylenda, C. A. (1991). A new approach to the study of tooth wear. *Journal of Dental Research*, **70**, 204–7.

Teaford, M. F. & Ungar, P. S. (2000). Diet and the evolution of the earliest human ancestors. *Proceedings of the National Academy of Sciences USA*, **97**, 13506–11.

Teaford, M. F. & Walker, A. (1983). Dental microwear in adult and still-born guinea pigs. *Archives of Oral Biology*, **28**, 1077–81.

(1984). Quantitative differences in dental microwear between primate species with different diets and a comment on the presumed diet of *Sivapithecus*. *American Journal of Physical Anthropology*, **64**, 191–200.

Ten Cate, A. R. (1985). *Oral Histology: Development, Structure and Function*. St Louis: Mosby.

Ten Cate, J. M. (1989). *Recent Advances in the Study of Dental Calculus*. Oxford: IRL Press at Oxford University Press.

Theilade, J., Fejerskov, O., & Hørsted, M. (1976). A transmission electron microscopic study of seven day old bacterial plaque in human tooth fissures. *Archives of Oral Biology*, **21**, 587–98.

Thesleff, I. & Åberg, T. (1999). Molecular regulation of tooth development. *Bone*, **25**, 123–5.

Thylstrup, A. & Fejerskov, O. (1994). *Textbook of Clinical Cariology*. Copenhagen: Munksgaard.

Tobias, P. V. (1967). *The Cranium and Maxillary Dentition of Australopithecus (Zinjanthropus) boisei*. Olduvai Gorge Volume II. Cambridge: Cambridge University Press.

Tong, H. (2001). Age profiles of rhino fauna from the Middle Pleistocene Nanjing Man Site, South China – explained by the rhino specimens of living species. *International Journal of Osteoarchaeology*, **11**, 231–7.

Townsend, G. C. (1980). Heritability of deciduous tooth size in Australian aboriginals. *American Journal of Physical Anthropology*, **53**, 297–300.

Townsend, G. C. & Alvesalo, L. (1985). Tooth size in 47 XYY males: evidence for a direct effect of the Y chromosome on growth. *Australian Dental Journal*, **30**, 268–72.

Townsend, G. C. & Brown, T. (1978). Heritability of permanent tooth size. *American Journal of Physical Anthropology*, **49**, 497–505.

(1979). Family studies of tooth size factors in the permanent dentition. *American Journal of Physical Anthropology*, **50**, 183–90.

(1980). Dental asymmetry in Australian aboriginals. *Human Biology*, **52**, 661–73.

Tupikova, N. V., Sidorova, G. A., & Konovalova, E. A. (1968). A method of age determination in *Clethrionomys*. *Acta Theriologica*, **13**, 99–115.

Turner II, C. G. (1987). Late Pleistocene and Holocene population history of east Asia based on dental variation. *American Journal of Physical Anthropology*, **73**, 305–21.

(1990). Major features of Sundadonty and Sinodonty, including suggestions about East Asian microevolution, population history, and late Pleistocene relationships with Australian aboriginals. *American Journal of Physical Anthropology*, **82**, 295–317.

Turner II, C. G. & Cadien, J. D. (1969). Dental chipping in Aleuts, Eskimos and Indians. *American Journal of Physical Anthropology*, **31**, 303–10.

Turner II, C. G., Nichol, C. R., & Scott, G. R. (1991). Scoring procedures for key morphological traits of the permanent dentition: the Arizona State University Dental Anthropology System. In Kelley, M. A. & Larsen, C. S. (eds.), *Advances in Dental Anthropology*. New York: Wiley-Liss, pp. 13–31.

Turner, A. (1984). Dental sex dimorphism in European lions (*Panthera leo L*) of the Upper Pleistocene: palaeolecological and palaeoethological implications. *Annales Zoologici Fennici*, **21**, 1–8.

Turner, A. & Antón, M. (1997). *The Big Cats and their Fossil Relatives*. New York: Columbia University Press.

Tveit, A. B., Espelid, I., & Fjelltveit, A. (1994). Clinical diagnosis of occlusal dentin caries. *Caries Research*, **28**, 368–72.

Ubelaker, D. H. (1978). *Human Skeletal Remains: Excavation, Analysis, Interpretation*. Chicago: Aldine.

(1989). *Human Skeletal Remains: Excavation, Analysis, Interpretation*. Washington DC: Taraxacum.

Uchiyama, J. (1999). Seasonality and age structure in an archaeological assemblage of Sika Deer (*Cervus nippon*). *International Journal of Osteoarchaeology*, **9**, 209–18.

Ungar, P. S. (1994). Incisor microwear of Sumatran anthropoid primates. *American Journal of Physical Anthropology*, **94**, 339–63.

(1998). Dental allometry, morphology and wear as evidence for diet in fossil primates. *Evolutionary Anthropology*, **6**, 205–17.

Ungar, P. S., Brown, C. A., Bergstrom, T. S., & Walker, A. (2003). Quantification of dental microwear by Tandem Scanning Confocal Microscopy and scale-sensitive fractal analyses. *Scanning*, **25**, 185–93.

Ungar, P. S. & Grine, F. E. (1991). Incisor size and wear in *Australopithecus africanus* and *Paranthropus robustus*. *Journal of Human Evolution*, **20**, 313–40.

Ungar, P. S. & Spencer, M. A. (1999). Incisor microwear, diet, and tooth use in three Amerindian populations. *American Journal of Physical Anthropology*, **109**, 387–96.

Ungar, P. S. & Teaford, M. F. (1996). Preliminary examination of non-occlusal dental microwear in anthropoids: implications for the study of fossil primates. *American Journal of Physical Anthropology*, **100**, 101–14.

Ungar, P. S. & Williamson, M. D. (2000). Exploring the effects of tooth wear on functional morphology: a preliminary study using dental topographic analysis. *Palaeontologia Electronica*, **3**, 1–18.

Unmack, K. & Rowles, S. L. (1963). Constituents of dental calculus from sheep. *Nature*, **197**, 486–7.

van Amerongen, J. P., Penning, C., Kidd, E. A. M., & Ten Cate, J. M. (1992). An in vitro assessment of the extent of caries under small occlusal cavities. *Caries Research*, **26**, 89–93.

van Bree, P. J. H. & Sinkeldam, E. J. (1969). Anomalies in the dentition of fox *Vulpes vulpes* (Linnaeus, 1758) from continental western Europe. *Bijdragen tot de Dierkunde*, **39**, 3–5.

van Bree, P. J. H., van Soest, R. W. M., & Strongman, L. (1974). Tooth wear as an indication of age in badgers (*Meles meles, L.*) and red foxes (*Vulpes vulpes, L.*). *Zeitschrift fur Saugetierk*, **39**, 243–8.

van der Merwe, M. (1997). Malocclusion in an African rodent. Is it necessarily fatal? *Journal of Zoology*, **243**, 689–94.

van Gerven, D. P. & Armelagos, G. J. (1983). 'Farewell to paleodemography?' Rumours of its death have been greatly exaggerated. *Journal of Human Evolution*, **12**, 353–60.

van Klinken, G. J. (1999). Bone collagen quality indicators for palaeodietary and radiocarbon measurements. *Journal of Archaeological Science*, **26**, 687–95.

van Nostrand, F. C. & Stephenson, A. B. (1964). Age determination for beavers by tooth development. *Journal of Wildlife Management*, **28**, 43–4.

van Valen, L. (1962). A study of fluctuating asymmetry. *Evolution*, **16**, 125–42.

Vasiliadis, L., Darling, A. I., & Levers, B. G. H. (1983a). The amount and distribution of sclerotic human root dentine. *Archives of Oral Biology*, **28**, 645–9.

(1983b). The histology of sclerotic human root dentine. *Archives of Oral Biology*, **28**, 693–700.

von den Driesch, A. (1976). *A Guide to the Measurement of Animal Bones from Archaeological Sites*. Peabody Museum Bulletin 1. Cambridge, MA: Harvard University.

von Ebner, V. (1902). Die Histologie der Zähne mit Einschluß der Histogenes. In Scheff, J. (ed.), *Handbuch der Zahnheilkunde*. Wien: A. Holder, pp. 243–302.

von Koenigswald, W. (1980). Schmeltzstruktur und Morphologie in den Molaren der Arvicolidae (Rodentia). *Abhandlungen der Senckenbergische Naturforschende Gesellschaft*, **239**, 1–129.

  (1982). Enamel structure in the molars of Arvicolinae (Rodentia, Mammalia), a key to functional morphology and phylogeny. In Kurtén, B. (ed.), *Teeth: Form, Function, and Evolution*. New York: Columbia University Press, pp. 109–22.

  (1997a). Brief survey of enamel diversity at the schmelzmuster level in Cenozoic placental mammals. In von Koenigswald, W. & Sander, P. M. (eds.), *Tooth Enamel Microstructure*. Rotterdam: A. A. Balkema, pp. 137–61.

  (1997b). Evolutionary trends in the differentiation of mammalian enamel ultrastructure. In von Koenigswald, W. & Sander, P. M. (eds.), *Tooth Enamel Microstructure*. Rotterdam: A. A. Balkema, pp. 203–36.

von Koenigswald, W. & Sander, P. M. (1997a). Glossary of terms used for enamel microstructures. In von Koenigswald, W. & Sander, P. M. (eds.), *Tooth Enamel Microstructure*. Rotterdam: A. A. Balkema, pp. 267–80.

  (1997b). Schmelzmuster differentiation in leading and trailing edges, a specific biomechanical adaptation in rodents. In von Koenigswald, W. & Sander, P. M. (eds.), *Tooth Enamel Microstructure*. Rotterdam: A. A. Balkema, 259–66.

von Koenigswald, W., Sander, P. M., Leite, M., Mörs, T., & Santel, W. (1994). Functional symmetries in the schmelzmuster and morphology in rootless rodent molars. *Zoological Journal of the Linnean Society*, **110**, 141–79.

Wahlert, J. H. (1968). Variability of rodent incisor enamel as viewed in thin section, and the microstructure of the enamel in fossil and recent rodent groups. *Breviora*, **303**, 1–18.

Waldron, H. A. (1983). On the post-mortem accumulation of lead by skeletal tissues. *Journal of Archaeological Science*, **10**, 35–40.

Walker, A., Hoeck, H. N., & Perez, L. (1978). Microwear of mammalian teeth as an indicator of diet. *Science*, **201**, 908–10.

Walker, P. L. (1996). Modern variation in tooth wear rates (abstract). *American Journal of Physical Anthropology*, Supplement **22**, 237.

Walker, P. L., Dean, G., & Shapiro, P. (1991). Estimating age from tooth wear in archaeological populations. In Kelley, M. A. & Larsen, C. S. (eds.), *Advances in Dental Anthropology*. New York: Wiley-Liss, pp. 169–78.

Walker, P. L. & Hagen, E. H. (1994). A topographical approach to dental microwear analysis. *American Journal of Physical Anthropology*, Supplement **18**, 203.

Walker, R. (2000). White-tailed deer (*Odocoileus virginianus*) mortality profiles: examples from the Southeastern United States. *Archaeozoologia*, **XI**, 175–86.

Wallace, A. R. (1876). *The Geographical Distribution of Mammals. With a study of the relations of living and extinct faunas as elucidating the past changes of the Earth's surface*. New York: Harper.

Wasserman, B. H., Moskow, B. S., & Rennert, M. C. (1970). Dental anatomy and coronal cementum in the Mongolian gerbil. *Journal of Periodontal Research*, **5**, 208–18.

Weatherell, J. A., Deutsch, D., Robinson, C., & Hallsworth, A. S. (1977). Assimilation of fluoride by enamel throughout the life of the tooth. *Caries Research*, **11**, 85–115.

Weaver, M. E. (1964). X-Ray diffraction study of calculus of the miniature pig. *Archives of Oral Biology*, **9**, 75–81.

Webb, S. D. (1974). Pleistocene llamas of Florida, with a brief review of the Lamini. In Webb, S. D. (ed.), *Pleistocene Mammals of Florida*. Gainesville: University Presses of Florida, pp. 170–213.

Weber, D. F. & Ashrafi, S. H. (1979). Structure of Retzius lines in partially demineralised human enamel. *Anatomical Record*, **194**, 563–70.

Weber, D. F. & Eisenmann, D. (1971). Microscopy of the neonatal line in developing human enamel. *American Journal of Anatomy*, **132**, 375–92.

Weber, D. F. & Glick, P. L. (1975). Correlative microscopy of enamel prism orientation. *American Journal of Anatomy*, **144**, 407–20.

Weidenreich, F. (1937). *The Dentition of Sinanthropus pekinensis: a Comparative Odontography of the Hominids*. Palaeontologica Sinica, New Series D, 1 (Whole Series 101). Peking.

Weinmann, J., Svoboda, J., & Woods, R. (1945). Hereditary disturbances of enamel formation and calcification. *Journal of the American Dental Association*, **32**, 397–418.

Weinreb, W. W. & Sharav, Y. (1964). Tooth development in sheep. *American Journal of Veterinary Research*, **25**, 891–908.

Werelds, R. J. (1961). Observations macroscopiques et microscopiques sur certains altérations postmortem des dents. *Bulletin du Groupement International pour les Recherches Scientifique en Stomatologie*, **4**, 7–60.

(1962). Nouvelles observations sur les dégredations post-mortem de la dentine et due cément des dents inhumées. *Bulletin du Groupement International pour les Recherches Scientifique en Stomatologie*, **5**, 559–91.

Whaites, E. (1992). *Essentials of Dental Radiography and Radiology*. Dental Series. Edinburgh: Churchill Livingstone.

Wheeler, J. C. (1982). Ageing llamas and alpacas by their teeth. *Llama World*, **1**, 12–17.

White, T. D. (1978). Early hominid enamel hypoplasia. *American Journal of Physical Anthropology*, **49**, 79–84.

Whittaker, D. K., Daniel, A. T., Williams, J. T., Rose, P., & Resteghini, R. (1985). Quantitative assessment of tooth wear, alveolar-crest height and continuing eruption in a Romano-British population. *Archives of Oral Biology*, **30**, 493–501.

Whittaker, D. K., Griffiths, S., Robson, A., Roger Davies, P., & Thomas, G. (1990). Continuing tooth eruption and alveolar crest height in an eighteenth-century population from Spitalfields, east London. *Archives of Oral Biology*, **35**, 81–5.

Whittaker, D. K., Parker, J. H., & Jenkins, C. (1982). Tooth attrition and continuing eruption in a Romano-British population. *Archives of Oral Biology*, **27**, 405–9.

Whittaker, D. K. & Richards, D. (1978). Scanning electron microscopy of the neonatal line in human enamel. *Archives of Oral Biology*, **23**, 45–50.

Whittaker, W. E. & Enloe, J. G. (2000). Bison dentition studies revisited: resolving ambiguity between archaeological and modern control samples. *Archaeozoologia*, **XI**, 113–20.

Wiggs, R. B. & Lobprise, H. B. (1997). *Veterinary Dentistry: Principles and Practice*. Philadelphia: Lippincott-Raven.

Willems, G., Moulin-Romsee, C., & Solheim, T. (2002). Non-destructive dental-age calculation methods in adults: intra- and inter-observer effects. *Forensic Science International*, **126**, 221–6.

Williams, R. A. D. & Elliott, J. C. (1989). *Basic and Applied Dental Biochemistry*. Dental Series. Edinburgh: Churchill Livingstone.

Wilson, D. F. & Schroff, F. R. (1970). The nature of the striae of Retzius as seen with the optical microscope. *Australian Dental Journal*, **15**, 3–24.

Wilson, M. (1974). The Casper local fauna and its fossil bison. In Frison, G. C. (ed.), *The Casper Site: a Hell Gap Bison Kill on the High Plains*. New York: Academic Press, pp. 125–71.

Winkler, L. A., Schwartz, J. H., & Swindler, D. R. (1996). Development of the orangutan permanent dentition: assessing patterns and varation in tooth development. *American Journal of Physical Anthropology*, **99**, 205–20.

Winter, G. B. & Brook, A. H. (1975). Enamel hypoplasia and anomalies of the enamel. *Dental Clinics of North America*, **19**, 3–24.

Wittwer-Backofen, U., Gampe, J., & Vaupel, J. W. (2004). Tooth cementum annulation for age estimation: results from a large known-age validation study. *American Journal of Physical Anthropology*, **123**, 119–29.

Wolpoff, M. H., Monge, J. M., & Lampl, M. (1988). Was Taung human or an ape? *Nature*, **335**, 501.

Wood, B. A. & Abbott, S. A. (1983). Analysis of the dental morphology of Plio-Pleistocene hominids. I. Mandibular molars: crown area measurements and morphological traits. *Journal of Anatomy*, **136**, 197–219.

Wood, B. A., Abbott, S. A., & Graham, S. H. (1983). Analysis of the dental morphology of Plio-Pleistocene hominids. II. Mandibular molars – study of cusp areas, fissure pattern and cross sectional shape of the crown. *Journal of Anatomy*, **137**, 287–314.

Wood, B. A. & Engelman, C. A. (1988). Analysis of the dental morphology of Plio-Pleistocene hominids. V. Maxillary postcanine tooth morphology. *Journal of Anatomy*, **161**, 1–35.

Wood, B. A. & Uytterschaut, H. (1987). Analysis of the dental morphology of Plio-Pleistocene hominids. III. Mandibular premolar crowns. *Journal of Anatomy*, **154**, 121–56.

Wood, N. K. & Goaz, P. W. (1997). *Differential Diagnosis of Oral and Maxillofacial Lesions*. St Louis: Mosby.

Worsaae, J. J. A. (1849) *Primeval Antiquities of Denmark* (trans. W. J. Thomas). London: J. H. Parker.

Wyckoff, R. W. G. (1972). *Biochemistry of Animal Fossils*. Bristol: Scientechnica.

y'Edynak, G. (1978). Culture, diet, and dental reduction in Mesolithic forager-fishers of Yugoslavia. *Current Anthropology*, **19**, 616–18.

(1989). Yugoslav Mesolithic dental reduction. *American Journal of Physical Anthropology*, **78**, 17–36.

(1992). Dental pathology: a factor in post-Pleistocene Yugoslav dental reduction. In Lukacs, J. R. (ed.), *Culture, Ecology and Dental Anthropology*. Journal of Human Ecology Special Issue 2. Delhi: Kamla-Raj Enterprises, pp. 133–44.

y'Edynak, G. & Fleisch, S. (1983). Microevolution and biological adaptability in the transition from food-collecting to food-producing in the Iron Gates of Yugoslavia. *Journal of Human Evolution*, **12**, 279–96.

Zeder, M. A. (1991). The equid remains from Tal-e Malyan, Southern Iran. In Meadow, R. H. & Uerpmann, H.-P. (eds.), *Equids in the Ancient World*. Beihefte zum Tübinger Atlas des Vorderen Orients, Reihe A (Naturwissenschaften) Nr 19/2. Wiesbaden: Dr Ludwig Reichert Verlag, pp. 366–412.

Zeiler, J. T. (1988). Age determination based on epiphyseal fusion in post-cranial bones and tooth wear in otters (*Lutra lutra*). *Journal of Archaeological Science*, **15**, 555–61.

Zsigmondy, O. (1893). On congenital defects of the enamel. *Dental Cosmos*, **35**, 709–17.

INDEX

aardvark 117
abrasion 214
Abrocomidae 75
abscesses 304, 307, 308, 309, 310, 313, 314
Acinonyx 57
Acomys 75, 243
actinomycosis 310
addax *Addax* 140
Aeretes 102
aetiology 286
Africa 7, 278
age estimation 207, 210, 211, 212, 216, 223–245, 299, 317
agenesis *see* congenital absence
Ailuropoda 60
Alcelaphus 140
Alces 138, 249
Allactaga 91
Allactagulus 91
Alopex 47
alpaca 143
Alticola 85
alveolae 9
alveolar bone 9, 311, 312
alveolar bone loss 311
alveolar crest 211, 311
alveolar process 10, 311, 314
ameloblast 155, 156, 176, 208, 273
ameloblastins 150
ameloblastoma 317
amelogenesis 155, 156
  imperfecta 168
amelogenins 149, 168
America 7, 275, 302
Ammospermophilus 102
Ammotragus 140
Andresen's lines 192, 247, 248
anodontia 281
Anomaluridae 74
anorganic preparation 199
anteaters 114
antelopes 138, 140
antemortem tooth loss 291, 306, 315
anterior 11
anterior teeth 11
Anthropoidea 42, 44
antibodies 304
antigens 303, 304
Antilocapra 140, 238
Antilocapridae 138, 140

Antilopinae 138, 140, 143
antimeres 266
antisymmetry 266
Antrozous 30
apatites 146, 148, 290
apes 44, 179, 225, 268, 281, 283, 287, 290, 293, 311
apex 9
apical 9
Aplodontia 100
Aplodontidae 74, 100, 102
Apodemus 75
Arctodus 60
Arctonyx 51
armadillos 111, 114
Artibeus 37
Artiodactyla 128, 158, 179, 229–239
Arvicanthis 75
Arvicola 85
Asellia 37
Asia 7, 275
asses 126, 279
ASUDAS 274
asymmetry 266, 267
  directional 266, 267, 273
  fluctuating 266, 267
Atlantoxerus 102
attrition *see* wear, and 189, 214, 231, 240, 241, 243, 255, 262, 284, 301
aoudad 140
Australia 301
australopithecine *Australopithecus* 167, 222, 269, 277

baboons 44, 167, 268, 273, 279
backscattered electrons 205
bacteria 286–287, 288, 290, 291, 295, 303, 307
badgers 13, 51, 242
Barbastella 30
Bassariscus 62
Bathygeridae 75
bats 29–42, 158, 165, 187, 287, 303
bears 17, 60, 251, 262, 268
beaver 18, 99, 243, 253
bell stage 208
beluga 71
Berardius 69
bharal 140
bilophodont 44, 124
birch mice 91
birefringence 204–205
bison *Bison* 140, 217, 239, 251

364

Blarina 25
Blarinella 25
body size 265
Bos 140
bovid Bovidae 18, 136, 140, 179, 217, 279, 281
Bovinae bovine 136, 140
brachydont 15, 74
Brachyones 79
Bradypodidae 111
broken mouth 307
brown striae of Retzius 161, 163, 164, 167, 192, 193, 204, 294
brushite 147, 148, 290
bruxism 221
Bubalus 140
buccal 10
buccolingual diameter 218, 260
   cervical 262
   maximum 262
bud stage 208, 266
buffalo 140
bunodont 15, 17, 19, 44, 128

calcite 148
calcium phosphates 146–148, 288
   amorphous 147, 184
calcospherites 185–187
calculus 148, 288–290
   subgingival 305
   supragingival 289
Callithricidae 43
Callorhinus 65, 245
Callosciurus 102
Calomyscus 78
camel 143, 239, 262
camelid Camelidae 18, 143–145, 311
Camelops 143
Camelus 143
canaliculae 194
canid Canidae 18, 47–51, 268, 281
canine 11, 12, 30, 45, 59, 67, 128–134, 135, 138, 237, 245, 268, 271, 282, 314
Canis 47, 241, 270–271
cap stage 208, 266
Capra 140, 179
Capreolus 138, 237
Capricornis 140
Caprinae caprine 138, 140, 143
Capromeryx 140
Capromyidae 75
Carabelli cusp 277
carbohydrates 291, 301
Cardiocranius 91
caribou 135, 218, 237
caries 159, 169, 189, 205, 287, 290, 303, 310, 317
   approximal 291, 293, 297, 302
   cement 293
   contact area 291
   coronal 292, 293
   dentine 291, 292, 293, 294, 295
   enamel 291, 294
   fissure 291, 293
   gross 293–294
   occlusal 291, 293, 297
   root surface 292–293, 295, 297
   wear and caries 293
carnassials 45, 46, 47, 51, 57, 60, 268, 271, 282
Carnivora carnivore 17, 45–63, 158, 163, 165, 177, 178, 187, 241–243, 268, 279, 280, 283, 303, 314
Carollia 37
Castor 99
Castoridae 74, 99, 100
Castoroides 99
cats 57, 241, 282, 289, 306, 310
cattle 140, 163, 165, 187, 194, 232–233, 253, 262, 266, 271–272, 290, 303, 311
Caviidae 75
CDJ 146, 196
Cebidae 43
CEJ 146, 292, 301
CEJ-AC distance 293, 312
cement 8, 10, 70, 193–198, 247, 248, 249, 250, 252, 255, 311, 314
   afibrillar 195
   extrinsic fibre 195
   formation of 194, 195
   intrinsic fibre 195
   layers of 196–198, 243, 245, 253
   mixed fibre 195
cement–dentine junction (*see* CDJ) 146
cement–enamel junction (*see* CEJ) 146
cementicles 195
cementoblasts 194, 212
cementocyte 193, 194, 195, 196, 250
   lacunae 194, 249
cementoma 317
Centurio 37
Cephalophinae 136
Ceratotherium 124
Cercopithecidae 44
Cervalces 138
cervical 9, 262, 266
cervid Cervidae 135, 138–140, 234–237, 279, 281
cervix 9
Cervus 138, 234–237, 250
Cetacea 69–73
chamois 140, 237
cheek teeth 11
cheetah 57
chewing 13
Chimerogale 25
chimpanzee 302, 306
Chinchillidae 75
chipmunks 102
Chiroderma 37
Chiroptera 29–42, 178
chiru 140
Choeroniscus 37
Chrysochloridae 20
cingulum 15, 30, 45, 67, 122, 132, 138, 262
cleaning of teeth 199

Clethrionomys 85
cloacae 310
clone theory 265–266
coatimundi 62
Coelodonta 124
Coelura 34
collagen 148–149, 184, 187, 192, 193, 195, 196, 204, 248, 304, 305
  fibres 148, 193–212
Condylura 34
Conepatus 51
congenital absence 272, 281
connate teeth 280
contact points 261, 262, 291
contour lines of Owen 192
coronal 9
cottontails 111
crevicular fluid 288
Cricetidae 78–85
Cricetinae cricetine 18, 78
Cricetulus 78
Cricetus 78
Crocidura 25
Crocuta 57, 179, 267
cross striations 159, 160, 164, 165, 192, 193, 204, 294
crossbite 283
crown 8
  area 264
  breadth 260
  index 264
  length 260
crypt 208, 211
Cryptotis 25
crystallites 147, 155, 184, 185
Ctenodactylidae 75, 95
Ctenodactylus 95
Ctenomyidae 75
Cuon 47
cusp 9, 277
Cuvieronius 118
Cynomys 102
cyst 308, 313, 316
Cystophora 65

Dasypodidae 111, 114
Dasyproctidae 75
Dasypus 114
Daubentoniidae 42
dead tract 189, 294, 310
death assemblages 225, 259
deciduous teeth 11, 12, 44, 45, 47, 58, 128, 135, 145, 198, 207, 211
decussation 158, 167, 177
deer 10, 18, 135, 138, 217, 234–237, 252, 262, 266, 290, 311
dehiscence 313
Delphinapterus 71
Delphinidae 71, 73, 248
Delphinoidea 71, 73
Delphinus 71

dental
  age 213, 216, 223
  arcade 10
  comb 42
  development 208–211, 225
  follicle 208, 212
  formula 12
  lamina 208
  papilla 208
denticles 189
dentine 3, 8, 149, 184–193, 208, 209, 210, 214, 215, 246, 247–249
  circumpulpal 185
  incremental structures 191, 245–253
  intertubular 185
  mantle 185, 187
  peritubular 185, 186, 187
  primary 185
  root dentine sclerosis/transparency 187, 254, 255
  secondary 185, 189, 214, 253, 255, 291, 294, 303, 310, 314
  tubules 185, 186, 189, 253, 254, 291, 294, 314
dentition 10, 11
desert dormouse 107
Desmana 25
desmans 20, 25
Desmodontidae 37, 40
Desmodus 37
developmental age 212, 213, 216
dhole 47
Diaemus 37
diagenetic foci 190
diastema 73, 115, 123, 126, 132, 135, 144, 282
diazones 178
Dicerorhinus 124
Diceros 124, 241
Dicrostonyx 85
Didelphidae, *Didelphis* 20
diet 153, 221, 222, 270, 284, 286, 288, 291, 301, 302, 303, 306
dilambdadont 17, 20, 30, 42
Dinaromys 85
Dinomyidae 75
Diphylla 37
Diplomesodon 25
Dipodidae 19, 91
Dipodillus 79, 243
Dipodomys 95
Diprotodont 19
Dipus 91
disease 6, 286
distal 10
DMF scores 295
DNA 151, 290
dogs 47, 241, 253, 258, 262, 270–271, 281, 282, 283, 290, 303, 306, 310
dolphins 71
domestication 270–272
dormice 19, 107
Dryomys 107
dugong *Dugong* 120, 245, 249, 268

Dugongidae 120–121
duikers 136

Echimyidae 75
ectoloph 126
Edentata 111–117
EDJ 146, 186, 291, 293
Eilodon 40
Elasmotherium 124
electron spin resonance (ESR) 154
elephant seal 65, 246, 249
elephant shrews 20, 26
Elephantidae 117–118, 120
elephants 3, 18, 118, 165, 180, 187, 189, 195, 214, 245, 268, 280
Elephantulus 26
Elephas 118, 245
Eliomys 107
elk 138
Ellobius 85
Emballonuridae 34
enamel 4, 8, 148, 155–184, 209, 214, 219, 248
    appositional 163
    defects *see* hypoplasia
    extensions 316
    imbricational 163
    knot 208, 209, 266
    matrix production 149, 155, 194
    maturation 149, 150, 155, 157, 163, 168, 194
    organ 208
    patterns 1, 2 and 3 157, 158
    pearls 316
    prism 156, 158, 159, 176–178, 199
    prism-free 157, 163
    prism sheath 150, 156
    protein 149–150
    radial 176
    tangential 177
    tubules 19, 158
    tufts 150
enamel–dentine junction (*see* EDJ) 146
enameloma 316
Enhydra 51
entoconid 15
environment 208, 210, 257, 264, 267, 273
Eothenomys 85
Eozapus 91
epidemiology 295–303
epithelium 10, 155, 208
Eptesicus 30
Equidae 123, 126, 128, 217, 279
Equus 126
Eremotherium 115
Erethizon 98
Erethizontidae 75, 98
Erignathus 65
Erinaceidae 20, 25
Erinaceus 25
eruption 195, 212, 213, 214, 216, 218, 231, 233–237, 238, 240, 241, 242, 292, 311, 315, 318
Eucatherium 140

Euchoreutes 91
Euderma 30
Eumetopias 65
Eumops 32
Europe 7
Eutamias 102
eutherian 12
exfoliation 12, 198, 211
externalindex 180
extinction 205
extrinsic fibres 194, 195, 196, 212, 248, 250, 253

facial 10
fats 288, 291
felid Felidae 57–60, 179, 268, 279
Felis 57
Feresa 71
fibroblast 194, 212, 305
field theory 265–266
fissure patterns 277
fluorapatite 146
fluorine 150, 158, 297
fluorosis 150, 169
flying foxes 40
flying squirrels 102
foxes 47, 242, 271, 281
fractures
    of jaws 314–315
    of teeth 293, 301, 303, 310, 313, 314
fruit bats 37, 40
functional age 216, 229
fur seal 65, 245

Galemys 25
Gazella 140
gazelles 140, 238, 253
gene pool 258
genets *Genetta* 51
genotype 257
genus 8
Geomyidae 74, 95–98
Geomys 95
Gerbillinae 79, 85
Gerbillus 79
gerbils 79, 243
giant deer 138
gingivae 10, 255, 289, 292, 303, 304, 307
gingival
    crevice *see* gingival sulcus
    cuff 211, 292, 304
    emergence 212, 223, 233
    sulcus 10, 303
gingivitis 305, 312
Giraffidae, giraffes 138
Glaucomys 102
GLG 247, 248, 249, 251
Gliridae 19, 107
Glirulus 107
Glis 107
Globicephala 71
Glossophaga 37

Glossotherium 117
glutton 51
glycoproteins 149, 288
Glyptodontidae 114–115
Glyptotherium 115
gnawing 4, 73
goats 140, 231–232
Gomphotheriidae 117, 118
gophers 95
goral 140
gorilla 268, 302
Grampus 71
Grant method 231, 233, 234, 319
granuloma 308, 313
Greenland 301
ground sloths 115–117
ground substance 149, 185, 192, 193, 195, 196, 248
growth 6, 207–214
growth layer groups (*see* GLG)
guanaco 143
Gulo 51, 242
gundis 95
Gustafson & Koch's chart 226
Gustafson's method 255

haematoxylin 196, 248, 249, 250
Halichoerus 65
hamsters 18, 78
hares 111, 280
hartebeest 140
heating 147, 148, 159, 190
hedgehogs 20, 25
Hemiauchenia 143
Hemiechinus 25
Hemitragus 140, 239
heritability 264, 273
Herpestes 51
Hesperomyinae 18, 78, 79
Heteromyidae 19, 74, 95
Hippopotamidae 132
Hippopotamus 132, 158, 179, 189, 268, 315
Hipposideros 37
Hippotraginae 138, 140
histology 146
Holarctic 4, 7, 69
hominid Hominidae 42, 44, 269
hominoid Hominoidea 275–279
Homo 13, 44, 167, 177, 269
homodont 69
homologous points 260
Homotherium 57
Hopewell-Smith's hyaline layer 187, 196
horse 18, 126, 163, 165, 179, 187, 194, 217, 240–241, 262, 279, 281, 282, 283, 292, 303, 307, 310, 311
Howship's lacunae 198
HSB (*see* Hunter-Schreger bands) 178
human 44, 163, 179, 187, 222, 223, 225–229, 252, 253, 254, 262, 263, 264, 265, 266, 268, 269, 270, 273, 274–281, 282, 287, 288, 289, 290, 291, 292, 293, 295, 297–301, 302, 306, 308, 310, 311, 312, 314, 317

Hunter-Schreger bands 178
Hyaena 57, 267
Hyaenidae 57, 179, 268
Hydrochaeridae, *Hydrochaeris* 75, 107
Hydrodamalus 120
hydroxyapatite 146, 155
Hylobatidae 44
Hylonycteris 37
Hyperacrius 85
hypercementosis 195
Hyperoodon 69
hypersensitivity 305
hypocalcification 168, 169
hypocone 15, 17
hypoconid 15
hypoconulid 15
hypodontia 281
hypolophid 15, 126
hypoplasia 168–176, 193, 210, 214, 264, 272, 315
hypsodont 15
Hyracoidea 122
hyraxes 122, 187
Hystricidae 74, 98, 99
hystricomorph Hystricomorpha 74, 75, 177, 180, 182
Hystrix 98

Ichneumia 51
identification 7, 145, 266
immunity 303–304
impaction 315
impressions 199
incisor 11, 12, 18, 19, 30, 37, 42, 73, 135, 180–183, 240, 297, 302, 307
inclination of HSB 180
incremental growth layers 247
Indriidae 42
inflammation 303–311
    acute 304, 308
    chronic 304, 308
inflammatory exudate 304, 305, 308
infolding 18
infundibulum 18, 115, 126, 132, 135, 144, 292, 303
inheritance 264, 272–274, 284
    Mendelian 258
    non-Mendelian/multifactorial 258, 273
inorganic component 146–148, 155, 184, 193
Insectivora insectivore 20–29, 148, 158, 165, 178, 187, 243, 245
interdental papilla 10, 211
interdental wall 311, 312
interglobular spaces 192, 248
internal enamel epithelium 208, 266, 273
intrinsic fibres 193, 194, 195, 196
involucrum 310
iron oxide 148
isotopes 152–153, 302
ivory 3

Jaculus 91
jerboas 19, 91

Index

jirds 79
jumping mice 85

kangaroo mice and rats 95
kinkajou 62
Kogia 70

labial 10
LAC/LAG 251
Lagenodelphis 71
Lagenorhynchus 71
lagomorph Lagomorpha 18, 110–111, 158, 180, 187, 315
Lagurus 85
laser scanner 275
Lasionycteris 30
Lasiurus 30
lead 151
lemmings 18, 85
Lemmus 85
Lemniscomys 75
Lemuridae 42
leopards 57, 241
Leporidae 111
Leptonycteris 37
Lepus 111
lesion 286, 291, 292, 293, 294, 295, 304, 305
life tables 223
lingual 10
lions 57, 241, 268
Lissodelphis 71
llama 143
long period lines 191
lophodont 15, 18, 19, 74, 75, 107, 122
lophs 9, 15, 118, 122, 126
Lorisidae 42
Loxodonta 118, 245
lumpy jaw 311
Lutra 51
Lycaon 47
lymphocyte 304, 305

Macaca macaques 44, 165, 179, 268
Macroscelididae 20, 26, 29
Macrotus 37
malocclusion 282
mamelon 17
mammoths 118, 163
Mammut 118
Mammuthus 118, 245
Mammutidae 117, 118
manatees 121
mandible 10
marmot *Marmota* 102, 243
marsupials 19–20, 158, 177
martens *Martes* 51, 242
Massoutheria 95
mastodon 118, 163, 290
matrix vesicles 185
maxilla 10

measurements 260–272
   interrelationships between 264–265
median sagittal plane 10
Megaceros 138
Megachiroptera 40, 42
Megalonychidae 115
Megalonyx 115
Megatheridae 115, 117
Meles 51, 242
Mellivora 51
Mephitis 51
Meriones 79, 243
mesenchyme 208
mesial 10
mesiodistal diameter 218, 260
   cervical diameter 262, 263
   maximum diameter 261, 262
Mesocricetus 78
Mesoplodon 69
mesostyle 15, 17
metacone 15, 17
metaconid 15, 46
metaconule 15
metaloph 124, 126
metalophid 15, 126
metastyle 15, 17
Metatheria 19
metrical variation 260
mice 18, 75, 78, 243, 279
Microchiroptera 29, 40
Microdipodops 95
micrometre 146
Micromys 75
microscope
   confocal 146, 159, 203, 252
   polarising 146, 202
   scanning electron 146, 205–206
   stereomicroscope 202
   transmitted light 202, 203–205, 294, 295
Microsorex 25
Microtinae microtine 19, 85
Microtus 85
microwear 219–223
   feature density 220
   pits 219, 220, 222
   scratches 219, 220, 221, 222
mineralisation 194, 196
Miniopteris 30
mink 51
Mirounga 65, 246
MMD 277
Moiré fringe contourography 275
molar 11, 12, 18, 45, 183, 184, 291, 297
mole rat 98
moles 20, 245
Molossidae 32, 34
Monachinae 67
Monachus 67
mongooses 51
monk seals 67
monkeys 43, 44, 179, 268, 287, 302

Monodon 71, 267
Monodontidae 71, 248
monotremes 19
moose 138, 249, 250, 290
Mormoopidae 34
Mormoops 34
morphogens 265, 266
Moschus 138
mountain beaver 100
multiserial 177
Muridae 18, 75–76
Murina 30
Mus 75, 243
Muscardinus 107
musk deer 138
musk ox 140
Mustela 51, 242, 253
Mustelidae 17, 51, 57, 179, 242, 268
MWS 231, 233, 234, 319
Mylodontidae 117
Mylohyus 129
Myomimus 107
myomorph Myomorpha 74, 180, 182, 183
Myopus 85
Myospalacinae 85
Myospalax 85
Myotis 30
Myotragus 140
Myrmecophagidae 114
Mysticeti 69

nanometre 146
Napaeozapus 91
narwhal 71, 248, 267, 268
Nasua 62
Natalidae 40
Natalus 51
Navahoceras 138
Neanderthals 269
Nearctic 7
necrosis 304, 308, 310
Nemorhaedus 140
Neofiber 85
Neomys 25
neonatal line 165, 166, 192, 246, 248
Neophocaena 71
neoplasm 317
Neotoma 78
Neotomodon 78
Nesokia 75, 243
Neurotrichus 21
non-metrical variation 260, 272–281
Nothrotheriops 115
Notiosorex 25
Nyctalus 30
Nyctereutes 47
Nycteridae 34
Nycteris 34
Nycticeius 30

occlusal 10
occlusion 213, 214, 281–284
Ochotona 111
Ochotonidae 111
Ochrotomys 78
octacalcium phosphate 147, 290
Octodontidae 75
Odobenidae 67, 69
Odobenus 67, 267
Odocoileus 138, 237, 250, 252
odontoblasts 208, 209
   processes 184, 190
odontocete Odontoceti 69–73, 158, 179, 187, 189, 192, 193, 195
odontoclasts 198
odontome 316–317
odontometry 260
okapi 138
Ondatra 85
Onychomys 78
openbite 283
opossums 20
orang-utan 302
Orcinus 71
Oreamnos 140
organic component 148–150, 155, 184, 193
Orycteropus 117
Oryctolagus 111
Oryx 140
Oryzomys 78
osteodentine 249
osteomyelitis 310
Otariidae 65, 179, 246
Otonycteris 30
otter 51, 242
overbite 283
overjet 283
Ovibos 140
Ovis 140, 179

Pachyuromys 79
Paguma 51
Palaearctic 7
Palaeollama 143
palatal 10
palm-civet 51
panda 60
pangolins 114
Panthera 57, 241
Pantholops 140, 281
Papio 44, 268
Pappogeomys 95
paracone 15, 17, 46
paraconid 15, 46
Paradipus 91
Paraechinus 25
paraloph 15
Paranthropus 167, 222, 269, 275, 277
Parascalops 20
parastyle 15, 17, 277
parazones 178

pauciserial 177
Payne method 231, 239
pc 63
PE 180
peccaries 129
Pecora *see* Ruminantia
Pedetidae 74
Pentalagus 111
periapical inflammation 195, 301, 307–310, 313
perikymata 163, 170
periodontal
  abscess 307, 313
  disease 292, 304–307, 311–313
  ligament 9, 126, 193, 194, 198, 212, 304, 305, 307, 308
  pocket 305, 307, 312
periodontitis 287, 291, 305–308, 311, 312
  acute 308
  periapical 308
periodontium 212
periosteum 308, 309, 310, 314
Perissodactyla 122–128, 158, 240–241
permanent teeth 11, 45, 207, 211
Perognathus 95
Peromyscus 78
Petaurista 102
Petromyidae 75
phagocytes 303, 304, 305
Phenacomys 85
phenotype 257
Phoca 65, 252
Phocaena 71
Phocaenidae 71
Phocaenoides 71
phocid Phocidae 65, 67, 179
Phocinae 65, 67
Phodopus 78
Pholidota 114
photogrammetry 275
Phyllostomatidae 37, 303
Physeter 70, 248, 267
Physeteridae 70, 71
phytoliths 219, 221, 290
PI 180
pigs 3, 128, 163, 165, 189, 233–234, 253, 268, 271, 272, 279, 281, 283, 315
pikas 111
pinniped Pinnipedia 63–69, 158, 178, 193, 249
Pipistrellus 30
Pitymys 85
plaque 2, 286–288, 291
  fluid 288
  subgingival 305
Platanistidae 69
Platygonus 129
Plecotus 30
pocket mice 95
Poecilictus 51
polar teeth 265, 266
polecats 51
polishing 199

polydontia, polygenesis 281
polyprotodont 19
Pongidae 44
population 257, 259, 269
porcupines 18, 98
porpoises 71
postcanine 19, 63, 65, 67, 69
Potos 62
prairie dog 102
Praomys 75
precement 194
predentine 185, 248
premaxilla 10
premolar 11, 12, 18, 30, 45
preservation 158–159, 189–190, 198
Primates 17, 42–44, 158, 163, 179, 187, 221, 222, 225–229, 262, 265, 268, 280, 313, 314
probable mutation effect 270
Proboscidea 117–120, 158, 180, 245
Procapra 140
Procavia 122
Procaviidae 122
Procyonidae 62, 63, 179
Prolagus 111
Prometheomys 85
pronghorn 138, 140, 238
Prosimii 42
protein 148, 149, 288, 291
protocone 15, 17
protoconid 15, 46
protoconule 15
protoloph 15, 124, 126
protostylid 277
Prototheria 19
Psammomys 79, 243
Pseudois 140
Pseudorca 71
Pteromys 102
Pteronotus 34
Pteropodidae 40, 42
Pteropus 40
pulp 8, 190, 291, 304
  chamber 8, 209, 248, 253, 307
  exposure 310
  stones (*see* denticles) 249
pulpitis 307, 314
Pygeretmus 91
pyrite 148

quadrants of dentition 12
Quadratic Crown Height Method 217
Quaternary 8

rabbits 111
racoon 62, 243
raccoon dog 47
radiolucency 291, 292, 308, 310, 312, 313
radio-opacity 292
Rangifer 138, 218
ratel 51
rats 18, 75, 78, 243

Rattus 75, 243
reduction in evolution 12, 269–270, 272, 283
reindeer 138
Reithrodontomys 78
replicas 199
resorption 198, 211, 255, 316
resting lines 248
reversal line 198
Rhinoceros 15, 18, 124
rhinoceroses Rhinocerotidae 122, 124, 126, 158, 165, 179, 241, 290
Rhinolophidae 37
Rhinolophus 37
Rhinopoma 34
Rhinopomatidae 34
Rhombomys 79
RI, refractive index 203, 294
ringtail 62
rodent Rodentia 12, 13, 18, 73–107, 148, 158, 177, 180–184, 187, 195, 243, 315
root dentine sclerosis/transparency *see* dentine
roots 8, 9, 245
Rousettus 40
ruminant Ruminantia 18, 132–143
Rupicapra 140, 237

sabre-tooths 57, 59
Saiga 140, 281
saliva 288, 289, 303
Salpingotus 91
samples 154–155, 259
Sangamona 138
Scalopus 20, 245
Scapanulus 20
Scaptonyx 20
schmelzmuster 178
Schour & Massler chart 225
Sciuridae 19, 102
sciuromorph Sciuromorpha 74, 180, 182
Sciurotamias 102
Sciurus 102, 243
Scotophilus 30
sea cows 120
sea lions 12, 65
seals 12, 65, 252
seasonality 245, 253
secodont 15
secondary electrons 205
sections 199–201
    buccolingual 201
    mesiodistal 201
    radial 201
    sagittal 201
    tangential 201
    thin 200
    transverse 201
sectorial premolar 44
Sekeetamys 79, 243
Selenarctos 60
selenodont 15, 18
Seleveniidae 107

Selevinia 107
SEM 146, 205
SEM-BSE 159, 165, 186, 190, 196, 205, 250, 252, 290, 294, 295
SEM-ET 159, 206
sequestrum 310
serow 140
sex determination 5, 237, 290
sexual dimorphism 128, 132, 138, 207, 257, 267–269, 272, 273
sheep 140, 221, 229–232, 252, 253, 283, 290, 303, 307, 311
short period lines 192
shovelling 275
shrew-moles 20
shrews 20, 25–26, 243
Sicista 91
significant 259
sinodont 278
sinus 309, 313
Sirenia 120–122, 180, 187
skeletal age 213
skunks 51
sloths 111
Smilodon 57
Solendontidae 20
Sorex 25, 243
Soricidae 20, 25–26
Soriculus 25
Spalacidae 98
Spalax 98
Spermophilopsis 102
Spermophilus 102
Spilogale 51
squirrels 19, 102, 243
starch 288, 291, 301
Stegodontidae 117
Stenella 71
Stenidae 71
Steno 71
Stephan curves 288
stoats 51
Stockoceros 140
strontium 151
Sturnira 37
Stylodipus 91
sugars 287, 288, 291, 301
Suidae 128
suiform Suiformes 17, 128–132
Suncus 25
sundadont 277
supernumerary teeth 281
Sus 128
Sylvilagus 111
Symbos 140
Synaptomys 85
syphilis 175

Tadarida 32
tahrs 140, 239
talonid 15, 17, 46

Index

Talpa 20, 245
Talpidae 20–25
Tamias 102
Tamiasciurus 102
Taphozous 34
tapirs, Tapiridae 122, 123, 124, 165, 290
Tapirus 123
Tarsiidae 42
Tatera 79
Tayassu 129
Tayassuidae 129, 132
Taxidea 51
temporomandibular joint (TMJ) 13
Tenrecidae tenrecs 20
Tetrameryx 140
Thalarctos 60
Thomomys 95
Thryonomidae 74
tiger 57
Tomes' granular layer 187
Tomes' process 155, 163, 174
tooth germs 208
toothbrush abrasion 219
trace elements 150–151
Tragulina 143
trauma 314–315
Tremarctos 60
Triaenops 37
tribology 219
tribosphenic 15, 19
Trichechidae 120, 122
Trichechus 121
trigon 15
trigonid 15, 17, 46
Trogopterus 102
Tubulidentata 117
tuftelins 150
Tupaiidae 42
turnover in tissues 207
Tursiops 71
tusks 67, 71, 117, 118, 120, 128, 129, 132, 140, 143, 144, 249, 268
Tylopoda 143–145

Ubelaker chart 225
underjet 283
ungulate 13, 163, 187, 195, 214, 251, 281, 313
uniserial 177
uranium 150–151
Urocyon 47
Urotrichus 21
Ursidae 60–62, 179, 268
Ursus 60, 268

variation 5, 257, 269, 272, 286
  continuous 258
  discontinuous 258
  quasicontinuous 273
vasodentine 189, 249
Vespertilio 30
Vespertilionidae 30
vestibular 10
vicuña 143
Viverridae 18, 51, 179, 268
vivianite 148
voles 18, 85, 243, 274, 279
von Ebner's lines 192, 247
von Korff fibres 185, 187
Vormela 51
Vulpes 47, 242

walrus 3, 67, 189, 249, 267, 268
wear 195, 214–223, 227–229, 231, 234–237, 241, 242, 243, 301, 302, 310, 311, 312
  abrasive 219, 220
  approximal attrition 214, 262, 284
  crown height recording 217, 218, 229, 237, 241
  erosive 219
  facets 214
  gradient 218
  image analysis recording 215
  occlusal attrition 214
  sliding 219
  see also abrasion, attrition
weasels 51
weddellite 148
whales 3, 69, 187
  beaked 69, 248
  bottlenosed 69
  killer 71–73
  pilot 71
  sperm 70, 248, 267
whitlockite 147, 148, 290
Wilson bands 165, 166
wolverine 51, 242
wolves *see* dogs

Xerus 102

zalambdadont 17, 20
Zalophus 65
Zapodidae 18, 74, 91
Zapus 91
zebras 126
Ziphiidae 69–70, 248
Ziphius 69
zokor 91
zooarchaeology 1